Multivariate Calibration

Multivariate Calibration

Harald Martens
Norwegian Computing Center,
N-0314 Oslo 3, Norway

Norwegian Food Research Institute,
Oslovegen 1, N-1430 Aas, Norway

and

Tormod Næs
Norwegian Food Research Institute,
Oslovegen 1, N-1430 Aas, Norway

JOHN WILEY & SONS

Chichester · New York · Brisbane · Toronto · Singapore

Copyright © 1989 by John Wiley & Sons Ltd.

Reprinted with corrections April 1991
First published as a paperback April 1991
Reprinted January 1993

Library of Congress Cataloging in Publication Data:

Martens, Harald.
    Multivariate calibration / Harald Martens and Tormod Naes.
        p.   cm.
    Bibliography: p.
    Includes index.
    ISBN 0 471 90979 3
    1. Calibration.   2. Chemistry, Analytic—Quantitative—Statistical
methods.   I. Naes, Tormod.   II. Title.
QD75.4.C34M37   1989
543'.07—dc20   89-14693
                                                                CIP

British Library Cataloguing in Publication Data:

Martens, Harald
    Multivariate calibration.
    1. Statistical mathematics
    I. Title   II. Naes, Tormod
    519.5

    ISBN 0 471 90979 3 ppc
    ISBN 0 471 93047 4 pbk

Printed and bound in Great Britain by
Biddles Ltd, Guildford and King's Lynn

To our wives Magni and Turid, and our children Silje, Johannes, Tone and Åsmund.

Contents

Preface

HOW TO USE THIS BOOK

The theory and practice of multivariate calibration has now come far enough that a unified treatment of the topic is needed. This book attempts to bridge a gap between the practice of calibration in chemistry and engineering and the statistical theory. It builds on the first author's experience in quantitative chemical analysis and instrument design and the second author's insight into statistical prediction.

Man's scientific knowledge about the universe consists of models—simplifying approximate descriptions of the real thing. No models are true, but some models are better than others—for a given purpose. When old models are taught as unassailable 'Physical Laws' in fixed language structures, they can change from a boon to a hindrance. The models and concepts developed before the laboratory computer and in isolated academic fields may have to be reformulated in order to be useful for today's and tomorrow's integrated high-speed quantitative analyses.

This book has been written to provide a unified mental and technical model framework for calibration. Hopefully, it can help chemists and technologists to go beyond certain limiting traditions in chemistry concerning selectivity, and in statistics concerning hypothesis testing. The book may also serve to illustrate to statisticians and mathematicians the wealth of background knowledge that practitioners apply when they work.

The aim of the book is to give a reasonably updated (1987/88) presentation of the main aspects of multivariate calibration. We have chosen to cover what we consider to be of most importance for practical use of multivariate calibration in research, development and routine operation. We do this by giving main attention to a few of the methods which we find of special interest. Some parts of the theory are treated rather lightly, but references are then given for further reading.

Each chapter and sub-chapter starts with a summary, and the book has an index and cross-reference list at the back.

The book is written at two different levels of detail. The book is written in order to give useful insight and methodology for practitioners in chemistry and technology. But it is also structured so that experienced chemometricians, statisticians and other data-analytic specialists can find more formal overviews, details and references. Level 2 of ambition is marked as 'Statistical Extensions'.

Level 1: Multivariate calibration for users

The main level of the book is structured as a tutorial on practical use of multivariate calibration, intended for graduate courses and self-study for chemists and technologists. The book requires little previous knowledge of statistics and mathematics at this level; the elementary matrix algebra and distributional theory required are explained in the book. At this level the book provides overviews and knowledge about a few central calibration methods.

Emphasis is given on teaching one particular class of calibration methods, the bilinear data compression methods. This is an approach applicable to a variety of calibration problems. But in order to use it effectively and safely, various aspects of calibration must be understood. The book's chapters can thus be described as follows.

1: WHY MULTIVARIATE CALIBRATION? An outline of why traditional univariate calibration often fails while multivariate calibration works. An opening for 'open-ended' empirical modelling instead of 'closed' causal modelling.
2: USEFUL STATISTICAL TOOLS. A brief summary of the problems involved in picking up the right information, and why.
3: HOW TO DO IT. Teaching how to use one good, general approach for multivariate calibration and prediction: bilinear regression plus an understanding of pragmatic multivariate calibration.
4: HOW TO MAKE SURE IT REALLY WORKS. Validation against too-cautious under-fitting and over-ambitious over-fitting.
5: HOW TO FIND EXTREME OUTLIERS OR ERRORS IN DATA. Types of anomalies and what to do with them.
6: HOW TO GET THE RIGHT TYPE OF INPUT DATA. Problem formulation and experimental design for calibration. Using what is already known, but preparing for the unknown.
7: HOW TO TAILOR YOUR DATA FOR OPTIMAL CALIBRATION. Pretreatment and linearization techniques.
8: A COMPLETE EXAMPLE. Design, pretreatment, calibration BASIC SETUP modelling and prediction.

Spectrophotometric examples are mainly used for illustration, ranging from straightforward causal modelling of well-behaved mixtures to empirical modelling of rather 'dirty' spectral data with overlapping peaks and turbidity.

Various other types of data suitable for multivariate calibration are also outlined.

Minimum curriculum:

A subset of the Level 1 material can be used as a curriculum for a minimum course in multivariate calibration. This should teach students of chemistry and engineering how to apply pragmatic multivariate calibration effectively and safely. The following selection provides the necessary motivation, methodology for calibration/validation/design/preprocessing and integrated understanding:

All of chapter 1
All of chapter 2 (but 2.1.1.13 postponed till section 3.3)
Sections 3.1, 3.3–3.4.5, 3.5.1–3.5.4.3
Sections 4.1–4.3.2.2, 4.5–4.6.1
Sections 5.1–5.2
All of chapter 6
Sections 7.1–7.4.2.3
All of chapter 8

Level 2: Quantitative chemometric methods for calibration

At this second level the book gives more theoretical and methodological detail. It is intended for statisticians and for specialists in chemometrics.

The text widens the perspective by comparing several alternative calibration methods. Many different aspects of model evaluation and optimization are described. Statistical properties and algorithm designs are more fully explained, and presently unresolved theoretical problems are pointed out. This part can be useful in method developments within calibration theory, software or advanced instrument design.

Computations

Most of the computations and illustrations in this book were done with the UNSCRAMBLER Version 2.0 on an IBM compatible PC.

Preface to Second Printing

Since the first printing of the book in 1989, the range of application of multivariate calibration and quantitative chemometrics has continued to grow.

A number of instrument companies now offer multivariate full-spectrum calibration methods, such as PLS regression, for their instruments. Several software packages for general data analysis have likewise been expanded recently to include quantitative chemometric modules.

This rapid development increases the need for statistically literate chemists and engineers, and for chemometrics oriented statisticians. It is important for users to understand multivariate calibration well enough to avoid using it as 'black box'. Hopefully this book can contribute in that respect.

For the present printing of this book, certain misprints have been corrected and minor changes have been carried out.

The first author has now changed address to Consensus Analysis AS, Ski Business Park, N-1400 Ski, Norway, to develop chemometrics software for instrumentation and process applications.

Acknowledgements

The authors would like to express their gratitude to a number of people who have contributed to the method developments and applications that led to this book being written. Included in this group are: Gerald Birth, Hans Rene Bjørsvik, Kim Esbensen, Sven-Åge Jensen, Paul Geladi, Jan Ruud Hansen, Inge Helland, William Hruschka, Tomas Isaksson, Bruce R. Kowalski, Magni Martens, Fred McClure, Roger Mossberg, Lars Munck, Karl H. Norris, Birthe Pedersen, Emil Spjøtvoll, Ed Stark, Bo Stenlöf, Rolf Sundberg, Veslemøy Tyssø, Svante Wold, Herman Wold and Tomas Öberg.

We also wish to thank the production staff at the Norwegian Food Research Institute, in particular Ragnhild Norang, Bjørg Narum Nilsen and Ulla Dyrnes.

During the writing of the book we have been financially supported by our employers. We hereby thank the Norwegian Food Research Institute—Kjell Ivar Hildrum and Einar Risvik in particular, as well as the Norwegian Computing Center—Eivind Damsleth and Hans Viggo Sæbø in particular, for their patience and generosity. In addition, Tecator AB is thanked for support and cooperation.

R. D. Kirby

1 Introduction to Multivariate Calibration

SUMMARY Multivariate calibration is a general selectivity and reliability enhancement tool. It is applicable to determination of major constituents as well as microcomponents and other qualities, and for a very wide range of instrument types. Successful examples range from spectrophotometric determination of the protein percentage in intact, whole wheat kernels, to chromatographic determination of dioxin in smoke at the nanogram range.

With multivariate calibration the need for sample preparation is greatly reduced. The reason is that selective input measurements are no longer needed—it is the output results that must be selective.

Multivariate calibration can thus stimulate the development of new analytical instruments. It can also increase the analytical capacity and reliability of traditional instruments. This extends the usefulness of quantitative chemical analysis in on-line industrial process control, analysis of intact biological or medical samples, low-cost pollution monitoring, etc.

This chapter provides a non-technical background. It shows why it can be useful to perform indirect measurements and calibrate these to yield valuable information, instead of always trying to measure the wanted information directly. It illustrates why selectivity problems often make multivariate procedures necessary and it motivates why the reader should make the effort of learning certain statistical techniques in order to ensure relevant, precise and reliable calibrations.

1.1 WHY MULTIVARIATE CALIBRATION?

SUMMARY   There is a need for improved quantitative information in science and technology. This requires transformation of measurements into informative results. Calibration is to establish this transformation.

1.1.1 ABSOLUTE VERSUS RELATIVE CALIBRATION

The word 'Calibrate' traditionally means to determine the inner diameter or capacity (the calibre) of a gun or some other cylinder, for instance using the traditional 'caliper' instrument. But if you want to determine the calibre of something, then you first have to learn how to relate your measurements X to some calibre reference Y: You have to calibrate your instrument.

In this case the calibration is a standardization to fixed scale, and is here termed absolute calibration. Such absolute calibrations should be traceable to legally accepted international standards, like the tuning of music instruments to a fixed frequency scale using a tuning fork.

However, in practical quantitative analysis, the absolute accuracy of the end results is often less important than their reliability and relevance.

It sometimes happens that absolute 'tuning fork' standards simply do not exist or are irrelevant for certain instruments. And more significantly, the tuning of the instrument strings does not ensure good music: absolute calibration of each individual variable may be irrelevant for the purpose of the analysis. One example of this is diffuse near infrared (NIR) spectroscopy, for which many of the techniques in this book originally were developed: The important thing is not to attain universally accepted absorbance readings at some individual wavelengths; the purpose is to predict protein content in a certain type of wheat samples, octane number in a certain type of gasoline etc. So while the NIR reflectance or transmission from the intact samples appear highly confusing, multivariate calibration converts several wavelengths into precise and relevant information.

Therefore in the present book the meaning of the word 'Calibrate' is generalized in the following way, here sometimes referred to as relative calibration:

To CALIBRATE is to use empirical data and prior knowledge for determining how to predict unknown quantitative information Y from available measurements X, via some mathematical transfer function.

Multivariate calibration then means determining how to use many measured variables $x_1, x_2, \ldots, x_K$ simultaneously for quantifying some target variable y. For instance, the X-variables could be chromatographic or spectroscopic measurements, and the target variable could be analyte concentration.

1.1.2 AN EXTREME CALIBRATION PROBLEM THAT CANNOT BE SOLVED BY TRADITIONAL METHODS

The following spectroscopic illustration summarizes the content of this book by demonstrating some advantages of multivariate calibration in a real, although somewhat exaggerated, example from chemical analysis.

Assume that you want to monitor the concentration of a chemical constituent in a complex liquid industrial process stream by high-speed light absorption spectroscopy (Figure 1.1a). The analyte in the example is actually the old litmus pH-indicator.

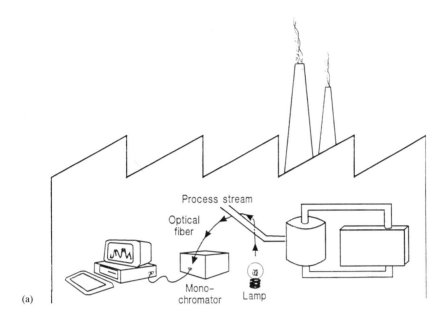

Figure 1.1 An example of selectivity problem: Spectroscopic quantification of litmus at unknown pH and unknown turbidity without sample preparation: a) Application potential: Remote high-speed analysis of a complex liquid process stream by fiber optic spectroscopy to determine the concentration of an analyte

Now you may have serious analytical problems:

The way that the analyte actually absorbs light in situ in the complex samples may be different from the spectrum of the analyte in pure form, if the constituent interacts with the solvent and with other constituents in the samples (analogy: the NIR spectrum of H_2O in wheat flour is different from that of pure H_2O.) So calibrating for the analyte in isolated, purified model systems may be of little use; it will have to be done empirically on realistic samples from the actual process.

But there may be other problems too: The analyte may not be stable and/or homogenous, and the measurements may be contaminated by interferences.

First of all, let us assume that there are natural pH variations in the process, and the in situ absorbance spectrum of the analyte changes with this pH variation (which of course litmus does).

Secondly, there are varying levels of particulate material in the liquid samples, causing strong turbidity changes in the samples to be analyzed (in this illustration unknown levels of ZnO powder were added).

And thirdly, there may be spectral interferences from other, more or less unidentified constituents and instrument variations.

Since your purpose for determining the analyte is on-line industrial process control, you have no time for cleaning and standardizing samples in the laboratory prior to the light absorption measurements. You may choose to measure the diffuse light transmittance T directly through the more or less turbid liquids, since these are fast, simple and precise measurements. Figure 1.1b shows the apparent absorbance spectra, $\log(1/T)$ of a set of such samples with varying analyte concentrations, varying pH and varying light scattering levels.

Figure 1.1c shows the 'best univariate calibration line' for concentration by traditional calibration procedures. The 'best' wavelength is the isospestic point of the constituent, (520 nm) marked by a vertical arrow in Figure 1.1b). The crosses represent the data of the 10 calibration samples used in determining this univariate calibration line; they show a rather unsatisfactory relationship between concentration of the analyte and the absorbance at its 'best' wavelength.

Under traditional circumstances chemists would conclude that these high-speed diffuse absorption data are unsatisfactory for determining the analyte in the given type of samples. This is illustrated for 'unknown' samples (A,B,C,D and E). The solid arrow marked A shows how the analyte concentration is subsequently predicted from the absorbance reading, via this best univariate calibration line. The concentration predictions are quite erroneous when compared to their correct values, as expected.

Figure 1.1d, in contrast, shows the corresponding prediction results obtained from the same data, but with multivariate calibration: The apparent absorbances from a number of different wavelengths were now combined in a statistical calibration model using 10 calibration samples as a training set. In this case the three normal samples' concentrations of the analyte were then correctly predicted, irrespective of level of pH and light scattering. This illustrates how multivariate calibration can greatly enhance the selectivity of analytical measurements.

The two letters D and E represent measurements under two (unknown to us) abnormal conditions: In one case the spectrophotometer instrument was not working properly and needed maintainance; in the other an unforeseen chemical constituent was present and interfered with the spectral reading. In both cases the obtained predicted analyte concentrations happened to come in a reasonable range, and might have passed unnoticed in spite of the gross errors involved.

But the multivariate 'disharmony analysis' identified both abnormal observations as outliers, and an error warning was automatically given by the software. Upon this prompt we compare the residual 'disharmony' spectrum of this outlier to the expected noise levels in normal samples, and get a quick indication of what the problems seem to be (Figure 1.1e). This illustrates how multivariate calibration can make quantitative analysis safer.

With multivariate calibration it is possible to build somewhat 'intelligent' analytical instruments that give quantitative, reliable determination of valuable information from high-speed, but highly non-selective input data. This can drastically simplify the analysis, since sample preparation can be minimized. At the same time it improves the reliability by adding automatic outlier checks.

This book ends in Chapter 8 with an in-depth treatment of this mentioned litmus example, using the calibration methodology treated in this book.

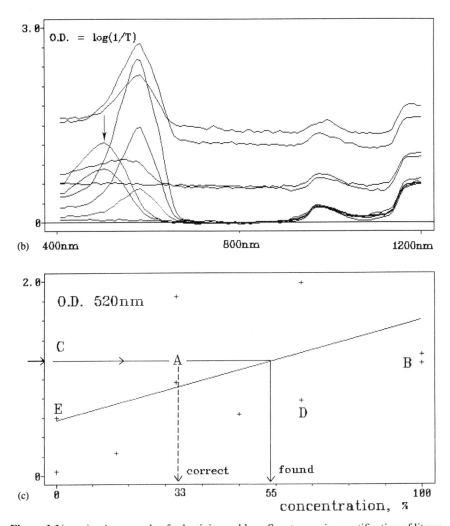

Figure 1.1(*cont.*)   An example of selectivity problem: Spectroscopic quantification of litmus at unknown pH and unknown turbidity without sample preparation: b) The spectra of optical density O.D. = log (1/diffuse transmittance) in the visible and near-infrared wavelength range (recorded at 100 wavelength channels), for 10 samples with varying litmus concentration, varying pH and varying turbidity. c) Univariate calibration: O.D. at 520 nm (the isospestic and hence 'best' wavelength) vs actual analyte concentration (measured by a slow off-line control method, in % of a reference solution containing 1.25 mg/ml litmus). Crosses: The 10 available calibration samples. Letters: Three unknown normal samples, A, B and C, and two unknown abnormal samples, D and E. The arrows show the prediction for sample A from measured signal at 520 nm (O.D. at $x_{\text{channel }15}$ = 1.2) via the univariate calibration line to predicted analyte concentration Y (55%). The true analyte concentration in A was 33%

6

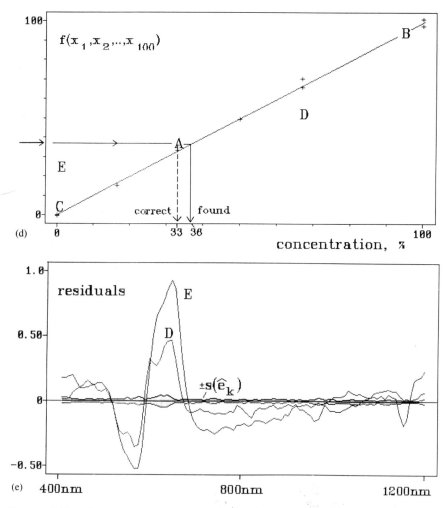

(d)

(e)

Figure 1.1(*cont.*) An example of selectivity problem: Spectroscopic quantification of litmus at unknown pH and unknown turbidity without sample preparation: d) Multivariate calibration: O.D. at 100 channels ($\mathbf{X} = (x_1, x_2, \ldots, x_k, \ldots x_{100})$) in the range 400–1200 nm combined mathematically, vs actual analyte concentration. Crosses: The 10 available calibration samples. Letters: Three unknown normal samples, A, B and C, and two unknown abnormal samples, D and E (which were now automatically detected as outliers from their O.D. spectra, see Figure 1.1e). e) Residual O.D. spectrum from the multivariate prediction of the two abnormal objects (D: an unexpected interferent, E: an unexpected instrument drift). The residuals are compared to the expected range of residuals for normal objects obtained from the calibration samples (shaded region). For a definition of $s(\widehat{\mathbf{e}}_k)$, see Chapter 5

1.1.3 ANALYSING INTACT SAMPLES

The power of multivariate calibration is fully documented by the success of diffuse NIR instruments (see e.g. Wetzel, 1983; Stark et al., 1986; Osborne and Fearn, 1986; and Williams and Norris, 1987). The NIR spectral region (700–3000 nm) has been referred to as 'nature's garbage dump', since so many chemical compounds show overlapping, broad spectra here. To make things worse, diffuse light spectroscopy of powders and slurries cannot be expected to follow traditional spectroscopic models like 'Beer's Law'. So quantitative analysis based on these non-selective spectra would have been impossible without multivariate data analysis. But with proper calibration, NIR spectroscopy is now rapidly replacing conventional chemical analytical techniques such as the Kjeldahl nitrogen method.

The sample preparation can be reduced to a minimum. In Figure 1.2a is given an illustration of results from NIR determinations of protein content in whole wheat kernels. The straight line illustrates where 100% perfect predictions should fall. The points corresponding to independent test samples are quite close to this line ($r = 0.97$, see section 2.1.2.2). The spectral measurements were in this example done in a Tecator Infratec instrument and the multivariate calibration was based on transmitted light intensity at a subset of 100 different wavelengths in the lower NIR wavelength region. For more details on the same example we refer to section 3.5.3.3. For comparison, a plot of $\log(1/T)$ 'the best' NIR wavelength versus protein concentration is given in Figure 1.2b

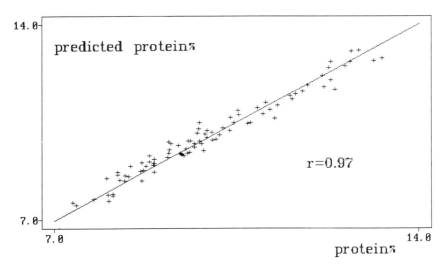

Figure 1.2 a) Plot of NIR determinations versus Kjeldahl measurements of protein% in whole wheat kernels. The multivariate calibration was done by PLS regression (see section 3.5). The plot shows both calibration and test samples. The prediction error (see section 4.3.1) for the test samples was equal to 0.31%. The correlation coefficient r (section 2.1.2) between y and $\hat{y}$ for the same test samples samples was equal to 0.97. For the calibration samples the correlation was 0.98

8

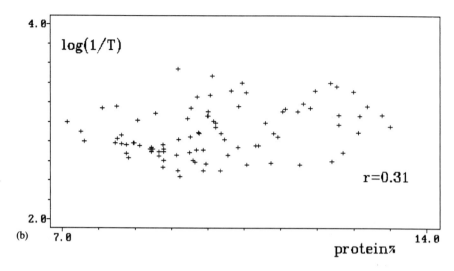

Figure 1.2(*cont.*) b) Plot of log(1/*T*) for 'the best' single NIR wavelength versus protein concentration for calibration and prediction samples. The correlation for the test samples was equal to 0.31. (The wavelength was selected as one with large regression coefficient value according to Figure 3.11t)

A pharmaceutical analogue to this example is the NIR analysis of intact tablets (Lodder and Hieftje, 1988). While NIR spectroscopy is the field where multivariate calibration has received the most attention till now, its applicability has been repeatedly demonstrated for other types of spectra, like IR (see Haaland and Thomas, 1988a,b; Nyden et al., 1988) and UV/vis fluorescence (Pedersen and Martens, 1989).

1.1.4 A NANO-GRAM EXAMPLE: CHROMATOGRAPHIC DIOXIN ANALYSIS

Many different measurement techniques can likewise benefit from multivariate calibration—other types of light spectroscopy, NMR, MS, accoustic spectrometry, image analysis, electrophoresis and chromatography to mention a few. The technique is applicable to macro-components (like protein in wheat in the example above) or micro-constituents (like environmental pollutants).

Here is one example of the latter: Dioxins are poisonous even at the extremely low levels present in smoke from municipal waste combustion and the dioxin determination is difficult at these low levels (Öberg and Bergström, 1987). Figure 1.3 shows how routine dioxin determination in flue-gas (smoke) from municipal waste combustion apparently can be simplified by multivariate calibration against a combination of more easily determined variables generated by the same combustion processes: The data points represent 33 flue gas samples from 11 different plants

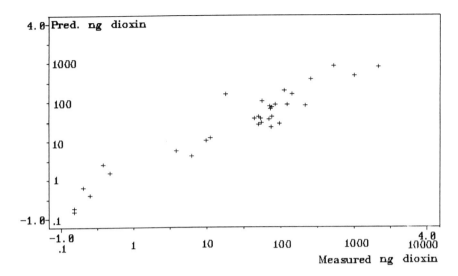

Figure 1.3 Multivariate calibration of an environmental toxin at the nanogram level: Abscissa: Total dioxin concentration, expressed as nm TCDD-equivalents per sample, determined by a complicated HPLC-GC-MS procedure in 33 samples from 11 municipal waste combustion factories. Ordinate: Total dioxin concentration correspondingly predicted as a function of 19 more easily measured chlorinated benzene and phenol isomers. The multivariate calibration function was determined by PLS (section 3.5) regression, using 2 PLS factors. The data, expressed on a logarithmic scale (-1 to 4), are taken from Öberg and Bergström (1988). The numbers 0.1 to 10 000 indicate concentration in ng

representing a variety of combustion conditions and flue gas cleaning systems (Öberg and Bergström, 1988).

The figure shows the relationship between Y = dioxin concentration (ng per sample) vs. a function of X = 19 different isomers of chlorinated benzenes and phenols. The Y-variable was obtained in the following way: Halogenated dioxins and dibenzofurans were determined in smoke extracts after HPLC clean-up, by high-resolution gas chromatography with selected-ion monitoring mass spectrometry, and summarized as ng tetrachlorinated dibenzo-p-dioxin (TCDD) equivalents per sample. The 19 X-variables in the smoke extracts are present at considerably higher concentrations and hence easier to determine; they were determined by a high-resolution gas chromatography, without any intermediate HPLC clean-up step.

The data were expressed in logarithmic form to calibrate over a wide enough range, and the calibration was done by PLS regression (section 3.5). The figure shows the results obtained for the calibration samples, but similar predictive ability was demonstrated using an independent test set.

Notice that the analyte Y in this example is not a single constituent—it is a summary of a group of similar chemical constituents. But similar results were obtained for the individual dioxin and dibenzofuran isomers.

So it appears that the important but very difficult determination of dioxins in waste combustion flue-gas can be replaced by simpler analyses: Other, more prevalent smoke constituents can be combined to predict the dioxins.

But this of course requires that the underlying processes that apparently generate both the dioxins Y and the other smoke constituents X remain the same in future waste incinerators as well.

Multivariate calibration has proven useful for many types of chromatographic and electrophoretic data, both for resolving partially overlapping peaks and for predicting more abstract qualities like sensory sweetness from combinations of chromatographic peaks.

1.1.5 RELEVANT MEASUREMENTS: RELIABLE PREDICTION OF NEEDED INFORMATION AT THE RIGHT TIME AT ACCEPTABLE COST

What makes measurements worth while? There must be a real need for the information that you try to obtain; it must be relevant for the given purpose. And this purpose is defined by the actual user of the information, not only by the specialist doing the measurements. The traditional barriers between scientific disciplines, each with its limited focus, therefore represent a problem. For instance:

* The detailed performance characteristics of electronic components are important to instrument makers, but they are of minor concern to the chemist who wants information about chemical concentrations.

* The identification and exact concentrations of every individual chemical constituents may be neither necessary nor sufficient for a product developer optimizing the flavour of a product. But the analytical chemist may think so.

Next, in order to make measurements worth while, they should provide the needed information with sufficient reliability in short enough time and at an acceptable cost. Like any other production process, production of analytical results ought to be optimized with respect to a conscious criterion that strikes a compromise between the desired and the attainable.

Within certain limits, analytical speed is often more important than precision, which again can be more important than accuracy. For instance, for monitoring industrial processes it is much more valuable to get imprecise measurements after a few seconds than to get highly accurate measurements the next day. In traditional academia the order of priority is often the reverse.

The expenses of analytical instruments can often be quickly paid back by the increased insight and quality control that they provide. But the choice of instrumentation and laboratory procedures must be geared to the analytical speed, sensitivity and precisions actually needed; analytical over-kill is a waste.

Sufficient methodology, software and application experience is now available for increased use of quantitative chemometrics in chemical analysis.

1.1.6 A WORLD UNDER INDIRECT OBSERVATION
REQUIRES CALIBRATION

We can seldom measure directly what we want to determine, and have to rely on indirect observation: We have to 'look where the light is'—to measure what we can measure, x. This means that we have to work upon limited rationality.

But complete knowledge about a system is not always necessary. What is necessary, however, is that we use what we already know, consciously seek new information, and are open for surprises. Then we can develop models—in the mind, and on paper or in the computer. With these models we do useful quantitative analyses and we learn as we go. So our models, our instruments and our problem formulations improve with time.

We often meet the dilemma that the more directly relevant a measurement is, the slower and/or less precise it is itself (Figure 1.4). The more complex the desired information is, the longer is often the step from measurement to information.

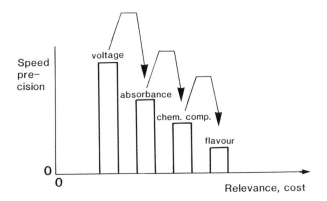

Figure 1.4 The dilemma of analytical instrumentation, and how it can be solved by calibration (arrows)

But the wanted information Y can be derived from faster or more precise measurements, X. For quantitative purposes this must be done by some mathematical transformation. Calibration concerns how to determine this empirical formula or function, $f(X)$ for a given type of variables (X,Y) and a given type of samples.

There will usually be measurement errors in our data, and our choice of transformation $f(\ )$ will usually be oversimplified, so

$$Y \approx f(X) \tag{1.1}$$

When we later predict information Y from measurements X, we have to use a concise relationship

$$\hat{Y} = f(X) \tag{1.2}$$

12

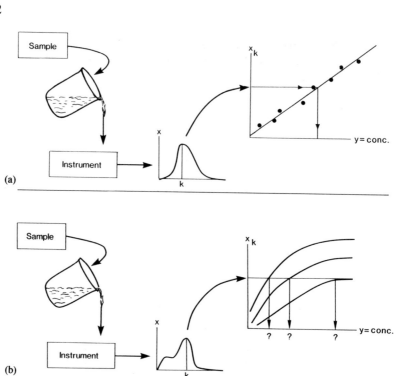

(a)

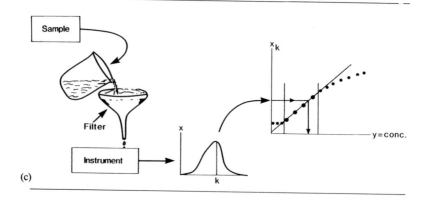

(b)

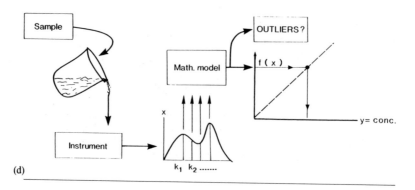

(c)

(d)

where the symbol $\widehat{Y}$ means 'predicted' or 'estimated' values of Y, and is pronounced 'hat', so $\widehat{Y}$ is called 'Y hat'. So we must bear in mind that the obtained values $\widehat{Y}$ are only approximations of the true unknown values, Y:

$$\widehat{Y} \approx Y \tag{1.3}$$

1.2 SELECTIVITY PROBLEMS

SUMMARY The problem of interference or lack of selectivity in analytical data is dicussed. Distinctions are made between interferences from other chemical constituents, from physical phenomena and the measurement itself. The need for *multivariate* calibration is stressed. The analogy between analytical interference and the need for multivariate calibration on one hand and listening to music on the other hand is stressed.

1.2.1 PROBLEMS: INTERFERENCE AND MISTAKES

In chemical analysis it is often difficult to obtain ideal measurements (Figure 1.5a), i.e. measurements that are selective for just the constituents that we want to determine. In addition to the usual more or less random measurment noise, the data may be affected by chemical and physical inteferences due to phenomena in the samples themselves, as well as by experimental interferences arising in the measurement process. Non-linearities often create additional problems: The instrument seldom responds linearly to changes in the constuent concentrations and to changes in the levels of interferents (Figure 1.5b).

Traditionally, interferences had to be removed physically to ensure selectivity, and in order to ensure linearity, only a narrow range of the instrument scales could be used (Figure 1.5c).

But in some cases this may be prohibitively expensive or physically impossible. And sometimes one is interested in quantifying several of the mutually interfering constituents, making the removal of interesting constituents illogical.

With multivariate calibration, interferences and individual non-linearities represent less problems. They are most easily dealt with if their influence on the

Figure 1.5 The ideal chemical measurement and the real world: a) A sample is analyzed in a high-precision instrument, producing a selective measurement that is linearly related to the concentration of the analyte. b) The instrumental measurement is not selective for the analyte, and the instrument response is non-linear. c) Traditional selectivity enhancement: Cleaning, standardizing and diluting each sample (symbolized by a filtering stage) prior to single-channel measurement x. The calibration is limited to the 'linear range'. Hopefully the data from new samples contain no unexpected trouble. d) Selectivity enhancement by multivariate calibration: The cleaning, standardizing and diluting is more or less replaced by mathematical modelling of multichannel measurements $x_1, x_2, x_3 \ldots$. This process removes interference effects, extends the linear range of the calibration and allows automatic outlier detection if new samples contain unexpected trouble

instrument response is known in advance. but with indirect multivariate calibration this is not necessary. Their effects are instead modelled mathematically, using data from representative calibration samples with sufficient variability (Figure 1.5d).

In some texts, e.g. Massart et al. (1988), there is made a distinction between so-called matrix effects and interferences. The matrix effect is defined as an effect changing the sensitivity of the sensors while an interference is an effect contributing to the response, but without influencing the sensitivity. In this book, however, both problem types are treated as interferences; purely multiplicative matrix effects etc. are addressed explicitly with regard to preprocessing (Chapter 7). The word *interference* here stands for any systematic effect on the spectrum caused by either chemical or physical phenomena.

1.2.1.1 Chemical interference in samples

Other constituents with overlapping spectra

'Chemical interferences' is here used for describing systematic errors in the quantitative determination of a certain analyte when these errors are caused by other chemical constituents or by chemically induced variations in the analyte's own instrument response.

Most chemical samples, at least those of biological type, are mixtures consisting of several different chemical constituents. Many of the constituents may be sufficiently similar to affect the same measurement type, although in somewhat diffferent ways. Therefore, it can be difficult to find simple measurements that are sufficiently selective for the analyte.

This is illustrated in Figures 1.6–1.10. Assume that we were interested in determining the concentration of a certain component in samples from an industrial extraction process (in the illustrating example: the blue-coloured compound litmus in aqueous solutions at pH 10), from visual-range transmission spectroscopy. Figure 1.6a gives the absorbance or Optical Density (O.D.) spectrum of the analyte under these conditions at various concentrations. The spectra were measured at 100 wavelength channels at 8 nm intervals in the 400–1200nm range through a single-strand two-way optical fiber, using a Guided Wave model 200 instrument. The original transmission (T) spectra have been linearized by the conventional O.D. = $\log(1/T)$.

In addition to the analyte absorbance peak, the smaller, but characteristic NIR absorbance peaks of the solvent H_2O around 900 and 1200 nm are visible; this wavelength range is included in the example because it can shed light on subsequent interferents.

Figure 1.6b shows the absorbance at the absorbance peak, 576nm, vs. the concentration of the constituent. (In the illustration the analyte concentration is given in percent of some reference concentration, to emphasize that useful and valid calibrations can be attained even when tractability to some standard unit like mg/ml is impossible.)

For simple, selective systems this conventional type of calibration curve

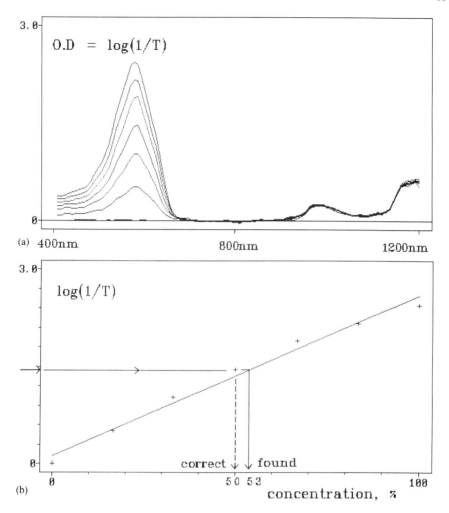

Figure 1.6 No interference problems: a) O.D. spectra (400–1200 nm) of the analyte (litmus) at various concentrations in pure aqueous solution at pH 10. b) Univariate calibration; O.D. at peak maximum, 576 nm (wavelength channel 22), vs analyte concentration (in percent of a reference concentration)

works well: The curve is obtained by plotting the absorbance of zero concentration and of one or more non-zero concentration curves against the concentration. Concentrations of future samples are determined (predicted) from their corresponding O.D. readings, e.g. O.D.=1.48 predicts 53 percent concentration of litmus (as opposed to the true value, 50 percent).

For this kind of well behaved data, univariate and multivariate calibration would give about the same prediction ability; the latter would probably give somewhat higher precision due to its averaging over many wavelengths. And of course the multivariate calibration gives outlier warnings, which is not the case for univariate calibration (Figure 1.1).

16

But what if an interferent were present in the samples? Figure 1.7a repeats some absorbance spectra of the analyte litmus together with some spectra for an unexpected chemical interferent with similar instrument responses (the pH-indicator Bromcresol Green in aqueous solutions at pH 10). The solvent peaks at 900 and 1200 nm are on the other hand unaffected by the variations in dissolved material.

As Figure 1.7b shows, the simple calibration curve based on the same single wavelength (576 nm) now gives quite erroneous analyte predictions for samples containing the unexpected interferents. For clear illustration, the true analyte concentration is zero for the samples containing the interferent. Similar selectivity

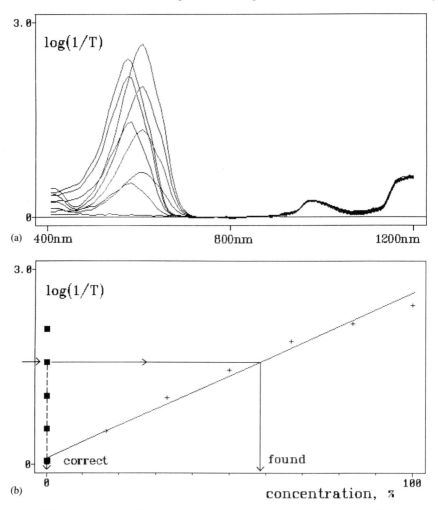

Figure 1.7 Chemical interference from other constituents: a) O.D. spectra of the samples with various analyte (litmus) concentrations and with various levels of an 'unidentified' interferent (actually pure solutions of Bromcresol Green) at pH 10. b) O.D. at 576 nm vs. analyte concentration (in percent). Crosses: the data from Figure 1.6. Squares: samples with interferent

errors would occur if the interferent had been present in mixtures with the analyte.

There is no other wavelength at which the litmus can be selectively determined with sufficiently strong signal. Thus, traditionally one would have to remove such interfering constituents prior to the spectral measurements, e.g. by some chromatographic cleaning step. But what if we do not have time for that? To get the desired selectivity in the results, we would then have to correct for the chemical interferences mathematically.

Multivariate calibration can easily handle chemical interferents like that in Figure 1.7, even if they are unidentified, with unknown spectra.

Variations in the analyte's spectrum

Figure 1.8 gives another example of chemical interference: The light absorbance spectrum of the analyte litmus in the visible range changes with the molecular state of the constituent—which is a function of pH. The figure gives the spectra for various concentrations of litmus at pH 4, 10 and some intermediate pH levels.

The figure shows that litmus has lower signal in its red-coloured state (pH 4) than in its blue-coloured state (pH 10). Figure 1.8b demonstrates the damaging effect this has for the univariate calibration at 576 nm.

Now, 576 nm is the absorbance maximum of blue litmus, and this of course is not optimal for dealing with the pH effect. Are there other wavelengths more suitable? In the range between 625 and 650 nm only the high-pH state absorbs, but reading absorbance at such a steep slope of a peak would make the calibration extremely sensitive to minor wavelength inaccuracies in the monochromator.

Since only two molecular states are evident in these data, it may be possible to find isosbestic wavelengths (e.g. 520 nm = x_{15}) where the two states of the analyte give the same signal. For simple, pure two-state systems this would give useful univariate calibrations. But as demonstrated previously in Figure 1.1c, it cannot accommodate more than one type of interference. For more complex problems it is of little help.

Traditionally one would have to standardize the sample condition, e.g. adjust the pH of every sample to some common level, prior to one-channel spectroscopy. With multivariate calibration, such systematic variation in the analyte signal is easily compensated for directly, so there is no need for extra sample preparation.

1.2.1.2 Physical interference in samples

By 'physical interferences' in the samples is meant systematic errors in the quantitative determination of a chemical constituent caused by physical effects rather than just the effect of other chemical constituents with similar instrument response.

Irrelevant physical phenomena in the samples can affect the measured signal strongly. To get the desired selectivity one must either keep such effects constant or compensate for them mathematically.

18

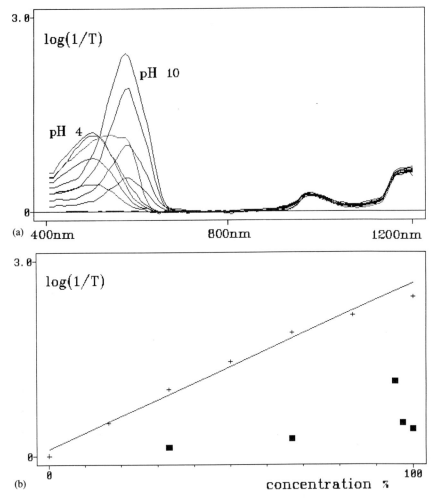

Figure 1.8 Chemical interference from a changing analyte: a) O.D. spectra of pure samples with different analyte concentration (litmus), and with different molecular states of the analyte, ranging from pH 4, (protonized, red colour), via intermediates to 10 (unprotonized, blue colour). b) O.D. at 576 nm vs analyte concentration (in percent). Crosses: the data from Figure 1.6. Squares: samples with varying pH

Figure 1.9 illustrates one common type of physical interference due to the samples, namely that of light scattering in light absorption spectroscopy: Figure 1.9a shows the spectra of apparent absorbance of the analyte litmus at various concentrations at pH 10 in the presence of various levels of turbidity (caused by three different levels of white ZnO powder dispersed in the aqueous solutions). Like the previous spectra in this example, these spectra were obtained by fiber-optic measurements of the now more or less opaque solutions.

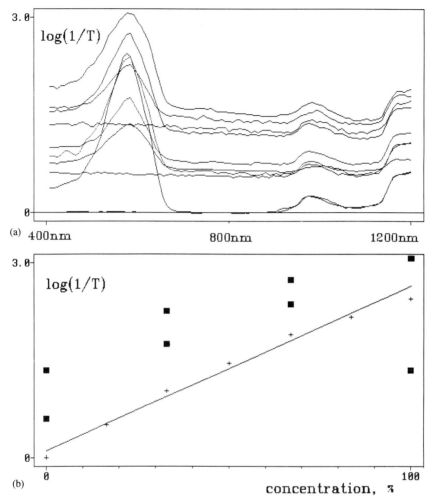

Figure 1.9 Physical interference from the sample: a) O.D. spectra of samples with equal analyte (litmus) concentrations and equal molecular state (pH 10) in aqueous solutions with different levels of light scattering (various amounts of white ZnO powder added). b) O.D. at 576 nm vs analyte concentration (in percent). Crosses: the data from Figure 1.6. Squares: samples with varying light scattering

The figure shows that whole absorbance spectra—including both the analyte peak, the baseline and the NIR solvent peaks, is greatly affected by the light scattering.

As Figure 1.9b illustrates, this has grave consequences for the univariate calibration of litmus. One could of course correct some of this by first subtracting the baseline offset in each spectrum in a way that would be a primitive multivariate calibration. But the light scattering effect is not strictly additive—as can be seen from some of the solvent peak curves, so it does not fully solve the problem.

Traditionally, turbidity would normally be removed by filtration or centrifugation prior to the one-channel measurement. With multivariate calibration the physical interferents can instead be corrected for mathematically, using e.g. the baseline and solvent peaks or multiplicative signal correction. So there is no need for sample purification prior to the measurements.

An example of a second type of physical interference due to the sample is the well known effect of temperature on the water absorption peaks in the near-infrared range (e.g. near 1940nm). Unless the temperature is the same for all samples, or the effect of temperature is somehow compensated for, the prediction of chemical composition (proteins, lipids, carbohydrates, water content, etc.) in foods and feeds from NIR reflectance spectra will be imprecise. In multivariate calibration it is possible to correct for temperature effects, either by using sample temperature as an extra predicting variable or by letting the calibration program model its effect as yet another 'unidentified interference harmony' from the spectra themselves.

1.2.1.3 Interferences from the measurement process itself

The interference from the analytical procedure itself concerns systematic errors due to the way the measurement is made. Variations in the way samples are prepared may introduce serious selectivity problems. And very few instruments respond solely to the quality of the samples analyzed.

For instance, in diffuse reflectance analysis of powdered samples, the grinding process prior to measurement may cause large particle size variations whose spectral effect is similar to the light scattering effect seen in Figure 1.9. In chromatrography, variations in temperature or pressure may change the patterns of overlap between peaks, and these effects may be amplified by an uncritical use of automatic peak integrators. Temperature variations during measurements can introduce systematic errors in many types of instrumentation, for instance in spectroscopy.

A rather extreme example of this is shown in Figure 1.10, where for illustration some litmus samples at pH 10 has been measured with an intentionally more or less maladjusted monochromator, simulating e.g. operator mistakes. The expected wavelength shifts are seen in Figure 1.10a. But when only one wavelength is used, the reading mistakenly looks like a normal reading at lower analyte concentrations. Such gross instrument errors should of course be avoided. But minor instrument variations can never be totally avoided, at least not without paying a price. With multivariate calibration it is possible to correct for at least minor instrument variations like these.

1.2.1.4 Mistakes and unexpected phenomena

There are a number of other error sources that cannot be foreseen at the time of the original calibration. Gradual or abrupt instrument drift over time is one such example. For instance, changing hardware components in the instruments, such as spectrophotometer lamps, can lead to effects that cannot be removed completely by standardizing the instrument at zero and 100% reading, e.g. due to nonlinear detector response curves. In complicated cases univariate calibration cannot correct

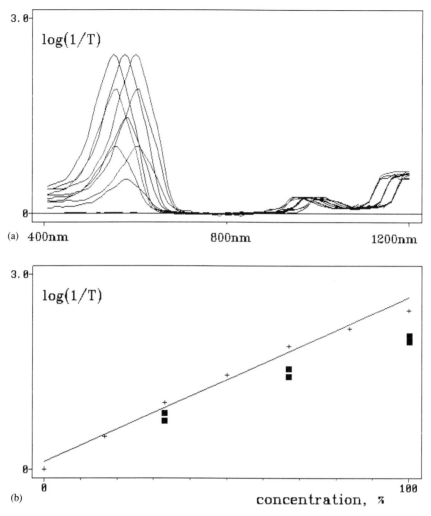

Figure 1.10 Physical interference from the measurement itself: a) O.D. spectra of pure samples with equal analyte (litmus) concentrations and equal molecular state (pH 10), but recorded with various degree of maladjustment of the monochromator. b) O.D. at 576 nm vs analyte concentration (in percent). Crosses: the data from Figure 1.6. Squares: samples with maladjusted monochromator

for such effects, but by multivariate calibration it is possible to build a 'bridge' across such discontinuities.

No matter how conscientious the analyst is, there is always a probability of unexpected errors, either due to an unexpected interferent, or due to unnoticed instrument trouble.

Problems of human mistakes and even fraud must also be taken into consideration, especially in routine analysis, where each sample cannot be given too much attention. Judging from many analysts' general over-optimistic assessment of their own accuracy compared to inter-laboratory round-robin tests, these and other errors can pass unnoticed, when univariate calibration is used.

With multivariate calibration many types of sample anomalies or instrument problems can automatically be detected as outliers. Human mistakes and fraud can also be revealed by these methods, although some people's stupidity and/or smartness can probably fool any mathematical tool.

1.2.2 CALIBRATING THE BODY'S SENSES

Rhythm and blues in the music from our instruments

Quantitative analysis with analytical instruments resembles the way we use our senses—our ears, eyes, mouth, nose, etc. Multivariate calibration is like learning to interpret our many sensory signals. Multivariate prediction is like using what we have learnt. This human analogy may help remove the mathematical mystique for some readers of this book. Let us first consider our multi-frequency instrument for mechanical vibration—our ears.

Analytical instruments are like music instruments: They must be tuned properly. And at a higher level: Learning to convert the data from analytical instruments into information is like learning to appreciate music.

Before every concert it is necessary to tune the musical instruments in an orchestra by standardization to the same scale (see section 1.1.1). For the gifted ear it is important that the orchestra is tuned to play in exactly the key prescribed by the composer. For the rest of us it is sufficient that the instruments are in tune with each other and stay that way all through the concert. Whether their A is played at 440 Hz or 442 Hz is of little concern to us. Absolute standardization is not always equally important.

Likewise in the laboratory: The absolute accuracy of the readings is often less essential than their precision and reproduceability.

In the laboratory we have the equivalent of single-string instruments (a pH-meter), multistring instruments (multi-channel, scanning spectrometers and chromatographs), and even whole orchestras of analytical instruments.

The simplest analytical instruments have only one channel, comparable to a single-string instrument. It is possible to play nicely on such primitive instruments, because our mind has a memory and can combine a sequence of sounds into patterns.

In all analytical work the data from the sequence of samples can be regarded as their melody—a sequence of sounds in a certain range and with a certain theme. The rate and pattern in which the samples are analyzed is their rhythm. The results from each sample are thus perceived in the context of the earlier results. From unexpected jumps in the melody or rhythms, abnormalities in samples can sometimes be detected.

Because we have this contextual memory, even the simple melody from a one-channel instrument does create a perception of music intended by a composer, instead of just a series of individual disconnected sounds.

There will always be a place for the elegance of the simple solo instrument. But new dimensions are added to our perception when two or more channels are played together, and here is where multivariate instrumentation comes in: Our mind combines the simultaneous sounds into harmonies and disharmonies (relative calibration, see section 1.1.1). Thus the twelve-string guitar has a richer sound than the four-string ukulele, and the symphony of an orchestra is yet richer.

In the chemical laboratory different analytical instruments ranging from the equivalent of a stretched rubber-band to a grand piano can be used, either solo or together as an 'orchestra'.

The simultaneous measurements on each sample are modelled in terms of harmonies and disharmonies. Certain mathematical relationships between sounds that we have learnt to appreciate we call harmonies, while unexpected, unfamiliar combinations are termed disharmony or dissonance. The harmonies describe the systematic interrelationships in the data; they may be well known in advance, or they may be established from new data just like new 'sounds' are created in music. After the permissible harmonies have been defined, the presence of additional sound combinations are perceived as disharmony, which indicates some type of anomaly.

Without multivariate data analytic tools, the scientist will perceive the stream of data from the multi-channel instruments as annoying cacophony rather than sweet music. Consequently, they will play fancy new analytical instruments the good old way—one string at a time. For instance, they may choose to use only one of the available wavelengths or chromatographic peaks, ignoring the rest in order to avoid getting flooded with data. This forces the user to ignore a lot of potentially important information.

Colour vision

Let us now consider the analogy of multi-wavelength spectrophotometers and our own eyes. Figure 1.11 illustrates the processes behind colour vision, according to the normally accepted general model (see e.g. Hunter, 1975; Valberg and Seim, 1983). When we look at a red cloth in sunshine, we apparently do much the same in our head as we would do with data from a multi-channel spectrophotometer.

When the sunlight hits the cloth, it penetrates a short way into the material before it is scattered back to the surface by the material's refractive inhomogeneity. While inside the cloth, part of the sunlight will be absorbed by the cloth pigments— primarily at non-red wavelengths, since this is a red cloth. The scattered light will then radiate diffusely from the cloth surface—some of it into our eyes and onto the retina where the photoreception takes place.

For people with normal colour vision the eye's retina has three different pigmented light-sensitive cones. These three nonlinear detectors have different sensitivities in the 380–760 nm visible range.

Thus in the eye the light intensity spectrum from the cloth is compressed into intensities of these three nerve signals. In the brain's 'computer' these three 'latent variables' are then combined into perceived colour. From this perceived colour

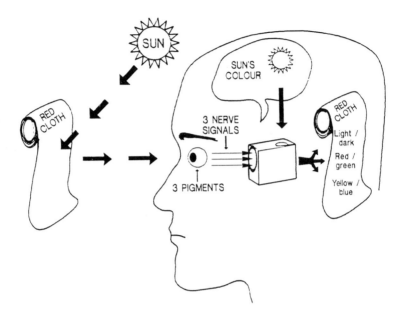

Figure 1.11 Colour vision: Compressing light at many different wavelengths via three nerve signals into a three-dimensional colour space of light/dark, red/green, yellow/blue, while correcting for the 'lamp's' spectrum

intensity we can for instance assess the concentration of red dye in the cloth, as well as higher level colour qualities like conformance to our preference.

For people with normal colour vision, 'colour' is perceived in three dimensions. This three-dimensional colour 'space' can be expressed by different words combinations, for instance 'lightness', 'hue' and 'saturation'. An equivalent mental coordinate system is illustrated in Figure 1.11: A light/dark axis, a red/green axis and a yellow/blue axis, corresponding to different geometric coordinate systems.

The way the brain's 'computer' apparently transforms the three nerve signals into the three-dimensional colour space is quite complicated. Among other things, the spectral distribution of the light source strongly affects the sample's light spectrum. So in order to assess the cloth's colour the brain automatically takes the colour of the sunlight into account in the transformation. This compensation is analogous to the way variations in the lamp intensity I_o are ordinarily corrected for in spectrophotometry: The sample's light intensity I is converted into transmission (T) or reflectance (R) by $T = I/I_o$ or $R = I/I_o$.

In general, the conversion of light intensity at many wavelengths via the three nerve signals into meaningful colour information is quite parallel to the main approach to multivariate mathematical modelling in this book: *The compression of input spectra via a few latent variables into the desired analyte concentrations and/or other, more abstract qualities.* The main difference is how the two calibration systems are established: In our colour vision system most of the calibration is defined by our genes, possibly with some early childhood learning added. But in order to use a new analytical instrument on complex samples, a mathematical calibration model must first be established, based on empirical data and theoretical considerations.

But there is another important differences too: Information that is lost in the eye's colour vision process is retained in multivariate spectrophotometry calibration and can be used for *automatic error warnings*. Cloths that contain quite different types of dye, and hence reflect light with intensity spectra of quite different shapes, can appear identical to the eye. This is because the eye's cone pigments only provide information integrated over the whole visible wavelength range. Thus if two different cloth dyes happen to trigger the three cone pigments in the same way, they will appear to have the same colour, at least under a given light source.

In contrast, when the data from a spectrophotometer are compressed mathematically into a few latent variables or 'main harmonies', each input spectrum is at the same time checked for possible 'disharmony'—abnormal lack-of-fit to the calibration model. Thus if we have calibrated to quantify a certain red dye A from measured reflectance spectra of cloths, then a cloth sample containing a red dye B with different absorbance spectrum will sound the alarm.

The eye can do more than colour assessments of homogenous materials. We can also see patterns and movements. Likewise, if we replace a spectrophotometer by a multi-channel video camera, multivariate calibration can help us determine spatial structures and dynamic changes with time.

The nose and tongue can likewise be compared to chromatographs or ion sensitive electrodes. In the same way that we use two eyes, two ears, a nose and a tongue, we can mathematically combine two or more multichannel instruments into very powerful analytical systems that can handle real world samples.

1.3   WHY STATISTICS?

SUMMARY   In order to predict information Y from measurements X, the prediction formula $\hat{Y} = f(X)$ must first be determined. A number of different multivariate calibration methods exist. For complex analytical problems with noise inherent in the system the parameters must be estimated statistically, based on realistic, empirical measurements from representative calibration samples.

1.3.1 CALIBRATION AND PREDICTION

Based on multivariate calibration, you can today predict the protein content in wheat grain from a few seconds' non-desctructive NIR spectroscopy (see e.g. section 1.1.3)—then you can eat or sow the grain, if you want. In contrast, the old Kjeldahl protein method required an accurate weighting and a slow and unpleasant chemical digestion, followed by critical distillation and spectroscopy stages.

But why should chemists and technologists worry about statistics when they calibrate their instruments? Because statistics has some useful concepts that are

26

often lacking in other sciences. With these concepts applied, the quality of analytical work can be improved.

Calibration in the wide sense concerns how to predict Y from X via some transfer function $f(\ )$:

$$\widehat{Y} = f(X) \tag{1.4}$$

Such techniques can be used for absolute calibration—tracing experimental readings back to some fixed standard, as well as for relative calibrations, converting measurements X to more useful information Y in general. Relative calibration can be used for determining chemical compositions as well as higher-order information such as toxicity.

But what is Y and what is X? In order to ensure clear conceptual 'grounding' of the statistical methods to be presented in this book, we now focus on calibration and

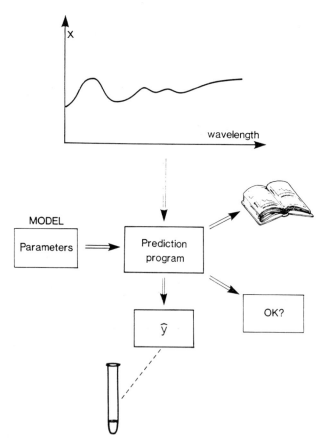

Figure 1.12 Prediction: converting instrument data X_i for sample i into predictions of more informative variables $\widehat{Y}_i$ using a previously established calibration model in a prediction computer program. With the prediction methods advocated in this book, the program can also provide outlier detection check (OK?) and new knowledge (symbolized by the book)

prediction in quantitative chemical analysis. In the following, the letter Y stands for concentrations to be determined and X stands for spectral measurements. Capital Y and X represent concentrations and instrument readings for a series of samples (matrices, see Section 2.1), lower-case letters y and x represent data from individual samples (vectors and scalars, see Section 2.1).

Prediction

Thus the general prediction equation can be written:

$$\widehat{Y} = f(X) \tag{1.5}$$

Figure 1.12 illustrates the basic operation in such prediction: X-data are read from an instrument, e.g. a scanning spectrophotometer or a chromatograph, and transformed into $\widehat{Y}$ by a prediction program via function $f(\ )$. This function has *some particular mathematical form* based on a *MODEL* formulation of the relation between Y and X.

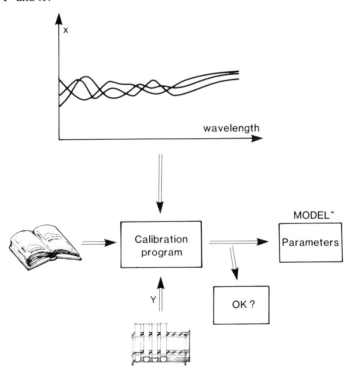

Figure 1.13 Calibration: Establishing a calibration model intended for later prediction of Y_i from X_i, by matching X- and Y- data from a set of calibration samples $i = 1,2,\ldots,I$ in some calibration program. Background knowledge (symbolized by the book) can be used as additional input information for experimental design, model formulation, a priori weights and estimation optimization. The modelling program provides internal consistency checks and new knowledge in addition to the final model

28

For instance, the calibration equation used in traditional univariate calibration is simply the linear predictor

$$\hat{y} = \hat{b}_0 + x\hat{b} \tag{1.6}$$

with mathematically determined constants $\hat{b}_0$ (offset) and $\hat{b}$ (slope). In conventional UV/vis. transmission spectroscopy x could be the absorbance at a certain wavelength, $b = 1/\epsilon$, (where ϵ = absorptivity) and $b_0 = 0$.

Calibration

We have to 'learn how to predict $\hat{Y}$ from X' before we can predict Y. In other words, the form of the mathematical function (or model), and the values of its parameters first have to be established or estimated. This is here called calibration, and needs statistical thinking.

Figure 1.13 shows how this usually requires data both from the instrument X and data Y. The calibration data Y may be known directly if the calibration samples represent artificial mixtures. But Y may also have been measured empirically by some reference procedure which is to be replaced by a more desirable method producing the X data.

1.3.2 WHY ESTIMATE?

In rare cases we know enough about the signal generation process to be able to set up a rather exact, complete mathematical description of the problems and thereby solve them. But in most cases our system knowledge is incomplete: We have to select some simplified mathematical model, and estimate its parameters empirically, in order to obtain an acceptable predictor $\hat{Y} = f(X)$.

Noise cannot be avoided

This estimated predictor model will be affected by various noise sources that always contaminate empirical measurements (Figure 1.14). Such noise, due to

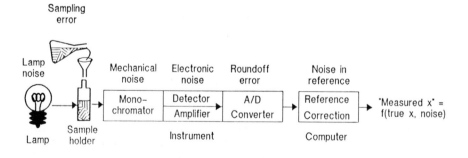

Figure 1.14 Sources of more or less random noise in quantitative instrumental analysis

uncontrollable effects, is usually considered to be 'random'. It is not truely random (as in the sense of sub-nuclear physics), because it is caused by something, but as in the flipping of coins or throwing of dice, these causal relationships are unknown to us. Even if we had known these initial states in great detail, their chaotic propagation through the measuring process would have made their resulting noise effect quite unpredictable. So 'random errors' is still a suitable description.

The noise concept can also be applied to the effect of uncontrollable sampling variability (Figure 1.15). In most analytical situations only sub-samples of material are actually being subjected to quantitative X and Y measurement. If this sub-sample happens to be quite peculiar compared to the bulk of the material which it should represent, this will also create a 'random measurement error' (see, for instance, Minkkinen (1987)).

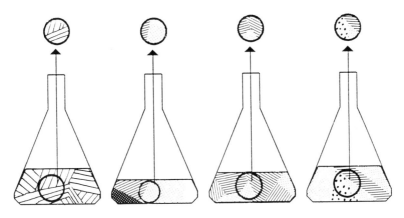

Figure 1.15   Random sampling noise

If several variables have been measured, the effect of apparent 'random errors' can often be greatly reduced by modelling systematic intercorrelation structures among the different variables. But there will always be an unexplainable *residual* left (see section 2.2). These uncertainties will affect the obtained calibration model, causing biases in its ability to predict Y from X in the future.

In order to understand how noise affects calibration and prediction, a minimum of statistical estimation theory is needed.

Models are just limited approximations

Models are usually not in themselves what the chemist or technologist wants information about. The world is under indirect observation and models are not true descriptions of the real world. Models are only simplified approximations of a selected part of the real world, seen from a particular point of view.

Therefore we have to choose which part of the real world we want to model for a given purpose. In order to do that, we must have an idea about how the world

30

varies, and we must have some concrete goal for our calibration modelling. Even for this purpose some statistical theory is required.

But statistics is not the favourite subject of most chemists and engineers. In the end of this first chapter we shall discuss why this is so, and indicate how this book attempts to overcome some of the inter-disciplinary barriers between chemistry and statistics.

1.4   STATISTICAL VS. CHEMICAL THINKING

The mind's limitations represent an obstacle in modern chemical analysis: Our instruments can create far more raw data than we can mentally digest. The human mind has an impressive storage and processing ability matched by no computer. Still, our mental modelling (see e.g. Johnson-Laird, 1983; Norman, 1983; Lakoff, 1987; Hampden-Turner, 1981) has clear limitations in certain areas where computers are superb.

Anyhow, the number of 'chunks' that the mind can handle rationally at one time appears to be rather limited. Therefore people—including chemists and statisticians—depend strongly on mental mechanisms for *simplification*, *ordering* and *explanation*.

Our need for simplification

Without selective filtering of the signals from our senses and selective suppression of our memories, our minds would be overwhelmed. But these simplifications sometimes create strange 'blind spots' in one's scientific field that are obvious when seen from other fields.

Consider for instance the inadequate handling of many modern laboratory data:

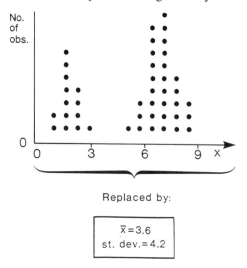

Figure 1.16   Summary statistics can hide the important information

Without statistical knowledge and multivariate software the chemist is forced to make subjective, incidental and suboptimal simplifications—like using only one wavelength from a scanning spectrophotometer. Most likely, the extra information lying in the other wavelengths in terms of potential selectivity enhancement and reliability checks will be in the blind spot—the chemist is not even aware that he is not aware of it.

Likewise in *bad* statistics—the important information in rather oddly distributed chemical data may be lost in the hands of some pocket calculator enthusiasts reducing data to meaningless average parameters, as illustrated in Figure 1.16.

Our need for order

Names make mental categories more easy to remember over time and more easy to communicate to others. So, empirical methods, limited deductions and practical simplifications receive names like 'Wurtz-Fittich reaction' in organic chemistry, 'Fisher's F-test for significance' in statistics and 'Beer's Law' in spectroscopy.

The naming makes classifications easier to remember. But over time such labels tend to stiffen into solid, categorical bricks in the foundation of our unquestioned 'established knowledge'. Sometimes the progress of science requires a new way of ordering our knowledge; the old categories and rules are too limiting. But the old order is often taken for granted, and that which threatens the old order tucked away in a blind spot.

For instance, all chemists know about the importance of interactions between various factors in chemical reactions. Still, many chemists design their experiments as if interactions did not exist: Following some unwritten ethical rule, one factor is varied at the time. In this way the experiment can be interpreted peacefully and in apparent compliance with well known 'laws' of elementary physical chemistry—but at the possible cost of critically reduced information. The fact that interactions are important are placed in the blind spot.

Even statisticians may have their blind spots. One of them concerns the use of established procedures like multiple linear regression (MLR). MLR and many other of today's standard statistical methods were originally developed way back before 1940 when it was expensive to make a new analytical technique, but rather cheap to ask a technician to go out and measure some more samples. Hence they often require the number of variables to be much lower than the number of samples, which is seldom the case in modern spectroscopy, chromatography etc.

Our need for 'causal' explanation

Without some type of mental pointer system between hierarchies of categories, our memory and thinking would be chaotic, unstructured. But these structuring mechanisms often appear as more or less incorrect *causal explanations*. This may sometimes lead scientists along 'blind alleys'.

Purely empirical relationships seem to be more foreign than apparently causal descriptions. It is harder to remember empirical results like 'variable X and Y

showed a linear correlation' than 'standard chemical knowledge' like 'According to Beer's Law the light absorbance is proportional to the concentration'. Beer's Law gives us both a quantitative description and a presumably causal explanation.

But Beer's Law is just an approximate model helpful for certain situations. Sometimes it may appear to receive moral authority as well: Many analytical chemists balk at the thought of doing reliable, precise quantitative spectrophotometry in turbid suspensions, because Beer's law is not fulfilled.

When it comes to multivariate calibration, chemists have the advantage over statisticians that they know their samples and instruments and can often make new measurements when needed. They also have a lot of background knowlege that is activated once the chemist detects trouble in the data. But chemists lack quantitative

Figure 1.17 An analytical chemist, being a specialist at analyzing individual samples, may think of the future unknown samples as a multitude of the type of samples being analyzed just now. The statistician knows very little about the analytical process, but worries about the distribution of future samples

concepts and methods for describing what is 'best in the long run' (Figure 1.17). Concepts like populations and representativity may be difficult to quantify for the chemist. Likewise the distinction between underlying model parameters which have to be estimated and the estimates themselves is evidently not clear for all chemists. In addition, chemists like to have things under control, and do not like to include unidentified systematic phenomena in their models. This limits their radius of action.

Statisticians, on the other hand, have a very good formal handle on distributions, and have valuable methods to assess uncertainties. But since they usually do not have a personal relationship to the samples and the instrumentation, they can become prisoners of the given data or locked into some type of standard statistical model. A danger in some statistical traditions is to focus too much on an elaborate mathematical model description of the problem, instead of driving for goal-oriented results.

Another danger for the statistician—as well as for others, is to forget key causal variables because they have not been measured. This was illustrated by Joiner (1981). Instead one may jump to conclusions based on the available variables—or even worse—based only on those variables incidentally not eliminated by some conventional automatic stepwise regression procedure.

Figure 1.18 shows a stylized example of how multivariate calibration can give causal understanding while not being bound by the traditional simplifying 'laws' like 'Beer's Law' etc: Two X-variables are plotted against each other, with the value of the analyte concentration Y mapped as numerical values in the plot. The trained chemist can see an interference pattern, a nonlinearity effect and an outlier from these calibration data. With such a calibration *map* established, the analyte concentration Y for new unknown samples of the same type can be predicted reasonably precisely from their (x_1, x_2) measurements. For simple purposes such

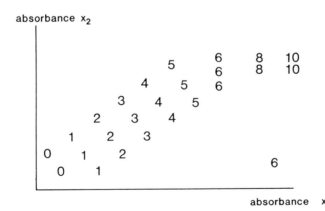

Figure 1.18 Mapping the absorbance at two different wavelengths. The numbers inside represent concentrations of the analyte constituent to be calibrated for. The figure shows a selectivity problem interference in the general direction 'northwest/southeast'), nonlinearity at high absorbances at wavelength 2, and an outlier in the lower 'southeast' corner

bivariate plots can be helpful; in more complex situations and for routine operation it is valuable to formulate the structure in the 'map' as a mathematical transfer function $\widehat{Y} = f(x_1, x_2, \ldots)$.

Quantitative chemometrics

During the last decade computers have become steadily better and cheaper, and the micro-processor is now supplying the chemist with enormous processing capacities at reasonable cost. Chemometrics (the use of mathematics and statistics on chemical data) has become an established scientific discipline.

Multivariate calibration is somewhat different from some of the other chemometric pattern recognition techniques, such as the classification and clustering methods. While the latter methods primarily are qualititative and often exploratory, calibration is quantitative and usually applied to data which are already known to have inter-relationships.

Multivariate calibration requires multivariate mathematical modelling, and the parameters of the models must be estimated statistically from empirical data. This book is an attempt to provide chemometric methods that are easy to understand and use for chemists, while also having good statistical properties. A combination of the chemist's and the statistician's point of view is presented: An empirical calibration approach is employed to concentrate the relevant information from several measured variables into a simplified 'map' that gives good statistical concentration *predictions* and at the same time allows the chemist to *interpret* the results.

See Kowalski (1984), Sharaf et al. (1986), Massart et al. (1988) and Meuzelaar and Isenhour (1987) for general overviews of the field of chemometrics.

2 Computational and Statistical Problems and Tools

2.1 COMPUTATIONAL AND STATISTICAL TOOLS

SUMMARY   The most basic computational and statistical tools used in the book are defined. *Key words*: Vector, matrix, matrix multiplication, diagonal matrix, unit vector, space, subspace , dimension, linear operator, trace, orthogonality, matrix inversion, projection, eigenvectors, eigenvalues, statistical distribution, expectation, variance, standard deviation, mean squared error, correlation, least squares.

2.1.1 SOME LINEAR ALGEBRA ILLUSTRATED

Describing multivariate calibration methods at sufficient precision and yet in a readable way requires matrix and vector notation. This section is written as to help the reader who is not familiar with matrix algebra. For readers of the more advanced level 2 a few more complicated tools will be required. For good reference books on matrix algebra useful in statistics we refer to Searle (1982) and Basilevski (1983). Readers who find the following tool-box too technical may read chapter 8 and browse through section 3.1–3.9 before returning to this chapter.

2.1.1.1 Scalars, matrices and vectors

In calibration we normally want to predict chemical concentrations (often denoted C) from instrument responses (often denoted R). But sometimes the order may be reversed, or we may want to predict e.g. flavour or toxicity from chemical concentrations.

   Therefore, we will here instead of the chemical notation (C=concentration, R=instrument response) change to the more general prediction notation (Y=predicted, X=predicting variables), corresponding to the general predictor formula $\hat{Y} = f(X)$. Normally $Y = C$ and $X = R$.

   Notice that we here consciously choose statistical prediction usage of (X, Y) instead of the (X, Y) convention used in some earlier calibration works (see e.g. Brown, 1982). The reason is that the main goal of calibration is future prediction, not the causal modelling as such, and we want to avoid a notation that subconsciously locks the thought into causal thinking. The question of causality is treated in more detail in section 2.2.

Notation is needed at various level of abstraction. Expressions at the purely conceptual level (as in 'X-variables') employ ordinary capital letters. It is important to distinguish between matrices, vectors and scalars; boldface letters are used explicitly for matrices and vectors. Capital boldface represents matrices and lowercase boldface represents vectors. Expressions inside brackets { }, ( ) or [ ] denote the contents of the matrices and vectors.

Frame 2.1 defines symbols and indices i, j and k used throughout this book.

Frame 2.1 Symbols and indices used

Notation convention:

$\mathbf{Y} = \{\hat{y}_{ij}\ i =, \ldots, I, j = 1, \ldots, J\}$
  = variables to be predicted $j = 1, 2, \ldots, J$
  (e.g. conc. of constituents)
  for objects $i = 1, 2, \ldots, I$.

$\mathbf{X} = \{\hat{x}_{ik}\ i = 1, \ldots, I, k = 1, \ldots, K\}$
  = predicting variables $k = 1, 2, \ldots, K$
  (e.g. absorbance at various wavelengths)
  for objects $i = 1, 2, \ldots, I$.

Example:

The concentration of a certain protein no. j measured in a certain protein-containing object number i can be regarded as a scalar y_{ij}

$\square$ scalar y_{ij}

The concentration of protein j measured in a set of objects $i=1,2,\ldots,I$ can be described as a vector $\mathbf{y}_{\hat{j}} = \{y_{\widehat{ij}}, i = 1, \ldots, I\}$,

$$I\ \square \qquad \text{Vector } \mathbf{y}_j = \begin{bmatrix} y_{ik} \\ \vdots \\ y_{Ij} \end{bmatrix}$$

The concentrations of proteins $j = 1, 2, \ldots, J$ measured in a set of objects $i = 1, 2, \ldots, I$ can be described as a matrix: $\mathbf{Y} = \{y_{ij} \; i = 1, 2, \ldots, I ; j = 1, 2, \ldots, J\}$.

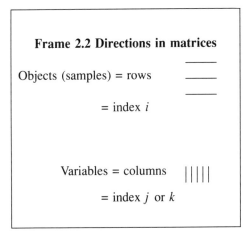

$$\text{Matrix } \mathbf{Y} = \begin{bmatrix} y_{11} \cdots y_{1J} \\ \cdot \quad \cdots \quad \cdot \\ y_{I1} \cdots y_{IJ} \end{bmatrix}$$

The data matrix convention outlined in Frame 2.2 is usually followed in this book. Notice that the word 'sample', in chemistry meaning a unit being analyzed, will normally be replaced by the word 'object'. This is done to avoid confusion with the statistical usage of 'sample', which implies a 'sampled sub-set of objects from a population'.

Frame 2.2 Directions in matrices

Objects (samples) = rows

$=$ index i

Variables = columns

$=$ index j or k

The symbol $'$ (in some other texts written as ^{T} or ^{t}, e.g. $\mathbf{x}^{\mathrm{T}}$) means 'transpose'. This is practical in order to make the matrix notation effective. For vectors it designates a row vector, as opposed to a column vector which are denoted without a transpose.

The transpose of a vector representing the 'spectrum' of an object i at variables $k = 1, 2, \ldots, K$ can be written as $\mathbf{x}_i{}' = (x_{i1}, \ldots, x_{ik})$ (a row) and illustrated as:

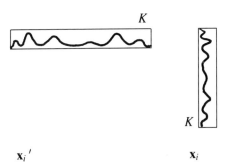

$\mathbf{x}_i{}'$ $\mathbf{x}_i$

In spectroscopy, $x_i{'}$ could e.g be absorbance response of object i at various wavelengths $k = 1, 2, \ldots, K$; in chromatrography it could represent areas or heights of peaks $k = 1, 2, \ldots, K$ of object i.

The transpose of a matrix representing e.g. the K amino acids of I objects corresponds to 'flipping the table around its diagonal' as illustrated below:

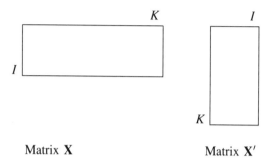

Matrix **X**   Matrix **X'**

2.1.1.2 Sums of matrices and multiplication with a scalar

A sum (or difference) of two vectors is defined as elementwise summation, e.g. for I-dimensional vectors **x** and **y** their sum **z** = **x** + **y** is defined by

$$z_i = x_i + y_i, \quad i = 1, 2, \ldots, I. \tag{2.1}$$

The sum (or difference) of two matrices is likewise performed for each separate element, e.g. for $I \times K$ matrices **X** and **Y** their sum **Z** is

$$z_{ik} = x_{ik} + y_{ik}$$

(for all i and k)

Of course this requires that the matrices or vectors added have the same dimension, and implies that the sum has the same dimension as the two.

A matrix and a vector can also be multiplied by a scalar and the definition of this is multiplication of each element by the scalar. This is written as **X** = $a*$**Z**, or only **X** = a**Z** (= **Z**a) and is illustrated as

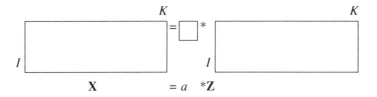

$$\mathbf{X} \qquad = a \quad *\mathbf{Z}$$

2.1.1.3 Matrix multiplication: The linear mixture model

Assume that we have measured chromatographically the amino acid composition profile $\mathbf{x}_i'$ of a certain object i, known to be a mixture of two well known proteins in unknown proportions, $j = 1$ and $j = 2$.

Assume that the instrument response is linear, the measurement noise negligible, and no other proteins or other interferences are present. We can then represent the profile $\mathbf{x}_i'$ by the linear mixture model:

$$\mathbf{x}_i' = \text{contribution}_1 + \text{contribution}_2$$

The amino acid contribution of each of the constituents can be expressed by its amino acid profile $\mathbf{k}_j'$, multiplied by its unknown concentration y_{ij}: The profile contribution of constituent j in object i is then equal to $(y_{ij}k_{1j}, \ldots, y_{ij}k_{Kj})$ which is written as $y_{ij}\mathbf{k}_j'$. This is multiplication of a scalar with a vector and is illustrated as:

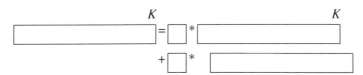

Thus we can sum the two contributions by a vector sum:

$$\mathbf{x}_i' = y_{i1}\mathbf{k}_1' + y_{i2}\mathbf{k}_2' \qquad (2.2)$$

illustrated as

Notice that k and $\mathbf{K}$ are used in two different ways in this book; namely as wavelength index $k = 1, 2, \ldots, K$ and as matrix of analyte spectra or profiles $\mathbf{K} = (\mathbf{k}_j, j = 1, 2, \ldots, J)$. Thus $\mathbf{K} = (k_{kj}, k = 1, 2, \ldots, K; j = 1, 2, \ldots, J)$. The latter is in accordance with chemometric notation tradition. In cases where confusion is possible, we explicitly state which of them we consider.

The previous formula can be written more elegantly as a matrix product: $\mathbf{x}_i' = \mathbf{y}_i'\mathbf{K}'$:

$$\mathbf{x}_i' \quad = \quad \mathbf{y}_i' \quad \mathbf{K}' \tag{2.3}$$

where J is the number of constituents (here: protein$_1$,protein$_2$), $\mathbf{y}_i'$ is a vector of their concentrations and $\mathbf{K}$ is a matrix of their amino acid profiles $\mathbf{k}_j . j = 1, 2, \ldots , J$.

Of course this can equivalently be expressed as $\mathbf{x}_i = \mathbf{K}\mathbf{y}_i$. Note the change of ordering of $\mathbf{K}$ and $\mathbf{y}$ in the transpose of the multiplication: $(\mathbf{y}_i'\mathbf{K}')' = \mathbf{K}\mathbf{y}_i$.

This shows how matrix multiplication, e.g. $\mathbf{T} = \mathbf{UV}$ (or $\mathbf{T} = \mathbf{U}*\mathbf{V}$) is done:

where

$$w = a * e + \ldots + b * g$$
$$x = a * f + \ldots + b * h$$
$$y = c * e + \ldots + d * g \tag{2.4}$$
$$z = c * f + \ldots + d * h$$

and ... represents the intermediate elements if there are more than two rows in the right-hand matrix of the multiplication (and hence more than two columns in the left-hand matrix of the multiplication).

Notice that the number of columns in $\mathbf{U}$ must equal the number of rows in $\mathbf{V}$!

We say that $\mathbf{V}$ is *premultiplied* by $\mathbf{U}$, and $\mathbf{U}$ is *post-multiplied* by $\mathbf{V}$, and $\mathbf{T}$ is their product.

2.1.1.4 Including offset in a linear model: The vector of 1's

What is the matrix representation of how the concentration of a single constituent y_{ij} ($J = 1$) varies with instrumental measurements x_{ik} at a single variable ($K = 1$) over objects $i = 1, 2, \ldots I$?

If we assume that this is a direct proportional relationship with no noise or curvature problems, then we can write it

$$y_i = x_{ik}b_k. \tag{2.5}$$

or more generally, $\mathbf{y} = \mathbf{x}_k b_k$; or simply $\mathbf{y} = \mathbf{x}b$, since we have only one X-variable:

$$\mathbf{y} \ = \ \mathbf{x} \quad b \tag{2.6}$$

If we also allow for an offset in the relationship which is often necessary, the relation is

$$\mathbf{y} \ = \ b_0 \ + \ \mathbf{x}b \tag{2.7}$$

To write this in matrix notation, vector $\mathbf{1}$ is a useful formal tool. Again with $K = 1$ we can suppress index k for simplicity:

$$\mathbf{y} = \mathbf{1}b_0 \ + \ \mathbf{x}b: \tag{2.8}$$

Putting vector $\mathbf{1}$ and vector $\mathbf{x}$ together in a matrix $\mathbf{X}$, and b_0 and b together in vector $\mathbf{b}$, we get:

$$\mathbf{y} \qquad \mathbf{X} \qquad \mathbf{b} \tag{2.9}$$

Hence, offsets can be included implicitly in linear matrix models thereby simplifying the notation. Likewise, if $\mathbf{y}$ depends on two or more X-variables ($K > 1$):

$$\mathbf{y} \ = \ b_0 \ + \ \mathbf{x}_1 b_1 \ + \ \ldots \ + \ \mathbf{x}_K b_K \tag{2.10}$$

which also can be written

$$\mathbf{y} \ = \ \mathbf{Xb} \tag{2.11}$$

where $\mathbf{y} = \{y_i\}$, $\mathbf{X}$ is composed of 1 and the columns $\mathbf{x}_1 \ldots \mathbf{x}_K$, and $\mathbf{b} = \{b_0.b_1.\ldots.b_K\}'$.

If we have two or more Y-variables ($J > 1$) the linear structure could be written

$$\mathbf{Y} \ = \ \mathbf{XB} \tag{2.12}$$

where $Y = \{y_{ij}\}$, X is as above and B contains the linear coefficients for all J components of Y relative to $x_k, k = 0, \ldots, K$.

2.1.1.5 Scaling rows or columns: The diagonal matrix

As we shall see below, it may sometimes be important to multiply every element in a row or column by the same number. A simple example: Transform the second column in X from, say, absorbance to milliabsorbance, without changing the first column. This type of transformation may be necessary in many of the calibration methods to be described in the book, in order to weight the information in one column so that it balances the information in another column in the same table (see Chapter 7). The technique of multiplying by a diagonal matrix can then be used:

In this example we can write: new $x_{ik} = $ (old $x_{ik})v_k$, $i = 1, 2, \ldots, I$ and $k = 1, 2, \ldots, K$, where $v_k = 1000$.

In matrix notation this means that we post-multiply the old $I \times K$ input matrix X_{inp} by the $K \times K$ diagonal matrix V, for instance:

$$\begin{bmatrix} 1 & e \\ 1 & f \\ 1 & g \\ 1 & h \end{bmatrix} = \begin{bmatrix} 1 & a \\ 1 & b \\ 1 & c \\ 1 & d \end{bmatrix} \begin{bmatrix} 1 & 0 \\ 0 & v_{22} \end{bmatrix}$$

$$\text{New } X = X_{inp} \; V \tag{2.13}$$

where $v_{22} = 1000$. Since all the off-diagonal elements of a diagonal matrix are zero, the usual effect of adding different columns or rows together in matrix multiplication is avoided. Hence, column 2 in X_{inp} is multiplied by 1000 and thus converted from absorbance units to milliabsorbance units, while column 1 is multiplied by 1 and hence unchanged.

Diagonal matrices can also be used for scaling rows instead of columns, by pre-multiplying the matrix to be scaled by a diagonal matrix.

2.1.1.6 The identity matrix I

The matrix equivalent of the number 1.0 is a special version of a diagonal matrix, termed the *identity* matrix and symbolized by the symbol I. It consists of 1 on every diagonal element and 0 on all off-diagonal elements. For instance, the 3x3 identity matrix is:

$$I = \begin{bmatrix} 1 & 0 & 0 \\ 0 & 1 & 0 \\ 0 & 0 & 1 \end{bmatrix}$$

Pre- or postmultiplying a matrix by I leaves the matrix unchanged, e.g.: $X = XI$ and $Y = IY$.

2.1.1.7 Space, subspace and linear operator

Geometrically, a vector with elements $k = 1.2.....K$ can be regarded as a line segment extending from an origin in a K-dimensional coordinate system (see Figure 2.1).

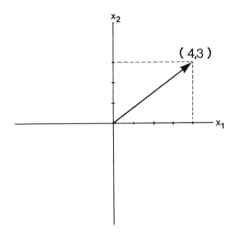

Figure 2.1 A vector is represented as a point or arrow in a two-way plot (x_1-x_2)

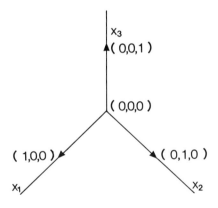

Figure 2.2 Illustration of *unit* vectors (1,0,0), (0,1,0) and (0,0,1) spanning the whole 3-dimensional space

By a Euclidean vector space or only *space* we mean the set of all possible linear combinations of some set of vectors. For instance, vectors (1,0,0)', (0,1,0)' and (0,0,1)' span a 3-dimensional space (Figure 2.2).

A linear combination is the same as a weighted sum. For instance, $x_1 w_1 + x_2 w_2 + ... + x_K w_K$ is a linear combination of the K vectors $x_k. k = 1.2.....K$; w_k here represent the weights (positive or negative).

By a *subspace* of a vector space we mean the set of all possible linear combinations of a particular subset of vectors from the full space. For instance, all *linear combinations* of the form $a(1, 2, 3)' + b(3, 2, 1)'$ where a and b are scalars form a subspace (a two-dimensional plane) in the above-mentioned 3-dimensional space.

A function $f(\ )$ from one vector space to another is termed a linear map or operator or simply linear function if it can be represented by a matrix, i.e. if it can be written as $f(\mathbf{x}) = \mathbf{B}\mathbf{x}$ for some matrix $\mathbf{B}$.

2.1.1.8 Linear independence, dimension and basis

If at least one of the vectors in a set of K vectors $k = 1, 2, \ldots, K$ can be written as a linear combination of the $K - 1$ others, then the K vectors in the set are *linearly dependent*. If it is not so, the vectors are *linearly independent*.

The two vectors (1,2,3) and (3,2,1) in the example in the preceding section are linearly independent, since (3,2,1) cannot just be written as (1,2,3)*c, where c is some constant. We say that the subspace spanned by them has *dimension* two. On the other hand, the vectors in the set (1,2,3), (3,2,1), (2,4,6) and (4,4,4) are linearly dependent. Here (2,4,6) = (1,2,3)*2 and (4,4,4)= (1,2,3) + (3,2,1).

More generally, the maximal number of linearly independent vectors in a space gives the *dimension* of the space. Orthogonal vectors (see 2.1.1.10) are good examples of linearly independent vectors.

Linearly independent vectors generating a space are called a *basis* for the space. Analogy: For charting a simplified flat landscape, e.g. a city plan, the directions north/south and east/west are a basis for the two-dimensional map space. Directions north/south and south/north would not serve as such a basis. But directions north/south and north-east/south-west would be another equivalent basis for the

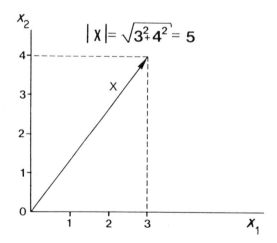

Figure 2.3 The length of a vector is defined according to the Pythagorean theorem

same map space, although with non-orthogonal axes. If so desired, height above sea level could be added to represent a third basis axis in a three-dimensional space.

2.1.1.9 Length (norm) of a vector

The length of vector $\mathbf{x} = (x_1, x_2, \ldots)'$ is defined as the square root of the sum of the squared elements in the vector:

$$|\mathbf{x}| = \sqrt{(x_1^2 + x_2^2 + \ldots)} \tag{2.14}$$

In matrix algebra this can be written: $|\mathbf{x}| = (\mathbf{x}'\mathbf{x})^{1/2}$.

Example: In the simplest case for the two-element vector $\mathbf{x} = (3.0, 4.0)$ this corresponds to the Pythagorean sentence (Figure 2.3).

2.1.1.10 Orthogonality between vectors

If two vectors $\mathbf{x} = (x_1, x_2, x_3)'$ and $\mathbf{z} = (z_1, z_2, z_3)'$ in 3-dimensional space (k = 3) are orthogonal to each other, the angle between them is 90°. Algebraically, their cross product sum (scalar-product of the vectors) is zero:

$$(x_1 z_1 + x_2 z_2 + x_3 z_3) = 0 \tag{2.15}$$

In matrix algebra this can for the two vectors $\mathbf{x}$ and $\mathbf{z}$ be written $\mathbf{x}'\mathbf{z} = 0$. For $K > 3$ this relation is taken as the definition of orthogonality. For an illustration, see Figure 2.4.

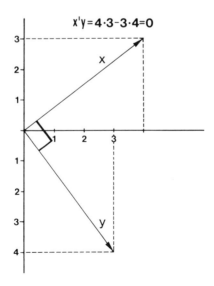

Figure 2.4 Illustration of orthogonality. The product of the orthogonal vectors is 0

Orthogonality between vectors is a very useful mathematical property that simplifies multivariate modelling algorithms, since one set of parameters or variables can be treated independently of other parameters. In calibration algorithms the orthogonality properites are important both between vectors of objects $i = 1, 2, \ldots, I$ and between vectors of variables $k = 1, 2, \ldots, K$.

2.1.1.11 Matrix inversion: The equivalent of division

The inverse of a single scalar number z can be written z^{-1}. *Matrix inversion* is a straight-forward extension of this. The inverse of a *square* matrix (a matrix with the same number of rows as columns) $\mathbf{Z}$ is written $\mathbf{Z}^{-1}$ and is defined such that the product equals the identity matrix of the same dimension, i.e.

$$\mathbf{ZZ}^{-1} = \mathbf{Z}^{-1}\mathbf{Z} = \mathbf{I} \qquad (2.16)$$

This multivariate inverse is useful in solving equations and in many statistical computations.

In general a matrix $\mathbf{Z}$ is only invertable if it has full 'rank'—i.e. the columns (or rows) are linearly independent. For 'ill conditioned' data matrices $\mathbf{Z}$ (see section 2.3.2) the inverse $\mathbf{Z}^{-1}$ may be impossible to compute, at least with sufficient reliability.

More specifically, the inverse does not exist if $\mathbf{Z}$ (dimension $J \times J$) contains less than J independent types of variation, e.g. because one column either is equal to another column, is equal to the difference between two other columns, or in general is equal to some linear combination of other columns in the table.

An example of a matrix that cannot be inverted is

$$\mathbf{Z} = \begin{bmatrix} 1 & 2 \\ 2 & 4 \end{bmatrix}$$

In this case the last column in $\mathbf{Z}$ is equal to the first one multiplied by 2, so they are not linearly independent. Trying to invert $\mathbf{Z}$ would be a multivariate equivalent to dividing a number by zero.

It often happens that the 'true' information in $\mathbf{Z}$ has this type of problem and cannot be inverted, but this is masked by random noise. Then an inverse can be computed technically, but it may be statistically meaningless to use it. In the present example, if the 'true' $\mathbf{Z}$ were the one given above, while a $\mathbf{Z}$ matrix somehow obtained from experimental data were

$$\mathbf{Z} = \begin{bmatrix} 1.001 & 1.99 \\ 1.99 & 4.00 \end{bmatrix}$$

then $\mathbf{Z}^{-1}$ would probably be meaningless (see section 2.3.1).

Generalized inverse

If the matrix to be inverted does not have full rank and therefore cannot be fully inverted, then a 'partial inversion' called the *generalized inverse* can sometimes be used instead.

A generalized inverse of a matrix $\mathbf{Z}$ is written $\mathbf{Z}^-$ and is defined by

$$\mathbf{Z} * \mathbf{Z}^- * \mathbf{Z} = \mathbf{Z} \qquad (2.17)$$

The generalized inverse is generally not unique and can be computed and expressed in several ways. The concept of generalized inverse may be useful in calibration when the input variables are collinear.

2.1.1.12 Projection

A mathematical concept which will be used repeatedly in this book in defining calibration methods is the word 'projection'.

In mathematical terms a *projection* of a vector $\mathbf{y}$ into a subspace spanned by linearly independent vectors $(\mathbf{x}_1, \ldots, \mathbf{x}_k) = \mathbf{X}$ is defined by the *linear operator* $\mathbf{X}(\mathbf{X'X})^{-1}\mathbf{X'y}$. We say that $\mathbf{y}$ is projected into the space spanned by $\mathbf{X}$.

Geometrically, a projection is described in Figure 2.5. As we see, the projection of $\mathbf{y}$ is found by drawing the line from $\mathbf{y}$ perpendicular to the space spanned by the two vectors $\mathbf{x}_1$ and $\mathbf{x}_2$.

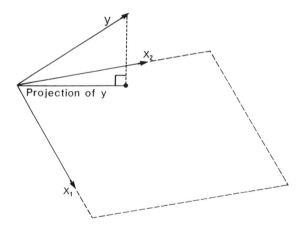

Figure 2.5 Projection of $\mathbf{y}$ down on the space spanned by the vectors $\mathbf{x}_1$ and $\mathbf{x}_2$

2.1.1.13 Eigenvectors, eigenvalues and singular value decomposition

These are fundamental concepts, although somewhat abstract at first sight. The eigenvector structure represents a stabilized basis for looking at experimental data, simpler and more informative than looking at the individual input variables or the individual objects.

Mathematically speaking, an *eigenvector* for a square matrix $\mathbf{Z}$ is defined as a vector $\mathbf{p}$ satifying

$$\mathbf{Zp} = \mathbf{p}\tau \tag{2.18}$$

for some scalar τ which is called the *eigenvalue* corresponding to the eigenvector **p**.

In this book we shall mainly consider eigenvectors of cross-products or so-called covariance-matrices, i.e. matrices of the form $\mathbf{X}'\mathbf{X}$ where $\mathbf{X}$ is a data matrix consisting of measurements on I objects and K variables. For such matrices there exist K distinct orthogonal eigenvectors and all eigenvalues will be larger than or equal to 0 (provided $I > K$).

In matrix notation the eigenvalues/eigenvectors of such a matrix satisfy

$$\mathbf{ZP} = \mathbf{P}(\mathrm{diag}(\tau_1, \tau_2, \ldots, \tau_K)) \tag{2.19}$$

where the eigenvectors in $\mathbf{P}$ are orthogonal to each other and of length 1, i.e.

$$\mathbf{P}'\mathbf{P} = \mathbf{I} = \mathbf{PP}' \tag{2.20}$$

Such an *orthogonal* matrix is sometimes termed *orthonormal*.

We can then express the matrix $\mathbf{Z}=\mathbf{X}'\mathbf{X}$ as product of its eigenstructure, i.e.

$$\mathbf{Z} = \mathbf{P}(\mathrm{diag}(\tau_1, \tau_2, \ldots, \tau_K))\mathbf{P}' \tag{2.21}$$

The inverse of $\mathbf{Z}$ can then be written

$$\mathbf{Z}^{-1} = \mathbf{P}(\mathrm{diag}(1/\tau_1, 1/\tau_2, \ldots, 1/\tau_K))\mathbf{P}' \tag{2.22}$$

if all the eigenvalues are larger than 0. If some eigenvalues, however, are equal to 0, then the inverse does not exist. A generalized inverse of some sort may then be required: e.g. the *Moore-Penrose* inverse (Searle, 1982)

$$\mathbf{Z}^- = \mathbf{P}(\mathrm{diag}(1/\tau_1 1/\tau_2, \ldots, 1/\tau_A, 0, 0, 0, 0 \ldots 0)\mathbf{P}' \tag{2.23}$$

where $A < K$.

The concepts of eigenvalues and eigenvectors are related to the *singular value decomposition* (SVD): A rectangular (but not necessarily quadratic) matrix $\mathbf{X}$ can be expressed as

$$\mathbf{X} = \mathbf{U}\sqrt{\mathbf{G}}\,\mathbf{P}' \tag{2.24}$$

where $\sqrt{\mathbf{G}} = \mathrm{diag}(\sqrt{\tau_a}, a = 1, 2, \ldots, A)$, with $\tau_a, a = 1, 2, \ldots, A$ being the positive eigenvalues for $\mathbf{XX}'$ and $\mathbf{X}'\mathbf{X}$ (which have the same eigenvalues), and $\mathbf{U}=(\mathbf{u}_a, a = 1, 2, \ldots, A)$ and $\mathbf{P} = (p_a, a = 1, 2, \ldots A)$ are the eigenvectors (corresponding to the positive eigenvalues) for $\mathbf{XX}'$ and $\mathbf{X}'\mathbf{X}$, respectively. Eigenvalues of value zero are hence ignored.

Singular value decomposition is again related to principal component analysis and the other calibration methods to be covered later. In general, when there are many inter-correlated X-variables in a calibration problem, the multivariate calibration techniques to be described ignore eigenvalues *equalling* zero, and also positive eigenvalues *near* zero, i.e. so small that they only reflect random measurement noise in the data. This will be described in more detail in Chapter 3.

2.1.2 STATISTICAL CONSIDERATIONS

Here we go through some basic statistical concepts necessary for reading the book. We refer to Box et al. (1978) for a more fundamental text on introductory statistics.

2.1.2.1 Population, distribution and parameters

The most fundamental concept of statistics is that of a population and its probability distribution. We cannot plan a calibration experiment without first having considered what population to consider. Neither can we assess how good or bad a calibration result is without such considerations. Still, chemists may have difficulties here, because statistical thinking is very different from the causal phenomenological orientation in chemistry.

Assume for instance that we shall determine the lipid content (Y) in foods, using a certain simplified analytical instrument X. Calibration for $\hat{Y} = f(X)$ is a question of local data approximation, and we cannot expect the same calibration formula to apply equally well for all kinds of food. So we define our target population to be e.g. ground beef sold to consumers. The fat percentage will certainly vary from batch to batch of ground beef, but there will still be a tendency of clustering at a certain region of the percentage axis. For instance, we know that 100% fat is an impossible value for meat. With a legal upper limit of, say, 20% fat, most of the objects will probably lie between 10% and 20% fat.

Imagine then that we can measure the fat percentage y for all individual objects in the actual *population*. There will then be a certain distribution of the percentages over a certain part of the axis. An example is shown in Figure 2.6 by a *histogram* or relative *frequency* diagram. The axis is divided into many equally wide intervals and the number of fat-values in each interval are counted. These numbers are divided by the total number of objects (and corrected for the length of the intervals if the

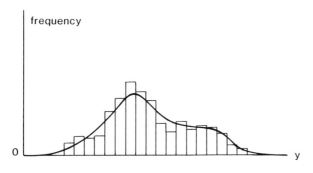

Figure 2.6 Histogram is illustrated by the bars. The smooth curve illustrates the underlying density curve

length is different from 1) and used as ordinates at the respective intervals.

When the population is large we can make the intervals smaller and smaller and still have many individuals in each interval. As the number increases we will approach a continuous curve like that presented in Figure 2.6. If this is continuous as in the figure it is called the density function of the population and represents the distribution of individuals.

If an individual is selected at random from the population and we measure the fat percentage y, the probability that y will fall in a certain interval corresponds to the area under the density function between the two end points of the interval. In this sense the distribution of fat percentage in the population as represented by the density tells us everything about probabilities connected to events for the y-variable. Therefore we call the continuous population distribution the *probability distribution* of the variable y.

When a set of objects (a 'sample' in the statistical meaning) of let us say 100 individuals are taken at random from a large population we have 100 y-measurements y_i; , $i = 1, 2, \ldots,$ 100.

In statistics, these 100 individuals would be called a *random* sample. Due to the different meaning of 'sample' in chemistry, we shall sometimes term it a random sample set. The histogram of the 100 intervals is not exactly equal to the population distribution, but the higher the number of objects is, the more similar they will be.

The *population* is thus an idealized concept of the total collection of unknown objects of the type or class in question. It usually cannot be assessed directly. The only thing we can do is to observe a subset of the population, and from these observations try to *estimate* the population or properties of the population as precisely as possible.

This distinction between the unknown population properties or *parameters* and the statistical *estimates* of these parameters is important in statistics. It allows a useful simplification in our *thinking*: First, an idealized *model* of how the unknown population is expected to behave is formulated, with respect to structures and to random variations. Only afterwards one worries about how to estimate the parameters in this chosen model. The important concept of a model is covered in more detail in section 2.2.

Statistical distributions are important in three connections: First of all, distributional considerations are necessary in order to plan calibration experiments. Secondly, the calibration modelling yields statistical summaries and these can be compared to more theoretical statistical distributions, for error warnings and validation. Thirdly, the optimal choice of calibration method depends on theoretical distribution assumptions about the input data (but in practice that is not very critical).

There are a number of statistical distribution types that are used as models for approximating the population distribution in practice. The most common of these is the normal distribution, illustrated in Figure 2.7 with its two parameters, the mean ($E(y)$) and the standard deviation σ (see section 2.1.2.2). A rule of thumb is given: The probability of an object falling within $\pm\sigma$ from the mean is about 67%, and within $\pm 2\sigma$ from the mean about 95%.

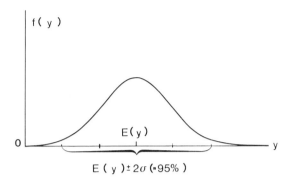

f(y)

E(y)

0

y

E (y)± 2σ (·95%)

Figure 2.7 The normal distribution with its expectation and approximate central 95% region illustrated

The normal distribution is often motivated by the 'central limit theorem'. This states that the sum of several different contributions will tend to be normally distributed, irrespective the probability distribution of the individual contributions. This is often the case e.g. for measurement errors in the long run.

All theoretical distributional assumptions should be checked in practice. For instance, in Figure 2.6 the distribution displays two maxima, indicating a mixed population. But generally the normal distribution is easy to use, is often reasonably realistic and sometimes leads to statistical results that are quite robust against departures from normality. Other important distribution types are the Student t-distribution and the Fisher F-distribution; they will be used for outlier tests in this book.

There are two conceptually different types of distributions—that of the actual phenomena in the samples to be measured (e.g. fat content and light scattering level) and that of the random measurement errors. The notation and context will reveal which of these distribution types is addressed in a given consideration.

(It should be noted that statistical hypothesis tests usually require assumptions about the actual distribution types, independence of residual elements, etc. In addition, they need estimates of degrees of freedom. In multivariate calibration these needs may sometimes be difficult to fulfill. With the methodology presented in this book such information is not even necessary to apply the methods and obtain good results. Therefore we will put less emphasis on what type of actual distribution the X and Y variables have and instead employ validation based on prediction ability and interpretability.)

2.1.2.2 Expectation, variance, standard deviation and correlation

In the following we shall go briefly through some standard statistical concepts and expressions concerning distributional parameters. Each parameter is first defined theoretically to show what it is intended to represent on a population basis. Then a practical method for estimating it from a subset of objects is given.

Expectation

The expectation of a variable y is defined as the average value in the full population from which y is selected. It is denoted by $E(y)$. In mathematical terms, this is the integral

$$E(y) = \int_{-\infty}^{\infty} y f(y) \, dy \qquad (2.25)$$

where $f(y)$ is the probability density function. Alternatively we can write it as

$$E(y) = \sum_{i=1}^{N} y_i/N \qquad (2.26)$$

where N denotes a large number of individuals in the population. In practice this parameter $E(y)$ cannot be measured, and we have to rely on a sub-set of objects. The natural estimate of $E(y)$ is of course the mean, $\bar{y} = \Sigma y/I$ where I is the size of the random sample subset. As this number I increases, the mean becomes a more precise estimate of the expectation.

Variance and standard deviation

The expectation and the mean are measures of location, i.e. measures of the 'centre' of the population and sample subset, respectively. In very many cases it is also natural to consider measures of dispersion, i.e. simple measures which tell us how wide the population is. Two such measures are variance and standard deviation where the latter is simply the square root of the former.

Statisticians usually prefer variance, partly because of the better computation properties (variances can be added). The standard deviation, however, is in the same unit as the measurements themselves and can be more easy to interpret; they have traditionally been preferred by chemists.

The theoretical (but unknown) population variance is usually denoted by σ^2 and is simply defined as the expected quadratic difference between the actual value and the expectation. In mathematical terms this means

$$\sigma^2 = E(y - E(y))^2 \qquad (2.27)$$

Standard deviation is simply defined as the square root of this and denoted σ. Sometimes a subscript etc. is used for indicating which variable it concerns, e.g. σ_y or $\sigma(x_k)$.

In a subset of I objects, the analogue is

$$s_y^2 = 1/(I-1) \sum_{i=1}^{I} (y_i - \bar{y})^2 \qquad (2.28)$$

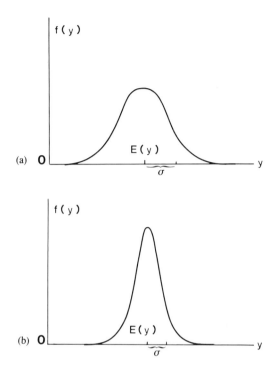

Figure 2.8 Two density curves with different standard deviation. The density with the broadest distribution is the one with the largest standard deviation

Notice that we divide by $I - 1$ instead of I. This is due to the more tractable statistical properties of s^2 as an estimate of the parameter (one degree-of-freedom is lost in the estimation of the mean, $\bar{y}$). The standard deviation of the sampled sub-set is defined as the square root of s_y^2 and termed s_y.

Figure 2.8 gives an illustration of two different populations with different standard deviation.

If variable y is e.g. an input variable (e.g. fat content), then variance σ_y^2 and its estimate s_y^2 concern the distribution of fat contents in the given type of products. Residuals can be likewise summarized: Let us let $\hat{f}$ denote the lack-of-fit between the true y and its prediction from x: $\hat{f} = y - g(x)$. Then we *may* make a priori assumptions about its variance $\sigma_{\hat{f}}^2$, and we can estimate it from empirical test data $\hat{f}$, as $s_{\hat{f}}^2$ (or $s(\hat{f})^2$).

Mean squared error

The theoretical mean squared error (MSE) of an estimate $\hat{y}$ for a value y is related to the variance and is defined as

$$\text{MSE}(\hat{y}) \;=\; E(y - \hat{y})^2 \tag{2.29}$$

This is interpreted as the expected squared deviation between estimated and underlying value. It is easily shown that the MSE for every value of y is equal to the variance plus the squared bias, i. e.

$$\text{MSE} \;=\; \sigma_{\hat{f}}^2 \;+\; \text{Bias}^2 \tag{2.30}$$

where Bias $= (E(\hat{y}) - y)$ (see section 4).

The parameters $\sigma_{\hat{f}}^2$, MSE and Bias are important parameters describing the performance of a calibration model for a certain purpose. In practice even MSE and Bias are estimated from the data; details are given in Chapter 4.

Covariance and correlation

The concepts *covariance* and *correlation* refer to the linear interrelation between two variables, say variables x and y. As above, these concepts are defined relative to a population, but have simple empirical estimates.

The covariance between x and y is defined by $E[(x - E(x))(y - E(y))]$ and is denoted by $\text{Cov}(x.y)$. As we see this is the expected product of the differences between x and y and their respective expectations. If large values of x occur together with large values of y and vice versa the covariance will be positive. If on the other hand large values of x occur together with small values of y and vice versa the covariance will be negative. If there is no relation between the two variables at all, i.e information about x gives no information about y, the covariance will be 0.

The covariance between x and y, however, is dependent of the scale or measurement unit of x and y. The correlation between x and y, theoretically defined by $\rho = \text{Cov}(x.y)/(\sigma_x \sigma_y)$, is developed just to cope with this. The correlation ρ will always be between -1 and 1 and be independent of the scale of x and y. If there is an exactly linear relation between x and y, then the correlation will be equal to -1 or 1. It will be 1 if large values of x correpond to large values of y and vice versa and it will be -1 if large values of x correpond to small values of y and vice versa. If the variables are independent, this 'true' correlation is equal to 0. The covariance is usually estimated by

$$\widehat{\text{Cov}(x,y)} \;=\; \sum_{i=1}^{I}(x_i - \bar{x})(y_i - \bar{y})/(I - 1) \tag{2.31}$$

The estimated correlation coefficient is usually denoted by r and is simply defined by

$$r_{x,y} \;=\; \widehat{\text{Cov}(x.y)}/(s_x s_y) \tag{2.32}$$

This r is, like ρ, a number between -1 and 1. Examples are given in Figure 2.9. The concept of covariance matrix will be met later in the book. It is defined as

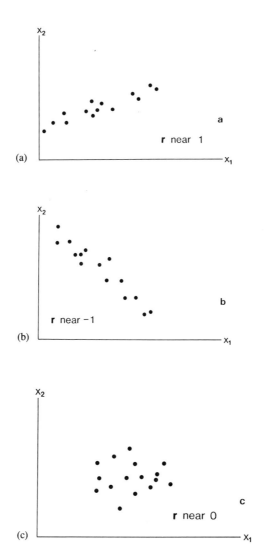

Figure 2.9 Correlation coefficients, r, of three different data sets

$$\text{Cov}(x) = \begin{bmatrix} \sigma^2(x_1) \cdot & & \text{Cov}(x_1, x_2) \cdots & & \cdot \\ \cdot & \cdot & & & \cdot \\ \cdot & & \sigma^2(x_k) & & \cdot \\ \cdot & \cdot & & & \cdot \\ \text{Cov}(x_1, x_K) & \cdot & \cdot & & \sigma^2(x_K) \end{bmatrix} \quad (2.33)$$

with variances on the diagonal and covariances as off-diagonal elements. The estimated version is defined similarly.

2.1.2.3 Least squares

The least squares (LS) principle will here be used in connection with estimation of parameters in a relationship between objects. But it likewise applies for relationships between variables, as later chapters will show.

Assume for simplicity that we have only one variable y (e.g. analyte concentration) and one variable x (e.g. an instrument measurement) and expect a reasonably linear relationship between the two. Since no models are perfect, we expect an error term f in the linear modelling—due to model errors and measurement noise in the data. We can then use the statistical model (see section 2.2):

$$y_i = b_0 + x_i b + f_i \quad i = 1, \ldots, I \tag{2.34}$$

where the model parameters b_0 and b and the residuals f_i are unknown quantities.

Assume further that we have obtained calibration data (x_i, y_i) from $i = 1, 2, \ldots, I$ independent objects (as opposed to several repeated measurements on just a few objects). We want to be able to predict y from x-measurements in future objects from the same population. Hence we need to determine the unknown population parameters, regression coefficients b_0 and b. To be able to assess the quality of the calibration we may also need an estimate of the residual variance σ_f^2.

A common and efficient way of doing this is to find the values of parameters b_0 and b that minimize the residual sum of squares of f, i.e.

$$\sum_{i=1}^{I} (y_i - (b_0 + x_i b))^2 \tag{2.35}$$

with respect to b_0 and b. The solutions are denoted $\hat{b}_0$ and $\hat{b}$ and are called the least squares (LS) estimators of the parameters b_0 and b. These estimators have a number of valuable statistical properties (see Weisberg, 1985, and section 3.6 in this book). For chemists not used to statistics it may suffice to realize that least squares minimization of the residuals

a) puts equal weight on positive and negative residuals
b) puts rather high weight on large individual residuals (since they are squared)

Geometrically the least squares principle can be illustrated as in Figure 2.10.

In matrix form for one or more X-variables the linear regression equation system can be written

$$\mathbf{y} = \mathbf{Xb} + \mathbf{f} \tag{2.36}$$

The least squares problem can then be formulated as minimizing the length of $\mathbf{f} = \mathbf{y} - \mathbf{Xb}$, i.e the scalar product $\mathbf{f'f} = (\mathbf{y} - \mathbf{Xb})'(\mathbf{y} - \mathbf{Xb})$. The solution to this is equal to

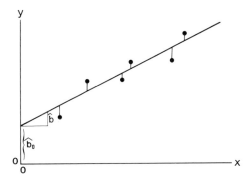

Figure 2.10 The least squares principle corresponds to finding the straight line which minimizes the sum of the squared lengths of the vertical lines

$$\widehat{\mathbf{b}} = (\mathbf{X'X})^{-1}\mathbf{X'y} \qquad (2.37)$$

As we see, the matrix $\mathbf{X'X}$ is inverted. This requires the columns of $\mathbf{X}$ to be linearly independent. If not, the LS estimator is not unique and must be formulated by a generalized inverse. If we plug $\widehat{\mathbf{b}}$ into the model equation we get

$$\widehat{\mathbf{y}} = \mathbf{X}\widehat{\mathbf{b}} = \mathbf{X}(\mathbf{X'X})^{-1}\mathbf{X'y} \qquad (2.38)$$

which is exactly the *projection* of $\mathbf{y}$ into the X-space. The term $\mathbf{X}\widehat{\mathbf{b}}$ is usually denoted by $\widehat{\mathbf{y}}$ and is used as a predictor or estimate of $\mathbf{y}$. In other words $\widehat{\mathbf{y}}$ is the vector in the X-space which is closest to the $\mathbf{y}$.

The estimator $\widehat{\mathbf{b}}$ has expectation equal to $\mathbf{b}$ (unbiased). The covariance matrix of $\widehat{\mathbf{b}}$ is useful for assessing various aspects of the calibration modelling (see later chapters). This covariance matrix can be written as

$$\sigma_f^2(\mathbf{X'X})^{-1} \qquad (2.39)$$

An estimate of σ_f^2 is obtained by $s_f^2 = (\mathbf{y} - \mathbf{X}\widehat{\mathbf{b}})'(\mathbf{y} - \mathbf{X}\widehat{\mathbf{b}})/(I - K)$, where I is the number of rows in $\mathbf{X}$ and $\mathbf{y}$, and K is the number of columns in $\mathbf{X}$. The estimate s_f^2 is also sometimes referred to as $\widehat{\sigma}_f^2$ (see chapter 4 for the dangers of overfitting).

If the columns in $\mathbf{y}$ and $\mathbf{X}$ represent input variables, like y=fat content and $(x_k. k = 1. 2. \ldots . K)$ = optical density at K different wavelengths, the LS solution in (2.37) is called the multiple linear regression (MLR) estimator of the regression coefficient spectrum $\mathbf{b} = (b_k. k = 1. 2. \ldots . K)$ for a single y-variable. This least squares application minimizes the squared residuals summed over *objects* $i = 1. 2. \ldots . I$.

Conversely, the same LS principle can be used for estimating e.g. constituent concentrations for a single object i, by fitting its data vector $\mathbf{x}_i$ to a matrix of constituent spectra; in that case the squared residuals are summed over *variables* $k = 1. 2. \ldots . K$ and the sum minimized. This is covered in more detail in section 3.6

concerning mixture modelling. In the bilinear calibration methods (sections 3.3–3.5) the principle of least squares is applied both over objects and over variables, iteratively (i.e. repeatedly).

Weighted least squares

The ordinary least squares (OLS or just LS) estimation principle is best suited when all the residuals whose sum of squares is minimized have the same expected uncertainty variance $\sigma(f_i)^2 = \sigma_f^2$ for $i = 1, 2, \ldots, I$.

If different observations are known to have different uncertainty variances, then the ordinary LS can be modified to weighted least squares (WLS). In this case a weighted sum of squared residuals is minimized. For instance, assume that the (x, y) data from different objects in the above example are the averages of replicates, and the number of such replicates differs between the objects $i = 1, 2, \ldots I$. Then the different objects' data $(x, y)_i$ must be expected to have different uncertainty variances. Instead of using the OLS principle we can use the WLS principle resulting in the expression

$$\hat{\mathbf{b}}_{\text{WLS}} = (\mathbf{X'V}^{-1}\mathbf{X})^{-1}\mathbf{X'V}^{-1}\mathbf{y} \tag{2.40}$$

where $\mathbf{V} = \text{diag}(\sigma(f_i)^2)$ is proportional to our a priori 'estimate' of the uncertainty variances for objects $i = 1, 2, \ldots, I$, e.g. the inverse of the number of replicates itself.

In practice, WLS can also be obtained in an easier way, by a priori weighting of the input data prior to OLS. For instance, when least squares fit is used for averaging residuals over two or more observations which are expected to have different noise levels $\sigma(f_i), i = 1, 2, \ldots, I$, then we can scale each input observation by $1/\sigma(f_i)$ in a preprocessing stage prior to OLS model fitting. Conversely, the topic of a priori weighting of variables is discussed in more detail in Chapter 7.

Generalized least squares

A further generalization of the weighted least squares principle is possible. This is applicable when the error covariance matrix for the K X-variables is not diagonal, e.g. when the different X-elements reflect some common systematic noise patterns. The estimator is called the generalized least squares (GLS) estimator and it can be written as

$$\hat{\mathbf{b}}_{\text{GLS}} = (\mathbf{X'\Sigma}^{-1}\mathbf{X})^{-1}\mathbf{X'\Sigma}^{-1}\mathbf{y} \tag{2.41}$$

where Σ is the covariance matrix of the noise vector f. The weighting with the inverse of Σ makes it possible to optimally balance the different types of information and noise sources in the system. Its covariance matrix is equal to

$$(\mathbf{X'\Sigma}^{-1}\mathbf{X})^{-1} \tag{2.42}$$

In the special case of diagonal covariance matrix Σ with equal elements, the

GLS estimator is identical to the OLS estimator, and if Σ is diagonal with unequal elements, the GLS estimator is equal to the WLS estimator.

However, the full covariance matrix is seldom known in advance, and has to be replaced by an estimate $\widehat{\Sigma}$ obtained from data. Depending on the problem, this estimation may be more or less difficult. And subsequently inverting this $\widehat{\Sigma}$ may also present a problem (see section 3.6.3). The explicit use of GLS itself as a calibration method is covered in section 3.6.

This chapter has given an introduction to some algebraic, geometric and statistical tools. Before applying them to calibration modelling, we first have to ask: What is a calibration model?

2.2   WHAT IS A CALIBRATION MODEL?

SUMMARY   The general structure of calibration models is given. Different specific alternatives are discussed, the forward regression model, the mixture model, the extended mixture model and regression on latent variables. It is emphasized that the statistical model used does not need to represent a causal structure of the data.

2.2.1 MODELS ARE USEFUL BUT NOT TRUE

Fundamentally, in order to convert measurements $\mathbf{X}$ to meaningful information $\mathbf{Y}$, we need a MODEL of how $\mathbf{X}$ and $\mathbf{Y}$ relate in the class of objects in question. Multivariate calibration requires a more conscious, flexible attitude to modelling than is common in some academic traditions. Before going on, let us therefore briefly go through the basic thinking in multivariate calibration:

Multivariate calibration involves models at several levels. Most fundamenal is our *mental model* of how we think the various variables in $\mathbf{X}$ and $\mathbf{Y}$ are phenomenologically related.

At an intermediate level some *abstract mathematical model* is chosen to represent the phenomena expected to be found between the X and Y-variables (the 'X–Y model'). The regression model $\mathbf{y}=\mathbf{Xb} + \mathbf{f}$ is an example of this. Similar models can be formulated between the X-variables themselves (the 'X–X model'). The mathematical model must fit to our prior mental model, but it does not have to be a detailed causal model! This abstract mathematical calibration model has unknown coefficients (model parameters, e.g. $\mathbf{b}$ in the above example).

During calibration these unknown model parameters are then estimated from calibration data, yielding the *resulting calibration model*. This resulting calibration model contains the desired predictor function $\widehat{\mathbf{Y}} = f(\mathbf{X})$ (e.g. $\widehat{\mathbf{y}} = \mathbf{x}'\mathbf{b}$). With uncertainty estimates, this can be termed the 'resulting X–Y calibration model'. It also contains an estimate of how the X-variables relate to each other—the 'resulting X–X calibration model'.

Finally, the resulting calibration model is used in prediction, and it is implicitly assumed that the prediction objects obey the same model structure as those used in calibration.

Of central importance is the fact that these model levels are interrelated: For instance, when we formulate our abstract mathematical models, we may realize that our a priori mental model was inconsistent or incomplete. Our empirical calibration modelling may show that both the mental and mathematical models have to be corrected. And future predictions may show that our resulting calibration model (and maybe even our mental model) may have to be updated.

In order to retain the dynamic power of integrated modelling without becoming stuck in unscientific subjectivism, three things are needed in calibration:

1) Clear goal formulation, including population considerations
2) Good validation procedures to ensure predictive ability
3) Good models, giving sufficient predictive ability and allowing interpretation and model criticism.

What is a good model, then?

A model is a simplified intermediate representation, intended to have, from a specific perspective and for a specific purpose, a structural or functional analogy to some phenomenon in the inaccessible, more complex reality. A good model in calibration is then a model that gives an adequate compromise between simplicity and completeness. The former implies sufficient interpretability; the latter implies sufficient realism and detailed description.

It is important to keep a certain mental distance to our scientific models. All models are wrong, but some models are very useful. Every model is a working hypothesis, based on a set of assumptions for a given purpose, and has its limitations. Unfortunately, such background information is often lost in the chain from problem specification, model formulation, via estimation to routine application. Standard statistical models can unjustly gain an elevated status that draws attention away from the real phenomena of interest. In chemical analysis, earlier explanations under certain conditions sometimes were elevated to seemingly universal status of 'Laws', with their limitations creating blind spots in the mind of the traditionalist (see section 1.4).

The modelling tools should be open-ended, i.e. if unexpected phenomena are registered, they should be revealed and accounted for in the modelling process. A model must not overshadow the real objects it is intended to approximate.

The underlying notion in this book is the following: It may often be dangerous to use possibly incomplete causal models which in their very structure and pretention lock the mind into possibly inadequate explanations ('hard modelling', like linear direct multicomponent unmixing). It is better to use pragmatic models that leave the mental inference to the subsequent graphical inspection phase ('soft modelling', like bilinear data approximation with extensive graphics and outlier warnings).

The optimal balance between theoretical model formulation and practical data analysis therefore depends on the extent of valid knowledge behind the model vs. the quality and relevance of the available data.

For further discussion on this topic we refer to Box et al. (1978).

Bearing in mind that our mathematical models are part of a larger, integrated model framework, and intended to be local approximations, not the 'truth', what does a mathematical calibration model consist of?

2.2.2 DATA = STRUCTURE + RESIDUALS

Roughly speaking, an abstract mathematical calibration model has a STRUCTURE part representing the systematic variation and a RESIDUAL part representing the difference between data and structure (more about specific assumptions are found in Chapter 3). For instance, the usual linear regression model is:

$$y = Xb + f \qquad (2.43)$$

In this model Xb represents the mathematical structural part and f represents the residual part consisting of presumably uncorrelated residuals (see section 2.1.2.3).

Some statisticians take primary interest in the distributional properties of the residual part, focusing on hypothesis tests, etc., while being less concerned with the interpretation of the structure part. Taken to the extreme, this approach is not so useful in multivariate calibration.

On the other hand, when chemists think of models, they will primarily focus on the structure part, trying to get as good mathematical formulation as possible, and subsequently as good interpretation of its estimate as possible. The residual part is regarded by a chemist as a bothersome nuisance that of course must be small but otherwise is of little interest. But if the chemist totally ignores the stochastic aspects of modelling, the probability of making optimal calibrations is low.

Good multivariate calibration modelling requires attention both to the structure and to the residual parts. A calibration model is founded on certain assumptions of chemical, mathematical and statistical nature. In general, some assumptions make a model better and more useful, while others make it worse. Therefore some underlying assumptions in calibration are now to be discussed.

We start our model considerations with a discussion of the common confusion between causality and prediction, before proceeding to the more statistical aspects.

2.2.3 CAUSALITY VS. PREDICTION

Our primary aim when we calibrate is to determine a function $f(\ )$ that allows quantitative predictions of Y from X-measurements:

$$\widehat{Y} = f(X). \qquad (2.44)$$

In order for this to work, the X and Y variables must display some reproducible intercorrelation in the given type of objects.

A scientist is always concerned about the causal basis for his or her observations. Calibration should be part of science's cyclic progress, alternating between theoretical modelling and empirical validation: We should always *try* to understand what the calibration data tell us.

But in multivariate calibration we do not *have to* understand everything *before* we start calibrating! Lack of causal insight can largely be compensated for empirically, by sensible choice of calibration object set. Further understanding can then be gained *during the calibration* and subsequent *predictions*, by studying the stuctures found in the data.

So the form of the calibration model inside the computer does not necessarily have to reflect a causal explanation of how the X and Y data are generated. There are several types of causal relationships on which the same predictive models $\widehat{Y} = f(X)$ can be based.

Let now U be an unmeasured—possibly unidentified—phenomenon of some sort, and let here the formula $Y \Leftarrow g(X)$ mean 'Y is caused by X via mechanism $g(\ )$'. Then the following causal relationship types can be met in chemical analysis (Frame 2.3):

Frame 2.3

Causal relationships $g(\ )$ for calibration

1) Selective forward causality: $Y \Leftarrow g(X)$ OK !
2) Selective reverse causality: $X \Leftarrow g(Y)$ OK !
3) Extended reverse causality: $X \Leftarrow g(Y, U)$ OK !
4) Indirect, common-cause: $(Y, X) \Leftarrow g(U)$ OK !
5) Incomplete causality: $Y \Leftarrow g(X, U)$ NO !!

For illustration, let us consider an example: analysis of toxic chemical compounds that absorb light in the ultraviolet (UV) wavelength range (see Figure 2.11).

1) Selective forward causality: Y is caused by X alone:

Toxicity, $Y \Leftarrow g$(conc. of toxins X)

(Predict toxicity from the concentrations of various toxins!)

2) Selective reverse causality: X is caused by Y alone:

UV absorbance, $X \Leftarrow g$(conc. of toxin Y)

(Predict toxin concentrations from UV absorbances!)

This is the causality assumed in many analytical instruments (see e.g. Solopchenko, 1987), but by no means the only one of interest.

3) Extended reverse causality: X is caused by Y and by unmeasured, possibly unidentified interferents U:

UV absorbance, $X \Leftarrow g$(conc. of toxin Y & conc. of interferents U)

(Predict toxin concentration from UV absorbance; use several UV wavelengths to solve the interference problems.)

4) Indirect, common cause: X and Y are both caused by unmeasured phenomena U:

$$\left\{ \begin{array}{c} \text{UV absorbance, } X \\ \& \\ \text{toxicity, } Y \end{array} \right\} \Leftarrow g(\text{conc. of toxin \& of } U)$$

(Predict toxicity directly from UV absorbance without going via explicit determinations of toxin and interferent concentrations. Use several UV wavelengths to solve the interference problem.)

5) The fifth causal relationship cannot give good predictions $\hat{Y}$ from X alone:

Incomplete causality: Y is caused by X and by unmeasured U:

$$\text{Toxicity, } Y \Leftarrow g(\text{conc. of toxin } X \text{ \& of anti- toxin } U)$$

(Predicting toxicity from concentration of toxin alone, in the presence of varying levels of anti-toxin? It cannot work, because the effects of the unknown anti-toxin concentration will not have been corrected for.)

If such an incomplete causal situation is encountered, one should look for other measurable X-variables that can reveal U in a multivariate calibration model.

This example illustrates how causal relationships and predictive direction are two quite different things. In this book is given calibration methodology that is applicable for causality relationships 1), 2), 3) and 4).

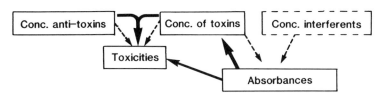

Figure 2.11 The toxin example illustrated. Dotted curves: causal directions. Solid arrows: examples of possible prediction directions

2.2.4 DIFFERENT CALIBRATION MODELS

Calibration aims at estimating a predictor formula $\hat{Y} = f(X)$, no matter what the causal relationships are. This requires a mathematical or statistical model, whose parameters are to be estimated. Frame 2.4 indicates a number of different model types that together with a certain estimation procedure (e.g. least squares) can be used to produce predictors of the type $\hat{Y} = f(X)$. More specific details of assumptions in the different cases are given in sections 3.3, 3.4, 3.5, 3.6 and 3.7.

Frame 2.4

Calibration models that can
give predictors $\widehat{Y} = f(X)$.
T represent systematic structure (possibly
unidentified) while E and F represent residuals

1)	Forward regression model	$Y = f(X)$	$+F$
2)	Mixture model	$X = h(Y)$	$+E$
3)	Extended mixture model	$X = h(Y, T)$	$+E$
4)	Regression	$T = h_1(X)$	
	on latent variables	$Y = h_2(T)$	$+F$
		$X = h_3(T)$	$+E$

The multiple linear regression model ($\mathbf{Y} = \mathbf{XB} + \mathbf{F}$) is the typical representative of type 1). The linear mixture model, termed Beer's 'Law' in spectroscopy ($\mathbf{X} = \mathbf{YK'} + \mathbf{E}$) represents type 2) and its extension (extended mixture model, see section 3.6.3) is a member of class 3). Regression on latent variables like partial least squares (PLS) regression is a representative of type 4) (section 3.5).

The forward regression model is a standard tool for statisticians. Multiple linear regression (MLR) in this model seeks to ascribe a unique aspect of the variability in the Y-data to each and every X-variable employed. But if two or more X-variables reflect the same basic phenomenon, this is impossible. Consequently, redundant variables have to be eliminated. This is often termed stepwise multiple linear regression (SMLR) in the literature (see e.g. section 3.7).

Chemists, on the other hand, may like the mixture model because it corresponds to their causal understanding (section 3.6). But this model does not allow for unidentified systematic phenomena caused by physical and chemical interferences, and therefore may perpetuate mistaken thinking.

Level 1 in this book focuses mainly on 4), the regression on latent variables T (section 3.3), since this is flexible and computationally simple. It uses the data well and allows 'smart' predictions that detect outliers. These methods also provide extensive possibilities for looking at the data and model from various points of view, both with respect to the variables and to the objects.

Level 2 covers other methods as well, in particular the extended mixture model (section 3.6.3), and discusses how the first two methods can be regarded as special cases of the last two.

2.3 STATISTICAL PROBLEMS IN CALIBRATION

SUMMARY This section is a presentation of the most important statistical problems in multivariate calibration that have to be solved. *Key words*:

random error, systematic error, collinearity, fitting criterion, experimental design, nonlinearities, outliers and updating.

In section 2.1 we presented the most basic mathematical and statistical tools which are used in this book and in this section we shall use these tools to discuss statistical problems in calibration. Solutions to the problems are presented in subsequent chapters.

2.3.1 THE EFFECT OF RANDOM ERROR

In section 1.3 we noted that *measurement errors* or *noise* cannot be completely avoided and in practice many types of noise have to be considered. They may be due to object heterogeneity, thermal noise in electronic components, uncontrolled interferences, human mistakes in chemical or instrumental analyses, computer round-off errors and incomplete data structure modelling. Although these may actually be deterministic, not random in the sub-nuclear sense of the word, we often assume these errors to be *randomly* distributed, because we do not understand them and cannot predict them.

The effect of random error is most easily illustrated by simple linear least squares regression.

Assume that we have to calibrate the relation between a chemical constituent y and a spectral measurement x and that the model

$$y = b_0 + xb + f \qquad (2.45)$$

holds for the relation between x and y (see section 2.1.2.3). Typical calibration data for this model are illustrated in Figure 2.12.

The estimators of b_0 and b are most easily found by LS and the solution is presented in section 2.1.2.3. Due to random errors f the estimates $\hat{b}_0$ and $\hat{b}$ will have a statistical uncertainty. Replicated computations on different data will give

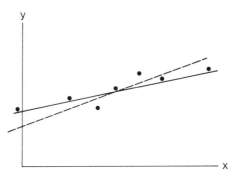

Figure 2.12 The solid line illustrates the line fitted to the 7 data points. The dotted line corresponds to the regression line when the two points furthest away from the centre are deleted. Even though all data points are generated by the same model, the two regression lines are quite different due to random noise

somewhat different estimates. The covariance matrix of $\widehat{b}_0$ and $\widehat{b}$ is for standard assumptions given in section 2.1.2.3.

In statistics one usually looks for estimators with as small uncertainty variation (variance) as possible. In fitting the model to data, the results depend on design of the experiment, on noise in the data (sampling noise, measurement noise), inadequacies in the mathematical model and on the number of objects used. In statistics, it is important to find methods that filter out as much as possible of the noise and get as precise estimates as possible.

In multivariate calibration it is useful to distinguish between independent errors for the different variables and systematic noise affecting all or at least some of them (see section 3.3 and section 3.6). If we measure e.g. absorbance at many

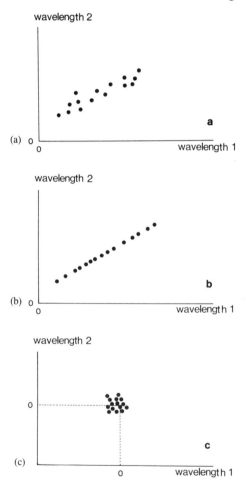

Figure 2.13 Repeated absorbance measurements at two wavelengths at constant analyte level, but varying levels of an unidentified interferent plus random noise contributions. The composite distribution in a) is a sum of the systematic distribution in b) plus the random noise in c)

wavelengths, only a part of the total measurement noise varies independently from wavelength to wavelength. The rest of the noise is due to presence of unidentified interferents with overlapping spectra affecting many wavelengths, interfering physical effects, etc.

This type of systematic, although unidentified errors gives rise to systematic *covariance* among the wavelengths, in addition to the variances of the individual variables (Figure 2.13). The systematic covariance can be picked up from the data and modelled together with the other, more desirable systematic variations, e.g. by T as described at the bottom of Frame 2.4.

For instance, in Figure 2.13(a) the absorbance at two wavelengths has been measured for different objects containing the *same* analyte concentration. Ideally all the data should have been equal. The variability is caused by two phenomena: Part of the variability is due to random, normally distributed thermal noise (c) from the electronic detector/amplifier system; this error is uncorrelated between the two wavelengths. But the other part (b) is an uncontrolled absorbance contribution from various levels of an unidentified interferent that also absorbs light at these wavelengths. By statistical techniques it is possible to model and separate this systematic noise from the totally unstructured noise.

2.3.2 COLLINEARITY

In mathematical terms, the X matrix is here called collinear (or multicollinear) if the columns in X are approximately or exactly linearly dependent. In other words, X is collinear if at least one of the X-variables can be written as an approximate or exact linear combination of the others. Remember that in case of exact linear dependence, the LS estimator is not unique and cannot be expressed by (2.37). In the following we will concentrate on the case where the equation (2.37) is valid.

From section 2.1.2.3 we know that $\text{cov}(\hat{b}) = \sigma_f^2 (X'X)^{-1}$. Using the eigenvector representation of $(X'X)^{-1}$ (see e.g. (2.22)) it can be shown that at least one diagonal (variance) element of this covariance matrix is large if X is collinear. This means that in case of collinearity some elements of the vector $\hat{b}$ from the LS fit have large variance. Thus, for the MLR solution, collinearity in X may have a detrimental effect on the stability of the coefficients of $\hat{b}$ and render them useless for causal interpretation.

Consider as an example the simple regression situation with two X-variables in Figure 2.14. In this case the correlation between the two X-variables is very high; variability along the dotted line in the $x_1 - x_2$ plane is very much larger than across this line. The collinearity problem in regression corresponds to the fact that it is difficult to estimate how y varies with this latter direction of small variation in X. If this minor direction in X is important for the prediction of Y (see section 3.4.6), then the collinearity represents a problem.

However, it is important to note that if this minor direction in X is more or less irrelevant for the prediction of Y, which is often the case in spectroscopy, then the collinearity is not a problem—provided that some other calibration method than MLR is used (Chapter 3). In Figure 2.14 it may be sufficient to base the

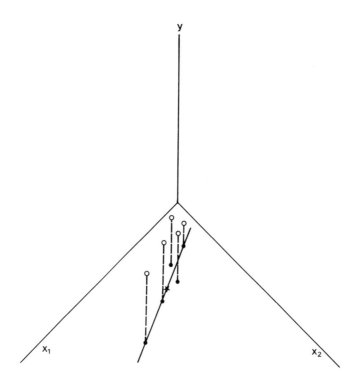

Figure 2.14 The two X-variables are collinear as illustrated by the high correlation between them. The star represents the centre in the $x_1 - x_2$ space

modelling and estimation on the first principal component in X (solid line in the X-plane, see chapter 3.4) and ignore the other X-direction. So using this principal component regression (PCR) or alternatively the partial least squares regression (PLSR) covered in sections 3.4 and 3.5, this collinearity between X-variables represents a stabilizing advantage rather than a problem! This will be discussed in more detail in sections 3.3–3.5.

However, other types of collinearity may be more important (see section 3.6). For instance we may want to quantify J different analytes simultaneously, from absorbance spectra of mixtures. The success of this will depend on how linearly independent their individual analyte spectra are, because the modelling will implicitly or explicitly consist of regression with the mixture spectra as regressands and the pure analyte spectra as regressors. If for instance two analytes have identical spectra, or if one analyte's spectrum resembles the sum of two other analytes' spectra, then the analysis will be unable to resolve these. The matrix of the pure analytes' spectra is then collinear; several of its columns are linearly dependent.

Thus it is valuable to have methods to assess the collinearity of a matrix. Several measures of collinearity exist, but some of them may be difficult to interpret. Two of the most important are variance inflation factor (VIF) and condition index (or number). The condition index is simply defined as the square root of the ratio of the largest and smallest eigenvalue of $\mathbf{X'X}$ and measures the ratio between the variation accounted for by the principal component (see section 3.4) with largest variance and the variation of the principal component with smallest variance. This will certainly be large when there is a collinearity between the x-variables, but it gives very little information about what type of collinearity or how important it is for the estimation.

The VIF exists for each parameter in the regression equation and measures the increase in variance compared to an orthogonal design (i.e. uncorrelated x-variables). In mathematical terms, the VIF for parameter k is defined as

$$\text{VIF}_k = 1/(1 - R^2_k) \tag{2.46}$$

where R^2_k is the squared correlation coefficient between x_k and $\hat{x}_k$ where $\hat{x}_k$ is the linear predictor of x_k based on some linear fit of x_k to the rest of the X-variables (see also section 3.7).

We refer to Belsley et al. (1980) and Stewart (1987) for further details on the use of the two measures.

There are a number of methods that can be used to solve the collinearity problem and in Chapter 3 we will go through a number of the most important candidates with main emphasis on so-called data compression methods. We will also discuss the advantages and disadvantages of the different methods and how to choose among them (Chapter 4).

2.3.3 DISTRIBUTION OF FUTURE OBJECTS WITH INTERFERENCES

Several statistical concepts of central importance in multivariate calibration may appear foreign to chemists at first sight.

One concept which has been covered in section 2.1 is the *population* of future objects. A related conceptual problem is the statistical *distribution* of this unknown population.

A third problem is that of modelling *unidentified* interference phenomena (Figure 2.13). A particularly foreign concept of course may be the statistical distribution of such unidentified interferences in future unknown objects!

These concepts may be essential for defining what is a 'good' calibration (see section 3.6.3).

2.3.4 FITTING MODELS TO DATA

In calibration we have to estimate the unknown parameters in the theoretical mathematical model in order to obtain the resulting X–Y and X–X calibration relationships.

There are many different statistical principles by which to fit models to experimental data; probably the most basic and important are the *least-squares* and *maximum likelihood (ML)* methods. By the *least squares* principle (see section 2.1.2.3) we simply minimize an average of the squared residuals, apparently without any serious assumptions about them needed.

In contrast, the *maximum likelihood* principle fits models to data based on more or less rigid statistical assumptions on the nature of the distributions of errors and true qualities in the data. Under certain assumptions the least squares and the maximum likelihood principles result in the same estimation method; otherwise they produce different statistical methodology.

In the present book the least squares principle is primarily used, because of its flexiblity and because it is intuitively easy to understand. It should be noted that both these techniques may be quite sensitive to serious *outliers*; one badly erroneous measurement may influence a calibration model as much as a number of good measurements! The topics of robustness and other warnings are therefore important.

2.3.5 OUTLIERS AND ROBUSTNESS

In practical applications of statistics one may always have observations showing some types of departure from the bulk of the data. Such observations are called *outliers* or abnormal observations. These can occur both in calibration and in prediction. In calibration such outlying observations may be detrimental to the quality of the predictor. In prediction it is important to have some type of error warning to avoid individual mistakes.

The outliers may occur for many different reasons. There may be printing error, chemical laboratory error, objects from another population, instrument error, etc.

Therefore outlier detection (and possibly rejection) is important. Many different tools for detecting outliers in regression type problems will be discussed in Chapter 5.

During the last decades *robust* statistical methods for automatic reduction of the influence of outliers have been developed (see e.g. Huber, 1981). Based on certain statistical assumptions about the distributions of possible outliers, the available data are 'massaged' to reduce their influence.

However, in calibration, an outlier may sometimes be highly informative and not at all erroneous: It may be an outlier only because it alone spans a certain type of important variability in the X-data. Therefore this book avoids 'robust statistics'. Instead, our approach is to alert the user for outliers, but let him or her decide what to do with them. This leaves the user more in control of the modelling.

Some authors also advocate outlier detection accompanied with robust statistics as a safe way to approach data, but this is not considered here.

2.3.6 SPANNING THE VARIABILITY SPACE

In order to be able to predict $\hat{Y} = f(X)$ sufficiently well, it is necessary to estimate $f(X)$ with sufficient reliability. This requires sufficiently informative calibration data. Therefore the choice of calibration objects can be of great importance.

The main points in experimental design for multivariate calibration is the spanning of all important types of variability by the calibration object set—both the analyte variations, the known interferences and the more or less unidentified interferences.

If nonlinearity problems or population heterogeneities are expected, various relevant combinations of these must also be spanned during calibration, in order to get good prediction ability.

In practice, it is important to have design strategies with a high probability of yielding the required information, even when our a priori understanding and theoretical model formulation is incomplete or wrong.

Strategies for calibration design are discussed in Chapter 6.

2.3.7 NONLINEARITIES

The choice of mathematical model should always be based on good fit to actual calibration data. Otherwise one may obtain unreliable mathematical prediction formulae that yield imprecise predictions.

In this book we will mainly cover linear regression methods. This means that we assume the relation between spectral and chemical variables, either separately or simultaneously, to be adequately approximated by a linear model (i.e. linear in the model parameters).

In some cases, however, nonlinearity is a problem. One may then select a model among a nonlinear class or simply use nonparametric regression. However, many nonlinear modelling tools are more complicated to use and/or specialized for very narrow applications.

Greater generality can be attained by meeting nonlinearity problems with a combination of the following three strategies:

1) Split the calibration population into several more narrow sub-populations so that linear approximations become more satisfactory. For instance, when calibrating for water pollution level from multichannel satellite images, calibrate for shallow lakes and deep lakes separately.
2) Preprocess the input data to make them more suitable for linear modelling. For instance, correct the satellite image data for atmospheric haze variations before calibrating.
3) If necessary, form nonlinear transformation of the input variables, and use them e.g. together with the original variables in the linear modelling. For instance, if 7 wavelength channels $x_k, k = 1, 2, \ldots, 7$ are available from the satellite, use also $x_k^2, k = 1, 2, \ldots, 7$ as in polynomial regression.

Linearity transformations are covered in Chapter 7. Nonlinear approaches to calibration modelling are discussed in section 3.7.

2.3.8 UNDERFITTING VERSUS OVERFITTING

When fitting an abstract mathematical model to data one has to decide on its degree of complexity, i.e. the number of independent parameters that have to be estimated. When more and more parameters are introduced into the model, the chance increases that the estimation process draws noise and other spurious phenomena from the calibration data into the resulting calibration model. The fit to the calibration data will look good, but when used in prediction the predictor may fail completely. This phenomenon is called overfitting.

The opposite phenomenon, namely applying a too small or unrealistic model is usually called underfitting. How to determine the optimal degree of complexity in a given calibration model, compromising between underfitting and overfitting, is a question of model validation.

In section 3.4 and Chapter 4 overfitting and underfitting are discussed in detail.

2.3.9 UPDATING

In calibration, an a priori understanding was expressed as: 1) choice of variables for measure (X and Y), 2) experimental design for selecting objects to measure, 3) a theoretical mathematical model, usually a 'soft' model. The parameters of this model were estimated from the obtained calibration data. There may be two reasons for wanting to update the resulting calibration model.

First of all the quality and quantity of available data may not have been sufficient at the time of calibration. As new control data are established, it may be desirable to merge these with the original calibration data and recalibrate in order to get a better estimation.

Secondly, there is the problem of drift. Analytical instruments may have a drift, and the population of objects to be analyzed (e.g. an industrial process) may change over time. The resulting calibration model may have to be updated, in order to re-establish its performance as local approximation.

In some simple situations the drift itself may be well understood and can be modelled and corrected for as function of time. In other situations the drift itself is of a simple form; then it is possible to correct for it by a simple adjustment, e.g. bias and slope correction, based on only a few new data. Otherwise, a more or less full recalibration may be required.

Further discussion on updating is given in Chapter 8 (see also sections 3.5 and 4.4).

3 Methods for Calibration

3.1 CLASSIFICATION OF CALIBRATION METHODS

SUMMARY In this section the following distinctions between calibration methods are discussed:

Univariate	— multivariate
Linear	— nonlinear
Selection	— weighting (full-spectrum)
Direct calibration	— indirect calibration
forward ('inverse')	— reverse ('classical')

As indicated in sections 2.2 and 2.3 there are a number of different calibration methods. Many of them will undoubtedly give similar results in many applications, but cases do exist where their performance can be quite different. The choice of method depends on the type of calibration problem and on what prior knowledge and data are available. Some methods, like the Partial Least Squares Regression (PLSR) are particularly versatile, giving satisfactory performance for a wide variety of situations.

The different methods can be classified in many different ways and here we give an overview of important classes of methods to be discussed later.

Univariate–multivariate

The most basic distinction is between univariate and multivariate calibration methods. The advantages of multivariate calibration were discussed in Chapter 1 and the majority of methods presented in the present chapter are multivariate. (It should be mentioned that univariate calibration is a special case of multivariate calibration.)

Linear–nonlinear

Another basic distinction is between linear and nonlinear calibration, i.e. methods that yield linear or nonlinear functions of the X-variables.

In more detail, by definition the linear calibration methods are those that can be written as

$$\hat{y} = \hat{b}_0 + \sum_{k=1}^{K} x_k \hat{b}_k \qquad (3.1)$$

where parameter estimates $\hat{b}_0$ and $\hat{b}_k$ are determined either from the data or from prior information about the X–Y and X–X relationships (see section 3.6). These methods are thus *linear in the parameters*, although they can accomodate several types of *nonlinear relationships in the data* (Chapter 7).

However, if a predictor of this type is not suitable, then it is possible to transform the spectral input variables by some *nonlinear preprocessing*, and then use these as predicting variables in the linear model (see section 2.3.7). The predictor is then of the type

$$\hat{y} = \hat{b}_0 + \sum_{k=1}^{K} n_k(x_k) \hat{b}_k \qquad (3.2)$$

where $n_k(x_k), k = 1, \ldots, K$ here are known nonlinear functions of the input variables x_k. This type of preprocessing is treated in Chapter 7.

Finally, more explicit *nonlinear modelling* methods can be used. These are discussed in section 3.7.

Selection–weighting (full-spectrum)

An important question in multivariate calibration is whether all available X-variables (e.g. wavelengths) should be used or only a few of them. Many of the calibration methods used in practice can only employ a small number of available predicting variables and have to ignore the rest, for purely mathematical reasons. In contrast, the full-spectrum methods given main attention in this book can use all the available wavelengths of relevance. In some cases a compromise between the two strategies is advantageous.

The distinction is particularly important when the X-variables are highly collinear. In this case full multiple linear regression may become useless, and one way to solve this is to decrease the number of wavelengths used as X-variables. The strategy advocated in this book is instead based on reducing the number of individual regressor variables by taking a few linear combinations of all available wavelengths (full-spectrum methods).

Direct–indirect

This distinction is related to the amount of prior information. Direct calibration means that the model parameters are known a priori and can be used to construct a predictor. An example of this is the direct 'unmixing' to resolve spectral overlaps etc.: When the X-responses of all the relevant constituents are known in advance, and the measurements are otherwise nicely behaved, then the linear mixture model can be solved directly (section 2.1.1.3). This can be the case in simple situations like transmission spectroscopy of transparent solutions containing a few constituents at low concentrations (section 3.6.1). But in many analytical situations such fully causal calibration is useless, because there are unidentified interferents, nonlinear instrumentation or the constituents interact with each other in mixtures and change their characteristics (see Chapter 1).

On the other hand, indirect calibration may need many empirical calibration data to allow statistical estimation of the unknown parameters in the calibration model. Intermediate possibilities also exist, where some parameters are known a priori and some are estimated from the calibration data. This distinction is considered in some detail in section 3.6.

Forward ('inverse')–reverse ('classical')

The present distinction concerns two ways of thinking about the calibration modelling; predictive vs causal modelling. Statistically, it is the question of regressing $\mathbf{Y}$ on $\mathbf{X}$ or regressing $\mathbf{X}$ on $\mathbf{Y}$.

The main purpose of calibration is prediction $\hat{\mathbf{Y}} = f(\mathbf{X})$, not causal modelling. This is therefore here called the 'forward' direction, from $\mathbf{X}$ to $\mathbf{Y}$.

A forward predictor can be obtained by regressing calibration data $\mathbf{Y}$ on $\mathbf{X}$, using the forward regression model

$$\mathbf{Y} = \mathbf{X}\mathbf{B} + \mathbf{F} \tag{3.3}$$

where $\mathbf{X}$ represents e.g. spectra, $\mathbf{Y}$ concentrations of one or more analytes, $\mathbf{B}$ are the corresponding regression coefficients and $\mathbf{F}$ their residuals (model errors, noise etc.). This yields a predictor of the multiple linear regression type

$$\hat{\mathbf{y}}_i{}' = \mathbf{x}_i{}'(\mathbf{X}'\mathbf{X})^{-1}\mathbf{X}'\mathbf{Y}$$

for future objects $i = 1, 2, \ldots$

Based on how the method was originally developed (see e.g. Krutchkoff, 1967), this type of calibration is sometimes referred to in statistics as 'inverse' calibration. This forward approach can be improved in various ways (sections 3.3–3.5), e.g. by regressing $\mathbf{Y}$ on a few linear *combinations* of the $\mathbf{X}$-variables, instead of on all the $\mathbf{X}$-variables themselves. For instance, the partial least squares regression (section 3.5) is such a modified forward method.

The forward regression model (3.3) resembles the causal structure in some calibration problems—for instance is toxicity $\mathbf{Y}$ caused by toxin concentrations $\mathbf{X}$: $\mathbf{Y} \Leftarrow g(\mathbf{X})$ (see section 2.2.3). But for most analytical applications, e.g. light spectroscopy, the causal structure is the reverse of model (3.3), i.e. the predicting variables $\mathbf{X}$ are caused by the analytes $\mathbf{Y}$: $\mathbf{X} \Leftarrow g(\mathbf{Y})$. It may then be appealing to develop the calibration model accordingly.

In the standard prediction notation chosen here, the linear (reverse) mixture model can for a set of objects be written as

$$\mathbf{X} = \mathbf{YK'} + \mathbf{E} \qquad (3.4)$$

where $\mathbf{X}$ represents spectra, $\mathbf{Y}$ represents the $\mathbf{J}$ analytes' concentrations, $\mathbf{K}=(\mathbf{k}_1, \ldots, \mathbf{k}_J)$ is the matrix of unit spectra (e.g. absorptivity) of the $\mathbf{J}$ analytes, and residual $\mathbf{E}$ represents model errors and random measurement noise.

If $\mathbf{K}$ is not fully known in advance then we first need an indirect estimate, $\hat{\mathbf{K}}$. This can be obtained, e.g. by first regressing $\mathbf{X}$ on $\mathbf{Y}$: $\hat{\mathbf{K}} = \mathbf{X'Y(Y'Y)}^{-1}$ in the calibration set.

Then a so-called 'reverse' (classical) calibration method is obtained, by replacing the $\mathbf{K}$ matrix in (3.4) and using the ordinary LS estimator for $\mathbf{y}$. This means that the reverse predictor for future objects $i = 1, 2, \ldots$ is

$$\hat{\mathbf{y}}_i{}' = \mathbf{x}_i{}'\hat{\mathbf{K}}(\hat{\mathbf{K}}'\hat{\mathbf{K}})^{-1} \qquad (3.5)$$

This method was the original and basic approach to calibration and is therefore sometimes refered to as 'classical' calibration. It may work well for many nicely behaving instruments applied to one-constituent problems and simple mixtures. But its very causal basis can be limiting: As it stands, (3.5) cannot accomodate $\mathbf{X}$-data which are affected by common selectivity problems like unidentified constituents, turbidity, temperature effects etc. This reverse modelling then has to be extended by more advanced statistical procedures, as described in section 3.6.

The terminology *reverse* and *forward* points to the way the actual regression modelling is done in the two cases, and does this in agreement with the main goal of the calibration, namely the 'forward' prediction. In the present book, the definition of what is $\mathbf{X}$ and what is $\mathbf{Y}$ has been taken from prediction theory, and not from earlier causal calibration traditions, which required the predictor to be written '$\hat{\mathbf{X}} = f(\mathbf{Y})$'.

The forward and reverse calibration methods give predictors with somewhat different statistical properties. But in general, if the available data are quite precise and linearly related, the difference is small. However, the calibration models obtained can also have quite different *interpretations*. The relative merits of forward and reverse calibration are discussed in sections 3.2 and 3.6.

Is it necessary to know all the methods?

The practical user does not have to master many different calibration methods (Frame 2.4). One calibration method, like the forward partial least squares regression (PLSR) (section 3.5) or PCR (section 3.6), can in fact handle most calibration problems. This is especially so if it is combined with sensible experimental design (Chapter 6), sensible preprocessing of the input data (chapter 7) and sensible validation and outlier detection (Chapters 4 and 5). Structurally the PLSR resembles the common-cause relationship, $(\mathbf{X}, \mathbf{Y}) \Leftarrow g(\mathbf{U})$, but can also be used for causal structures like $\mathbf{Y} \Leftarrow g(\mathbf{X})$ and $\mathbf{X} \Leftarrow g(\mathbf{Y})$ (Frame 2.3).

However, in order to understand the fundamental calibration problems that have to be solved, and to master the philosophy behind such a flexible calibration method, the user may benefit from studying various other methods as well.

A brief overview over various multivariate data–analytic methods is given in Martens et al.(1983c). The next section illustrates some basic issues in calibration, in terms of univariate calibration. Subsequent sections compare and describe a few main calibration methods in more detail.

3.2   UNIVARIATE CALIBRATION

SUMMARY   Univariate calibration is reviewed. It is focussed on the linear 'classical' (reverse) and 'inverse' (forward) methods, but nonlinear methods are also briefly described. The main results on prediction ability and functional relationship are given.

According to the terminology, univariate calibration refers to construction of a relation between two variables x and y such that x can be used to predict y. This type of calibration has been in common use for many years in chemical analysis and there are certainly well-defined and simple cases where it can be valuable.

This section is devoted to a presentation of established univariate methods and of their statistical properties. In addition to a review of the statistical theory on univariate calibration, the section can be useful as an introduction to the more complicated multivariate methods to be treated later.

Assume as above that we have a set of $i = 1, 2, \ldots, I$ calibration objects, i.e. I observations of the 2-element vector (x_i, y_i). The means, variances and covariance in the calibration objects are in terms of these data defined as

$$\bar{x} = \sum_{i=1}^{I} x_i / I \tag{3.6}$$

$$\bar{y} = \sum_{i=1}^{I} y_i / I \tag{3.7}$$

$$s_x^2 = \sum_{i=1}^{I} (x_i - \bar{x})^2 / (I - 1) \tag{3.8}$$

$$s_y^2 = \sum_{i=1}^{I} (y_i - \bar{y})^2 / (I - 1) \tag{3.9}$$

$$\widehat{\text{Cov}(x, y)} = \sum_{i=1}^{I} (x_i - \bar{x})(y_i - \bar{y}) / (I - 1) \tag{3.10}$$

We shall first consider linear relations in which X-data are assumed generated according to the linear mixture model, i.e.

$$x = k_0 + yk + e \qquad (3.11)$$

where in the case of spectroscopy the intercept k_0 would be a baséline offset (background effect), k the absorptivity constant and error e random measurement noise due to the instrument. The error terms are assumed independent with expectation equal to 0 and variance equal to σ_e^2 (see section 2.1.2.3). This model was originally presented by Eisenhart (1939) and used extensively in chemistry thereafter. In applying this linear, reverse data model we assume that measurements of y are error free and the linear model is an adequate approximation of the actual X–Y structure (see Lwin and Spiegelman, 1986; Yum, 1987).

As explained in the previous section, the somewhat awkward X–Y notation in this causal model is the consequence of choosing the more natural predictor notation $\hat{Y} = f(X)$ and not the other way around.

A typical scatterplot of data generated according to this model is shown in Figure 3.1. The data are given in Table 3.1.

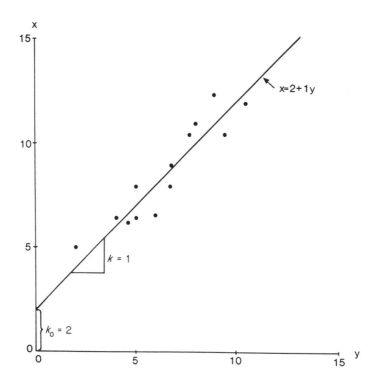

Figure 3.1 Typical calibration data generated by model (3.11)

Table 3.1 Data for Figure 3.1

nr.	Target y	Measurement x
1	2	5
2	4	6.5
3	4.5	6.25
4	5	6.5
5	5	8
6	6	6.5
7	6.5	8
8	6.5	9
9	7.75	10.5
10	8	11
11	9	12.5
12	9.5	10.5
13	10.5	12

'Classical' versus 'inverse' calibration

The calibration for data generated according to model (3.11) can be done in many ways. The main body of the literature, however, is devoted to comparisons of the *linear 'inverse'* and *'classical'* approach. Since the 'inverse' vs. 'classical' terminology is so established in the univariate calibration literature, it will be used here instead of the forward vs. reverse terminology used elsewhere in this book (see section 3.1).

In 1967 Krutchkoff started the discussion of the relative merits of these methods by suggesting to use the 'inverse' predictor

$$\hat{y}_{\text{inv}} = \bar{y} + \frac{\widehat{\text{Cov}(x,y)}}{s_x^{\;2}}(x - \bar{x}) \qquad (3.12)$$

obtained by regressing y on x over the calibration objects. This was certainly against established practice which insisted that the best approach was to use the fitted linear regression line of x on y, i.e. the equation

$$x = \bar{x} + \frac{\widehat{\text{Cov}(x,y)}}{s_y^{\;2}}(y - \bar{y}) \qquad (3.13)$$

which is just an estimate of equation (3.11) with the usual least squares estimators of the parameters, i.e.

$$\hat{k} = \frac{\widehat{\text{Cov}(x,y)}}{s_y^{\;2}}, \; \hat{k}_0 = \bar{x} - \bar{y}\hat{k} \qquad (3.14)$$

The predictor obtained by reversing this causal model is equal to

$$\widehat{y}_{cl} = \bar{y} + \frac{s_y^2}{\widehat{\text{Cov}(x,y)}}(x - \bar{x}) \tag{3.15}$$

We refer to Figure 3.2 for an illustration where the calibration is done by the data in Table 3.1. The three prediction objects are generated from the same model as the calibration objects and are for illustration selected at different levels of y.

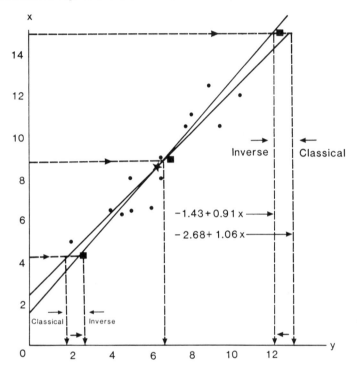

Figure 3.2 Illustration of prediction by the 'inverse' and the 'classical' predictor. There are 13 calibration objects (circles) and 3 test objects (squares) not present in the calibration. Notice the shrinking of the 'inverse' predictor towards the mean $(\bar{x}, \bar{y})$ in the calibration set

Functional relationship between $\widehat{y}_{inv}$ and $\widehat{y}_{cl}$

The relationship between the two predictors is developed in e.g. Hoadley (1970) and can according to Næs (1985a) be written as

$$(\widehat{y}_{inv} - \bar{y}) = \frac{\widehat{k}^2 s_y^2}{\widehat{\sigma}_e^2(I-2)/(I-1) + \widehat{k}^2 s_y^2}(\widehat{y}_{cl} - \bar{y}) \tag{3.16}$$

where $\widehat{k}$ and $\widehat{\sigma}_e^2$ are the usual estimates of k and σ_e^2 obtained from the calibration objects (see section 2.1.2.3). We see that

$$\hat{k}^2 s_y^2 / (\hat{\sigma}_e^2 (I - 2)/(I - 1) + \hat{k}^2 s_y^2) \tag{3.17}$$

is always less than 1 and it follows that

$$|\hat{y}_{\text{inv}} - \bar{y}| < |\hat{y}_{\text{cl}} - \bar{y}| \tag{3.18}$$

i.e. the 'inverse' predictor represents a shrinking of the 'classical' predictor towards the centre of the calibration objects, $\bar{y}$. From the relation (3.16) we also see that the difference between the two predictors (apart from $(I - 2)/(I - 1)$ which is close to 1 for moderate and large sample size) is only dependent on the signal to noise ratio

$$\frac{\hat{k}^2 s_y^2}{\hat{\sigma}_e^2} = \frac{\widehat{\text{Var}(yk)}}{\text{Var}(\hat{e})} \tag{3.19}$$

When this ratio is large the two predictors are indistinguishable. In other words, if $\hat{\sigma}_e^2$ is small compared to the numerator $\hat{k}^2 s_y^2$, $\hat{y}_{\text{inv}}$ is almost equal to $\hat{y}_{\text{cl}}$.

The two predictors based on the data in Figure 3.1 are given in Figure 3.2. The relation between them is equal to

$$(\hat{y}_{\text{inv}} - 6.48) = 0.86(\hat{y}_{\text{cl}} - 6.48) \tag{3.20}$$

so the ratio in (3.17) is here 0.86. Figure 3.3 shows the lack of fit $(y - \hat{y})$ for 13 calibration objects, as a function of the true y-level. The shrinking effect can be

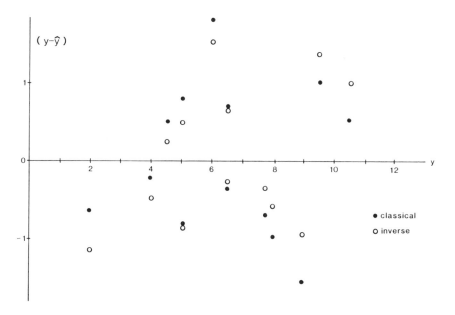

Figure 3.3 Residuals $(y - \hat{y})$ from the 13 calibration objects, based on the 'inverse' and 'classical' prediction. The open circles correspond to the 'inverse' predictor

82

seen: The inverse predictor gives best fit at medium Y-levels, while the classical predictor is the better at the highest and lowest Y-levels. The effect in (3.18) is also clear in Figure 3.2 for the two test objects in the ends of the calibration region.

As will be clear below, the 'inverse' calibration (regressing Y on X) is intended to give better predictive ability (lower mean square error, MSE) than the 'classical' (regressing X on Y), on the average for the whole unknown calibration population (see e.g. Næs and Martens, 1984). The 'inverse' predictions are stabilized against random noise in the calibration data by exerting a slight 'pull' of all predictions towards the estimated population center. Consequently, for objects with higher-than-average y-levels the prediction $\hat{y}_{inv}$ is systematically too low, and for objects with lower-than-average y-levels $\hat{y}_{inv}$ is systematically too high.

However, the importance of such 'inverse' (forward) vs. 'classical' (reverse) calibration should not be exaggerated: If the data fit well to the linear calibration model, the two calibration approaches give about the same result.

In the following a more detailed summary is given of the statistical literature concerning univariate calibration.

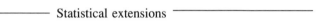

Statistical extensions

Statistical properties of $\hat{y}_{inv}$ and $\hat{y}_{cl}$

Krutchkoff (1967) presented a Monte Carlo study which was in favour of the 'inverse' predictor with respect to a modified version of the MSE. The modification was a truncation of $\hat{y}$ and was necessary since MSE($\hat{y}_{cl}$) was theoretically infinite. He concluded that the 'inverse' predictor was uniformly superior to the 'classical' predictor with respect to this criterion. This conclusion was controversial because it was against established practice and several authors joined the discussion.

Williams (1969) argued that since MSE of the 'classical' approach was infinite Krutchkoff's study was of little value. Others, however, argued that this objection was of little practical importance, since a truncation would always be done in practice.

Williams also argued that the 'inverse' predictor was difficult to justify from sound statistical principles in so-called *controlled* calibration where the calibration objects are selected due to a conscious design strategy for the constituents. Viewpoints on this were presented by Hoadley (1970), Brown (1979) and Lwin and Maritz (1982) who derived the 'inverse' predictor from different established statistical principles or criteria. Hoadley and Brown showed by assuming that future y's come from a natural population that the 'inverse' predictor could be justified as a Bayesian estimator for y. Hoadley used a full-Bayesian approach based on distributions of all parameters involved, while Brown assumed only distribution on y. The results in Brown (1979) showed that the 'inverse' predictor is especially well suited for situations where the calibration objects are representative for the population of objects (*natural* calibration). The 'inverse' method can of course also be justified as the ordinary regression estimator if the ordinary regression model

$$y = b_0 + xb_1 + f \tag{3.21}$$

(with ordinary regression assumptions on the f's) treated in e.g. section 2.1.3 is reasonable.

Faced with Williams' critique of MSE as criterion, Halperin (1970) proposed to use other comparison criteria than MSE. Asymptotic MSE (MSE as N is equal to infinity) and the asymptotic version of the measure

$$P(|\hat{y}_{inv} - y| < |\hat{y}_{cl} - y|) \tag{3.22}$$

for different values of y were analysed (here P denotes probability). If this latter quantity is larger than 1/2, the 'inverse' predictor is said to be closer to the correct value than the 'classical' method. (Notice that studying the asymptotic behaviour of the predictors in this way is identical to studying the *direct* calibration analogues of the $\hat{y}_{inv}$ and $\hat{y}_{cl}$). For both criteria it turned out that the 'inverse' method was best for Y-values in a region near the centre $\bar{y}$ of the calibration objects, while the 'classical' predictor was best outside this region. This is in harmony with the fact that the 'inverse' method can be developed as a Bayesian estimator of y. Since Krutchkoff only predicted y values in a limited interval, this also explains why he could conlude that the 'inverse' method is uniformly superior to the 'classical'.

In more detail, the y-region where the 'inverse' method is asymptotically the (MSE) best is given by the interval

$$\left(\bar{y} - \sqrt{(2s_y^2 + \sigma_e^2/k^2)}, \bar{y} + \sqrt{(2s_y^2 + \sigma_e^2/k^2)}\right) \tag{3.23}$$

Notice that the length of the interval is always larger than $2\sqrt{(2s_y^2)}$. The asymptotic risk functions or expected quadratic error for the predictors can be found in Halperin and two examples with different value of σ_e are plotted in Figure 3.4. As we see the region where the 'inverse' method is better than the 'classical' is quite large compared to the standard deviation s_y, but the 'inverse' method gives only small improvement over the 'classical' method when the σ_e is small, i.e when the X-instrument is very precise.

Similar problems as those treated in Halperin (1970) are considered in Shukla (1972) and Oman (1985). All these results show that the 'inverse' method is best suited for interpolation, while the 'classical' method is the best in extrapolation. In addition, in natural calibration, the average prediction ability is best for the 'inverse' method.

Such properties may be important for various reasons. In control applications for instance one is often interested in precise predictions near a certain limit Y-value, for instance an upper or lower legal limit of the analyte, y_{limit}. In such cases one can select a calibration set with its

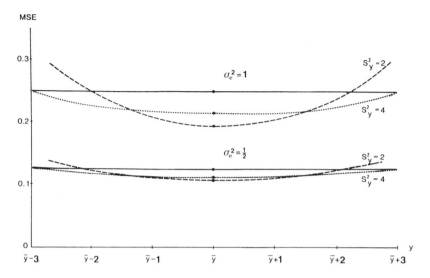

Figure 3.4 Illustration of asymptotic risk functions for the two predictors for different choices of parameters σ_e^2 and s_y^2 in the model $x = 0.5y + e$. The straight lines corresponds to the MSE of the 'classical' method, while the curves represent risk functions of 'inverse' predictors. Notice that the 'classical' method is independent of the variance of y, and independent of the actual y-value. Compared to this, the prediction error of the 'inverse' methods is lower near $\bar{y}$ and higher far from $\bar{y}$, in particular when the lack-of-fit residuals are large (high σ_e^2) and the span of the calibration is narrow (low s_y^2)

mean close to y_{limit} and with 'inverse' calibration obtain precise results near this centre.

In other cases one may want a uniformly good predictive ability over a wide range of y-levels. This could be the case if payment for some produce is to be based on $\hat{y}$; it would not be correct to under-pay the good producers and over-pay the bad producers, so the inverse predictor is not fully satisfactory. One can then bring in a wider calibration material. Using some 'classical' calibration method (or if the predictor from an 'inverse' method were transformed in analogy to (3.16)), this could make the resulting predictor applicable over a wider range, with especially good extrapolation properties.

Empirical procedures of obtaining estimates of MSE in practice are presented in Chapter 4.

Other univariate predictors

The 'inverse' and 'classical' approaches are based on regression of Y on X or X on Y respectively. It should be noted that intermediate solutions like the principal component curve (the eigenvector of e.g. standardized (x, y), see sections 2.1 and 3.4) or curves based on errors-in-variables models (Lwin and Spiegelman, 1986) also can be used. Another compromise between the 'inverse' and 'classical' approach is proposed in Næs (1985) and is motivated by the risk functions such as those in Figure 3.4. The predictor is equal to

$$\hat{y}_{cl} \text{ if } \hat{y}_{cl} \text{ is outside range } \hat{F}$$
$$\hat{y}_{inv} \text{ if } \hat{y}_{cl} \text{ is inside range } \hat{F} \tag{3.24}$$

where $\hat{F}$ is an estimate of the interval around $\bar{y}$ (3.23) in which the inverse predictor is best. This predictor is thus nonlinear in x.

Other nonlinear approaches to univariate calibration can be found in Lawton et al. (1972), Lwin and Maritz (1980) and Knafl et al. (1984). Lawton et al. (1972) used an approach based on modelling x as a nonlinear function of y while Knafl et al. (1984) used an nonparametric regression method to estimate the relation between x and y, i.e. the function g in the relation $x = g(y) + e$. For determination of unknown y the estimated g-relation has to be inverted.

Both approaches are then of the 'classical' type, i.e the measurement process is modelled, estimated and inverted for prediction. The approach in Lwin and Maritz (1980) is more like the 'inverse' method in the sense that they assume that y has a distribution and the predictor is developed as an empirical Bayes estimator of y.

Confidence intervals

Confidence intervals for y have been studied by several authors. The most famous is probably the interval presented by Scheffé (1973) (see also Spiegelman, 1980). He assumed that x can be written as a monotone parametric function of y (linear in the parameters) plus noise. He then derived a confidence interval for y by a conditional sampling approach requiring the specification of two significance levels and producing statements like 'I am 95% sure that 90% of the intervals I make are true'. A similar approach is considered by Oden (1973). Confidence intervals based on $\hat{y}_{inv}$ and $\hat{y}_{cl}$ can be found in Hoadley (1970). Nonparametric intervals can be found in Knafl et al. (1984).

 End of statistical extensions

3.3   DATA COMPRESSION: FROM MANY X-VARIABLES TO A FEW FACTORS T

SUMMARY   The general structure of multivariate data compression methods is given. The philosophy behind the strategy is discussed and the bilinear methods

(PCR and PLS) are presented as important methods in the class. The relations to other methods such as stepwise linear regression (SMLR) and Fourier regression are also briefly discussed.

3.3.1 THE IMPORTANCE OF DATA REDUCTION

The following problems are quite common when we want to predict Y from X:

1) *Lack of selectivity*: No single X-variable is sufficient to predict Y.
2) *Collinearity*: There may be redundancy and hence intercorrelations (collinearity) between the X-variables.
3) *Lack of knowledge*: Our a priori understanding of the mechanisms behind the data may be incomplete or wrong.

The reason for problem 1) may be that Y is caused by several phenomena represented by different X-variables or that the X-measurements are affected by several interferents other than analyte Y (see Frame 2.3). To attain selectivity through multivariate calibration we need to use several X-variables $x_1, x_2, \ldots, x_K$.

Problem 2) may arise because we do want a certain redundancy between the variables, in order to stabilize the predictions against noise in X and in order to get calibration models that do not leave out important information and that allow outlier detection. Thereby we cannot use calibration methods that assume each X-variable to have unique information about Y, like the traditional multiple linear regression (MLR). Some 'rank reduction' is needed for the calibration.

Problem 3) stops us from using detailed causal modelling in the calibration. For instance, we may not know a priori the X-response of every constituent and interference phenomenon as they occur *in situ*, including their possible interactions. Or the instrument response may be difficult to linearize completely. Then the linear mixture model is dangerous to use: Unless it is extended to account for the unidentified covariance structure involved, it will give wrong predictions. But with such statistically advanced extensions (section 3.6.3) the resulting calibration model may be difficult to master for the chemist.

There is a need for flexible calibration methods that handle the three problems above in an easy way which gives both good predictions and increased understanding.

3.3.2 DATA COMPRESSION METHODS

Let us now consider a general framework for 'rank reduction' or 'data compression'.

The basic structure of this approach is that the information in the many observed (manifest) variables $\mathbf{x}' = (x_1, x_2, \ldots, x_k)$ is concentrated onto a few underlying ('latent') variables, called components, scores, regression factors or just *factors*, $t_1, t_2, \ldots t_A$, i.e. (see Frame 2.4, model 4)

$$(t_1, \ldots, t_A)' = h_1[(x_1, \ldots, x_k)'] \tag{3.25}$$

and that these factors are used as regressors in the regression equation with $y = (y_1, \ldots, y_J)$ as regressand i.e.

$$(y_1, \ldots, y_J)' = h_2[(t_1, \ldots, t_A)'] + \mathbf{f}' \tag{3.26}$$

Here $\mathbf{f}$ represents those contributions to y which cannot be explained by the factors $\mathbf{t} = (t_1, \ldots t_A)$. The A factors in $\mathbf{t}$ are thought of as representing the systematic variation in the X-spectra that is important for predicting y. Together the two functions h_1 and h_2 form the desired predictor $\hat{y} = f(\mathbf{x})$, with $f(\mathbf{x}) = h_2(h_1(\mathbf{x}))$.

The data compression of many X-variables into a few T-variables simplifies the statistical calibration, by reducing the number of model parameters that have to be estimated in $f(\mathbf{x})$ in the X–Y regression. It can also simplify the interpretation of the results by revealing the main X–X relationships; the first few factors $t_1, \ldots, t_A$ together can give us a few good two-dimensional 'windows' into the K-dimensional x-space.

It is common practice to use linear modelling to approximate the relationships in the data. Linear approximations are usually developed around some 'typical' sample quality $(\mathbf{x}_*', \mathbf{y}_*')$, estimated e.g. by the means of the calibration data, $\bar{\mathbf{x}}'$ and $\bar{\mathbf{y}}'$.

Let $\mathbf{X}$ and $\mathbf{Y}$ then represent the centered input data, i.e. $\mathbf{X} = \mathbf{X}_{\text{input}} - \mathbf{1}\bar{\mathbf{x}}'$ and $\mathbf{Y} = \mathbf{Y}_{\text{input}} - \mathbf{1}\bar{\mathbf{y}}'$ (later $\mathbf{X}_0$ and $\mathbf{Y}_0$). The linear data compression model can then be written:

$$\mathbf{T} = \mathbf{XV} \tag{3.27}$$

$$\mathbf{Y} = \mathbf{TQ}' + \mathbf{F} \tag{3.28}$$

After determination of $\mathbf{V}$ and estimation of $\mathbf{Q}$, we then for the centered $\mathbf{Y}$ obtain the predictor $\hat{\mathbf{Y}} = \mathbf{XV}\hat{\mathbf{Q}}$, based on the centered $\mathbf{X}$. Modified to uncentered $\mathbf{X}$ and $\mathbf{Y}$ this yields

$$\hat{\mathbf{Y}} = \mathbf{1}\hat{\mathbf{b}}_0' + \mathbf{X}\hat{\mathbf{B}} \tag{3.29}$$

where

$$\hat{\mathbf{B}} = \mathbf{V}\hat{\mathbf{Q}}' \tag{3.30}$$

and

$$\mathbf{b}_0' = \bar{\mathbf{y}}' - \bar{\mathbf{x}}'\hat{\mathbf{B}} \tag{3.31}$$

It may be noted that contrary to the full MLR predictor, the individual regression coefficients b_{kj} are not 'independently' estimated for the data compression methods. The reason is that the collinearity problem in $\mathbf{X}$ was solved by pulling out a lower number of combined or 'compressed' regressors $\mathbf{T}$ from $\mathbf{X}$; thus the regression coefficients $\hat{b}_{kj}$ are here functions of the model parameter estimates ($\mathbf{V}$ and $\hat{\mathbf{Q}}$).

It should, however, be noted that all the linear data compression methods considered here converge to the full-rank MLR when the number of factors, A, equals the number of X-variables, K. But then they have lost their ability to solve collinearity problems and to avoid overfitting (Chapter 4), so the optimal number of factors to be used is generally lower than K.

Many different methods (i.e. definitions of **V**) are available for this type of calibration modelling on combined X-variables **T**. The performance of the methods can be quite different under different conditions. The choice should in principle depend on the prior knowledge about the calibration problem, about e.g. the underlying causal phenomena, smoothness of the X–X relations, noise levels etc.

We shall here focus on particularly flexible methods that require very little prior knowledge about causal relationships, sometimes called 'soft modelling' or bilinear methods (BLM). These methods estimate **V** from the calibration data themselves, thereby making the data themselves responsible for extracting the relevant information. The two basic methods in this class are principal component regression (PCR) and partial least squares regression (PLSR) which are the topics in the two subsequent sections.

But for the sake of overview, some other regression methods that are normally expressed in other terms (stepwise multiple linear regression, Fourier regression etc.) can also be formulated inside this data compression framework. These methods will be treated later (section 3.7), but their place in this framework is illustrated in Frame 3.1, together with the bilinear methods.

We emphasize again (see also section 3.1) that due to the linear regression of **y** on functions, T, of the matrix **X**, the data compression methods naturally belong to the forward ('inverse') class of calibration methods. Therefore they share the theoretical properties of the EBLP method covered in section 3.6.3. We refer to that section for some basic properties of forward calibration methods.

Frame 3.1.

Data compression methods

Each factor $a = 1, 2, \ldots, A$ in (3.27) is defined by a vector $\mathbf{v}_a$, which is obtained in different ways for different methods:

1. *Stepwise multiple linear regression (see section 3.7.4.1)*

   Each vector $\mathbf{v}_a$ consists of all zeroes except for one selected element with the value 1 for X-variable no. a, e.g. $\mathbf{v}_a = (..0...1...0..)'$.
   Various strategies (forward or backward, best-of-all-combinations etc. are used to select which X-variables to include. The X- and Y-data only are used in this selection process (symbolized by dotted rectangles).
   With **T=XV**, this means that the matrix of regression factors $\mathbf{T} = \{\mathbf{t}_a, a = 1, 2, \ldots, A\}$ simply consists of A individual X-variables, selected to represent all the relevant information in X.

   (continued on next page)

Frame 3.1 *continued*

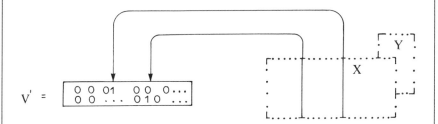

$$V' = \begin{array}{cccc} 0 \ 0 \ 0 1 & 0 \ 0 \ 0 \cdots \\ 0 \ 0 \ \cdots & 0 \ 1 \ 0 \ \cdots \end{array}$$

2. *Hruschka regression (see section 3.7.1.2):*

Each vector v_a represents the unique information in the spectrum of a certain object, x_i (a row in **X**).

The method selects the combination of objects $a = i_1, i_2, \ldots, i_A$ that together produces scores **T=XV** which correlate the best with the variation in **Y**. The Y-data may be used in this selection process.

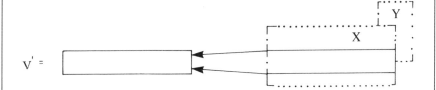

3. *Fourier regression (see section 3.7.1.1)*

Each v_a consists of smooth, mathematically pre-defined 'spectra' consisting of low- and medium-frequency sine and/or cosine functions. Each factor score t_a is thus a linear combination of the K X-variables. Both **Y** and **X** can be used for selecting the best combination of trigonometric functions.

(*continued on next page*)

90

Frame 3.1 *continued*

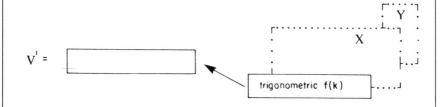

4. *Bilinear modelling (sections 3.3.3, 3.4 and 3.5)*

Each $\mathbf{v}_a$ consists of coefficients estimated from the calibration data themselves according to some optimization criterion.
For PLS (section 3.5) the A factors together summarize the information in $\mathbf{X}$ that is relevant for modelling $\mathbf{Y}$; hence both $\mathbf{X}$ and $\mathbf{Y}$ are used in the estimation of $\mathbf{V}$.
In PCR, in contrast, only the X-data are used in estimating $\mathbf{V}$; the Y-data are just used for selecting which principal component factors from $\mathbf{X}$ to use in the final model.

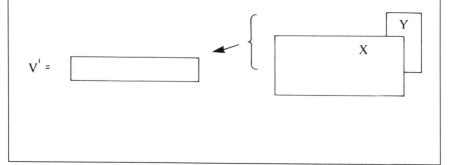

3.3.3 BILINEAR MODELLING (BLM)

Like the other data compression methods, the bilinear methods are usually applied as forward ('inverse') calibration methods (see sections 3.1, 3.2 and 3.6): $\mathbf{Y}$ is modelled as a function of $\mathbf{X}$ (actually of $\mathbf{T}=\mathbf{XV}$) and not the other way around.

However, contrary to the other methods in this framework (Frame 3.1), the bilinear methods imply some type of (least squares) *estimation* of the elements in $\mathbf{V}$ from the calibration data themselves, instead of selection only (SMLR and Hruschka regression) or fixed mathematical transformation (Fourier regression). To emphasize this estimation we shall here use the symbol $\hat{\mathbf{V}}$ and $\hat{\mathbf{T}}$ for the estimates of $\mathbf{V}$ and $\mathbf{T}$ respectively. The underlying $\mathbf{V}$ and $\mathbf{T}$ can then be thought of as the values of $\hat{\mathbf{V}}$ and $\hat{\mathbf{T}}$ for a large (infinite) number of calibration objects from the actual population (see e.g. Næs and Martens, 1985). The name '*bilinear* modelling' comes from the way $\mathbf{X}$ itself is approximated by a model $\mathbf{X} = h_3(\mathbf{T}) + \mathbf{E}$ that is the product of *two* sets of linear parameters to be estimated, termed scores ($\mathbf{T}$) and loadings ($\mathbf{P}$, see equation (3.32)) plus noise $\mathbf{E}$.

A priori linearization and weighting of the input variables

Prior to application of bilinear calibration models, some preprocessing may have to be considered. The relation between variables should be sufficiently linear and/or the calibration population should be sufficiently narrow to ensure reasonable bilinear modelling (see Chapter 7). In addition, the different steps in BLM-estimation are based on some type of least squares fit, and it may therefore be important that the X-variables are properly weighted (scaled) relative to one another prior to the modelling. Similarly, if two or more Y-variables are to be calibrated for simultaneously in PLS2 (see section 3.5.4), the Y-variables must also be weighted relative to one another prior to the modelling.

We here assume that our variables $\mathbf{X}_{\text{input}}$ and $\mathbf{Y}_{\text{input}}$ have already been properly linearized and weighted to ensure comparable noise levels (see Chapter 7).

Calibration model

In terms of centered $\mathbf{X}$ and $\mathbf{Y}$ variables (see section 3.3.2) the full bilinear calibration model can be written

$$\mathbf{X} = \mathbf{TP'} + \mathbf{E} \qquad (3.32)$$

$$\mathbf{Y} = \mathbf{TQ'} + \mathbf{F} \qquad (3.33)$$

where, as before

$$\mathbf{T} = \mathbf{XV} \qquad (3.34)$$

Loading matrix $\mathbf{P}$ represents the regression coefficients of $\mathbf{X}$ on $\mathbf{T}$ in the same way as $\mathbf{Q}$ represents the regression coefficients of $\mathbf{Y}$ on $\mathbf{T}$.

The residuals $\mathbf{E}$ and $\mathbf{F}$ represent the unique variation in $\mathbf{X}$ and $\mathbf{Y}$ that is not explained by the A-factor bilinear structure. These residuals can be due to measurement noise, operator mistakes, nonlinearities etc. As will be shown in Chapter 5, residuals may be very important for diagnostic checks of the model.

Conceptually, the bilinear calibration model $(\mathbf{X}, \mathbf{Y}) = h(\mathbf{T}) + (\mathbf{E}, \mathbf{F})$ resembles the causal common-cause model structure $(\mathbf{X}, \mathbf{Y}) \Leftarrow g(\mathbf{U})$ (Frame 2.3), but it also provides suitable representation of other types of data generated by e.g. the reverse mixture model $X \Leftarrow g(Y)$, the extended reverse mixture model $X \Leftarrow g(Y,U)$ or the forward causal structure $Y \Leftarrow g(X)$. However, it should be remembered that no calibration method can handle the incomplete forward structure $Y \Leftarrow g(X,U)$, since the effect of the unknown phenomena U cannot be modelled from X.

Mathematically, the bilinear (BLM) model $(X,Y)=T(P',Q')+(E,F)$ can be regarded as an intermediate between the extended reverse mixture model $(X = YK' + TP' + E)$ and the forward MLR model $(Y = XB + F)$ (Frame 2.4). Unlike the mixture models, the bilinear model does not assume each of the *individual* unknown model parameters (in this case the various elements t_{ia}, p_{ka}, q_{ja} and v_{ka}) to represent specific chemical or physical realities. But on the other hand the bilinear models both *describe* and *compensate* for the selectivity problems, in contrast to the forward MLR, which only compensates for the selectivity problems without describing them.

The underlying assumption in BLM is that the non-random structure in the many X (and Y) variables are caused by *something*, although just *what* these fewer phenomena are, may be more or less unknown to us. Together the bilinear model factors are intended to *map* these phenomena—to the extent that they can be resolved from the available calibration data. The phenomena can then be studied graphically in the resulting compressed bilinear factor model.

Consider for instance the modelling of additive mixtures with structure $X \approx YK'$ by the more empirical bilinear modelling $X \approx TP'$ (using e.g. centered X- and Y-variables). The causally interpretable structure YK' is then spanned by the bilinear model, i.e. $YK' \approx TP'$: The $A = J$ factor score columns in T span the same between-object variabilities as the constituent concentrations, the J columns in Y. Hence in this case $Y \approx TZ$, where Z is here some unknown matrix that depends on the choice of input data. Likewise, since $ZZ^{-1} = I$, the factor loading spectra, the A rows in P', span the same spectral variabilities as the constituent spectra, the J rows in K': $K' \approx Z^{-1}P'$ (see e.g. section 3.7.2.3).

So for data from a set of additive mixtures of $J = 2$ independent and distinguishable analytes with no other interferents, plots of the obtained $\hat{t}_2$ vs $\hat{t}_1$ and $\hat{p}_2$ vs $\hat{p}_1$ would reveal the relevant structure of Y and K, respectively. If X in addition to the $J = 2$ analytes were affected by a third additive phenomenon (e.g. some unidentified physical interference), we would expect to find $A = 3$ bilinear factors in X, and we could study the analytes and the unidentified interferent in two-dimensional plots of factors 1 vs 2, 1 vs 3 and 2 vs 3, or a three-dimensional plot of factors 1 vs 2 vs 3. This will be illustrated more extensively in successive sections.

However, it should be noted that the individual bilinear factors $t_a p_a'$, $a = 1, 2, \ldots A$, cannot necessarily be interpreted directly as individual constituent contributions $y_j k_j'$. That would require the rotation matrix Z to come out as a diagonal matrix, and that requires very special experimental designs.

When a high number of factors $a = 1, 2, \ldots, A$ is required to give adequate modelling, it may be difficult to do graphical interpretation without subsequent rotation and summary techniques. The psychometric factor analysis literature (e.g.

Harman, 1967) give further details on factor rotation. This topic is, however, not of central importance in multivariate calibration; in many applications the first few bilinear factors carry the most important causal information, while later factors primarily serve to correct for minor phenomena like nonlinearities etc.

Despite the rotational ambiguities, the rotation flexibility is also a clear advantage: The bilinear methods thereby avoid the dangers of unwarranted causal modelling when our a priori causal understanding and mathematical model formulation is incomplete or erroneous. Instead, they give *pragmatic* data compression that when used interactively with proper graphics and validation allows the user to attain good predictive ability *and* good causal insight at the same time.

Estimating the bilinear model parameters

Frame 3.2 summarizes the calibration and prediction based on bilinear regression.

Frame 3.2

Bilinear calibration methods for centered $\mathbf{X}$ and $\mathbf{Y}$

Calibration:

$$\hat{\mathbf{V}} = f(\mathbf{X}, \mathbf{Y}) \qquad \text{determine } \hat{\mathbf{V}}$$
$$\hat{\mathbf{T}} = \mathbf{X}\hat{\mathbf{V}} \qquad \text{compute } \hat{\mathbf{T}}$$
$$\mathbf{X} = \hat{\mathbf{T}}\mathbf{P}' + \mathbf{E}$$
$$\mathbf{Y} = \hat{\mathbf{T}}\mathbf{Q}' + \mathbf{F} \qquad \left.\right\} \text{compute } \hat{\mathbf{P}}, \hat{\mathbf{Q}}, \hat{\mathbf{E}} \text{ and } \hat{\mathbf{F}}$$

Full prediction:

$$\hat{\mathbf{t}}_i' = \mathbf{x}_i'\hat{\mathbf{V}} \qquad \text{compute } \hat{\mathbf{t}}_i$$
$$\hat{\mathbf{y}}_i' = \hat{\mathbf{t}}_i'\hat{\mathbf{Q}}' \qquad \text{compute } \hat{\mathbf{y}}_i$$
$$\hat{\mathbf{e}}_i' = \mathbf{x}_i' - \hat{\mathbf{t}}_i'\hat{\mathbf{P}}' \qquad \text{compute residuals } \hat{\mathbf{e}}_i$$

Given the centered input data $\mathbf{X}, \mathbf{Y}$, how are the parameters $\mathbf{V}, \mathbf{T}, \mathbf{P}$ and $\mathbf{Q}$ in the bilinear calibration model estimated and used?

First of all, the $\hat{\mathbf{V}}$ matrix is determined to optimize a certain criterion, which *characterizes* the method. The scores $\hat{\mathbf{T}}$ are then found as $\hat{\mathbf{T}} = \mathbf{X}\hat{\mathbf{V}}$. subsequently, the loadings $\mathbf{P}$ and $\mathbf{Q}$ are estimated by multiple linear regression of each individual variable $\mathbf{x}_k$ on the obtained factors $\hat{\mathbf{T}} = \{\mathbf{t}_a, a = 1, 2, \ldots, A\}$. In matrix notation this can be written:

$$\hat{\mathbf{P}}' = (\hat{\mathbf{T}}'\hat{\mathbf{T}})^{-1}\hat{\mathbf{T}}'\mathbf{X} \tag{3.35}$$

$$\hat{\mathbf{Q}}' = (\hat{\mathbf{T}}'\hat{\mathbf{T}})^{-1}\hat{\mathbf{T}}'\mathbf{Y} \tag{3.36}$$

The residuals can be obtained as:

$$\widehat{\mathbf{E}} = \mathbf{X} - \widehat{\mathbf{T}}\widehat{\mathbf{P}}' \tag{3.37}$$

and

$$\widehat{\mathbf{F}} = \mathbf{Y} - \widehat{\mathbf{T}}\widehat{\mathbf{Q}}' \tag{3.38}$$

The maximum number of factors $A_{\max}$ usually equals $\min(K, I)$, ($\widehat{\mathbf{E}}$ then equals zero). But in most cases this solution, which corresponds to the MLR (if $I > K$), is too high. How to determine the dimension of the model, i.e the optimal number A to be used, is of fundamental importance in all types of data compression. This model assessment is generally based on predictive validation; the model should only include factors that improve the prediction of $\mathbf{Y}$ in independent test objects. But graphical interpretability is also important. For instance, the model should not include factors that are obviously due to noise in the data, as judged by graphical inspection of its loadings and scores. More detail is given in Chapter 4.

Prediction in bilinear models

Subsequent prediction of unknown y_i from x_i-measurements for a new sample can be done in two ways; full prediction (Frame 3.2) and short prediction. They give identical predictions $\widehat{y}_{ij}$, but otherwise different amounts of information.

Common to both methods is of course that the input measurements $\mathbf{x}$ and the output predictions $\widehat{y}_{ij}$ must be transformed according to the linearization and a priori weights used prior to calibration.

Full prediction   The most informative prediction technique is to compute $\widehat{y}_i$ via new scores $\widehat{\mathbf{t}}_i' = (\widehat{t}_{i1}, \ldots, \widehat{t}_{iA})$. This means that each input data vector $\mathbf{x}_i$ is first corrected for the mean as in the calibration. Next, the centered $\mathbf{X}$-vector is multiplied by $\widehat{\mathbf{V}}$ to obtain the scores or factors $\widehat{\mathbf{t}}_i' = (\widehat{t}_{i1}, \ldots, \widehat{t}_{iA})$

Figure 3.5 Arrow diagram of different approaches to linear multivariate calibration modelling. One Y-variable is to be determined from four X-variables. The X-variables are assumed to reflect two types of systematic variation, caused by the analyte ($\mathbf{Y}$) and one minor unidentified interferent ($\mathbf{U}$). a) Causal mixture modelling: The analyte concentration $\mathbf{Y}$ is supposed to model the X-variations, and the interference $\mathbf{U}$ is erroneously forgotten. This leads to mistaken causal interpretation of the calibration model and to decreased predictive ability due to the unmodelled interference. b) Multiple linear regression: The four X-variables are expected to have four more or less independent types of information about $\mathbf{Y}$. Collinearity problems arise, since there in fact are only two ($\mathbf{Y}$ and $\mathbf{U}$) plus possibly some noise. This leads to interpretation problems and decreased predictive ability due to over-fitting. c) Principal component regression: The X–X covariance structure is found to give rise to two main

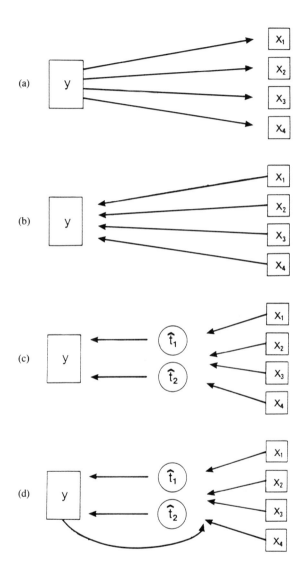

regression factors $\widehat{\mathbf{t}}_1$ and $\widehat{\mathbf{t}}_2$ from **X**. Together, these two factors give good description of both **X** and **Y**. The unidentified interference **U** is detected and possibly identified from graphical inspection of the bilinear model. d) Partial least squares regression: The X–Y and X–X covariance structures are found to give rise to one major and one minor regression factors $\widehat{\mathbf{t}}_1$ and $\widehat{\mathbf{t}}_2$ from **X**. As for PCR, the two factors together give a good description of both **X** and **Y**. But factor $\mathbf{t}_1$ alone accounts for most of this Y-modelling, so a simpler calibration model could be attained

$$\hat{\mathbf{t}}_i' = (\mathbf{x}_i' - \bar{\mathbf{x}}')\hat{\mathbf{V}} \qquad (3.39)$$

These scores are then multiplied by $\hat{\mathbf{Q}}$, and the center of $\mathbf{y}$ is added, yielding the predictor:

$$\hat{\mathbf{y}}_i' = \bar{\mathbf{y}}' + \hat{\mathbf{t}}_i'\hat{\mathbf{Q}}' \qquad (3.40)$$

Notice that scores of the unknown object, $\hat{\mathbf{t}}_i$, can be plotted together with the corresponding scores from the calibration samples, to show how the new X-data compare to the X-variabilities seen earlier. Statistical outlier tests can be performed, on $\hat{\mathbf{t}}_i$ as well as on the residual spectrum (see e.g. Chapter 5):

$$\hat{\mathbf{e}}_i' = \mathbf{x}_i' - \bar{\mathbf{x}}' - \hat{\mathbf{t}}_i'\hat{\mathbf{P}}' \qquad (3.41)$$

This residual spectrum can then be studied graphically and submitted to various types of pattern recognition techniques for more extended interpretation, if so desired.

Short prediction   Alternatively, the bilinear predictor of the uncentered $\mathbf{y}_i$ based on the uncentered $\mathbf{x}_i$ can also be expressed in terms of a linear function of $\mathbf{x}_i$ in the form

$$\hat{\mathbf{y}}_i' = \hat{\mathbf{b}}_0' + \mathbf{x}_i'\hat{\mathbf{B}}, \qquad (3.42)$$

where

$$\hat{\mathbf{B}} = \hat{\mathbf{V}}\hat{\mathbf{Q}}' \qquad (3.43)$$

and

$$\hat{\mathbf{b}}_0' = \bar{\mathbf{y}}' - \bar{\mathbf{x}}'\hat{\mathbf{B}} \qquad (3.44)$$

Contrary to the full prediction, this short prediction does not provide an explicit 'modelling' of what is going on in the new X-data in terms of the expected analyte and interference variations and possible unexpected phenomena. The short prediction only seeks to predict $\mathbf{Y}$ from $\mathbf{X}$ in such a way that the expected interferences are compensated for. In this respect it resembles conventional MLR, although with a stabilization against collinearity in $\mathbf{X}$.

PLS-PCR as bilinear methods

In this book two bilinear calibration methods will be covered, the PCR and the PLS regression. The difference between them is illustrated in Figure 3.5, which also compares them conceptually to the two most extreme linear calibration methods, the forward MLR and the reverse mixture modelling.

Another well known method closely linked to the bilinear framework is the latent root regression treated in e.g. Webster et al. (1974). This method is not considered in this book.

3.4 PRINCIPAL COMPONENT REGRESSION (PCR)

SUMMARY The concept of principal components of $\mathbf{X}$ is defined. Principal component regression (PCR) is presented as regression of $\mathbf{Y}$ on selected principal components of $\mathbf{X}$. This is illustrated with two simple examples. Properties of PCR are given together with a discussion on selection of eigenvectors.

The name principal component regression (PCR) stems from the fact that we use principal component analysis (PCA, Hotelling, 1933; Anderson, 1958; Mardia et al., 1980; Rao, 1965) of $\mathbf{X}$ to determine weights $\hat{\mathbf{V}}$ to be used in (3.27). This means that we compute the so-called principal components of $\mathbf{X}$ and use only a few of them in the regression equation (3.28) (Gunst and Mason, 1979; Joliffe, 1986). Some details on how to select the best principal components for regression is discussed in section 3.4.6. Mandel (1982) gives further detail on PCR.

In PCR the weight matrix $\hat{\mathbf{V}}$ and the loading matrix $\hat{\mathbf{P}}$ are identical; we shall use letter $\hat{\mathbf{P}}$ to represent both $\hat{\mathbf{P}}$ and $\hat{\mathbf{V}}$.

3.4.1 PRINCIPAL COMPONENT ANALYSIS (PCA): COMPRESS X TO ITS MOST DOMINANT FACTORS $\hat{\mathbf{T}}$

The most frequent application of PCA is in cases where the X-variables are expected to be collinear. There can be many reasons for collinearity: The number of analytes and interference is lower than the number of X-variables—the spectral responses for some analytes or interference may resemble each other—the experimental design is such that the level of certain analytes or interferents correlate with each other in the calibration set.

This collinearity means that the matrix $\mathbf{X}$ will have some dominating types of variability that carry most of the available information. Redundancy and smaller noise variabilities can then be removed.

The purpose of PCA is then to express the main information in the variables $\mathbf{X} = \{\mathbf{x}_k, k = 1, 2, \ldots, K\}$ by a lower number of variables $\hat{\mathbf{T}} = \{\hat{\mathbf{t}}_1, \ldots, \hat{\mathbf{t}}_A\}$ $(A < K)$, the so-called principal components of $\mathbf{X}$. To give best results, the different X-variables in the calibration objects should be scaled to similar noise levels and centered prior to this PCA data approximation (see sections 3.3.3 and 7.1)

The PCA is illustrated conceptually in Figure 3.6. Figure 3.6a shows the distribution of a certain calibration set; the points represent objects $i = 1, 2, \ldots, I$ for which variables x_1 and x_2 have been measured (they represent the K axes when K different X-variables have been measured). These data points are equally well described in the x_1, x_2 space and in the space spanned by any other pair of orthogonal axes, for instance the two principal components $\hat{\mathbf{t}}_1$ (running through the longest axis in the ellipsoid-shaped 'cloud' of data points) and $\hat{\mathbf{t}}_2$ (running

98

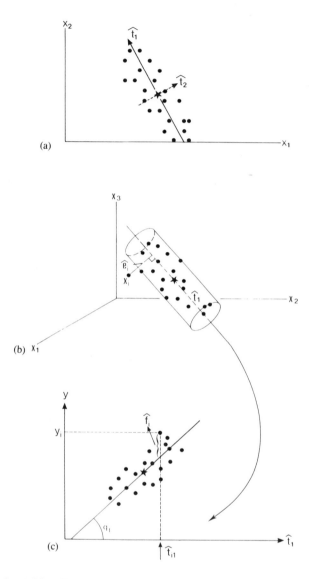

Figure 3.6 a) PCA illustrated for two X-variables with one dominating component. The star represents the mean of the data set, $\overline{\mathbf{x}}$. The solid arrow points in the directions of the first principal component, as defined by its loading $\widehat{\mathbf{p}}_1$. The projection of the objects $i=1,2,\ldots,I$ onto the factor gives their factor scores $\widehat{t}_{i1}, i=1,2,\ldots,I$. The second principal component is given by the dotted arrow. b) Plot of three X-variables for another data set with one major principal component. The 'beer-can' around the principal component in the X-space forms the basis for outlier tests (Chapter 5). An object's data vector $\mathbf{x}_i$ has a residual vector $\widehat{\mathbf{e}}_i$ perpendicular to the principal components. c) The plot shows how score vector $\widehat{\mathbf{t}}_1$ is used as regressor with $\mathbf{y}$ as regressand (PCR), yielding the Y-loading $\widehat{q}_1$. The vertical deviation from an object's data $(y_i, \widehat{t}_i)$ to the regression line is called Y-residual $\widehat{\mathbf{f}}_i$

perpendicular to the $\hat{\mathbf{t}}_1$ direction at the center of the calibration set). Figure 3.6b shows a plot of another data set in 3-dimensional space with one major principal component. In Figure 3.6c is illustrated how this $\hat{t}_{i1}$ can be used in regression.

For further illustration we refer to section 3.4.4.

More formally, the loading $\hat{\mathbf{p}}_1$ is defined as the normalized (length = 1) vector that maximizes the empirical variance of $\hat{\mathbf{p}}_1'\mathbf{x}$, or in other words maximizes $\hat{\mathbf{p}}_1'\mathbf{X}'\mathbf{X}\hat{\mathbf{p}}_1 = \hat{\mathbf{t}}_1'\hat{\mathbf{t}}_1$. The next factor's loading, $\hat{\mathbf{p}}_2$, is defined as the vector maximizing the same quantity, i.e. the scalar $\hat{\mathbf{p}}_2'\mathbf{X}'\mathbf{X}\hat{\mathbf{p}}_2 = \hat{\mathbf{t}}_2'\hat{\mathbf{t}}_2$ under the constraint that the $\hat{\mathbf{t}}_1$ and $\hat{\mathbf{t}}_2$ are orthogonal, i.e $\hat{\mathbf{t}}_1'\hat{\mathbf{t}}_2 = 0$. The procedure continues this way under the constraint that new factors' scores are uncorrelated orthogonal with those of the previous factors.

It is good calibration practice to extract more factors than expected to be needed, and afterwards eliminate those that appear to be irrelevant or unreliable. This permits possible surprises—there may be unexpected phenomena in the X-data, and they may then be revealed and corrected for.

The eigenvalues, which show how much variability each factor removes from $\mathbf{X}$, are here denoted by $\hat{\tau}_1.\hat{\tau}_a \ldots \hat{\tau}_K$. These sums-of-squares of the principal component scores $\hat{\mathbf{t}}_1. \ldots \hat{\mathbf{t}}_K$ are computed as $\hat{\tau}_a = \hat{\mathbf{t}}_a'\hat{\mathbf{t}}_a. a = 1.2. \ldots . k$.

It can be shown that in addition to orthogonal scores one also successively obtains ortogonal loading vectors $\hat{\mathbf{p}}_1. \ldots \hat{\mathbf{p}}_K$. In matrix language the orthogonality properties of the loadings and scores can be written as

$$\hat{\mathbf{P}}'\hat{\mathbf{P}} = \mathbf{I} \tag{3.45}$$

$$\hat{\mathbf{T}}'\hat{\mathbf{T}} = \mathrm{diag}(\hat{\tau}_a) \tag{3.46}$$

For centered X-variables it can be shown that the $\hat{\mathbf{p}}_a. a = 1.2. \ldots . A$ vectors are eigenvectors of $\mathbf{X}'\mathbf{X}$ with the $\hat{\tau}$'s as eigenvalues. This means that all $\hat{\mathbf{p}}$'s satisfy the equation

$$\mathbf{X}'\mathbf{X}\hat{\mathbf{p}}_a = \hat{\mathbf{p}}_a\hat{\tau}_a \tag{3.47}$$

Likewise, it can be shown that the scores $\hat{\mathbf{t}}_a. a = 1.2. \ldots . A$ represent the corresponding eigenvectors of $\mathbf{X}\mathbf{X}'$, scaled to length $\sqrt{\hat{\tau}_a}$.

It can also be shown that if all the $A = K$ eigenvectors have been extracted (some of them possibly with eigenvalues of or near zero), then $\mathbf{X}$ can be written as $\mathbf{X} = \hat{\mathbf{T}}\hat{\mathbf{P}}'$. Scaling the principal components scores for each factor to length one and denoting the resulting matrix $\hat{\mathbf{U}}$, we can write the $\mathbf{X}$ matrix as $\hat{\mathbf{U}} diag(\sqrt{\hat{\tau}_a})\hat{\mathbf{P}}'$ which is exactly the singualar value decomposition of $\mathbf{X}$ defined in equation (2.24). With only the first few principal components collected in $\hat{\mathbf{T}}$ (i.e. $A < K$), the data matrix $\mathbf{X}$ is instead *approximated* by $\hat{\mathbf{T}}\hat{\mathbf{P}}'$, i.e. $\mathbf{X} = \hat{\mathbf{T}}\hat{\mathbf{P}}' + \hat{\mathbf{E}}$ where $\hat{\mathbf{E}}$ represents X-residuals as in model (3.32).

An interesting property of PCA in this model is that $\hat{\mathbf{P}}$ can be found by least squares fitting of $\mathbf{X}$ to $\hat{\mathbf{T}}$ in the same way as $\hat{\mathbf{T}}$ can be found by least squares fitting of $\mathbf{X}$ to $\hat{\mathbf{P}}$. We have $\hat{\mathbf{P}}' = (\hat{\mathbf{T}}'\hat{\mathbf{T}})^{-1}\hat{\mathbf{T}}'\mathbf{X}$ and $\hat{\mathbf{T}} = \mathbf{X}\hat{\mathbf{P}}(\hat{\mathbf{P}}'\hat{\mathbf{P}})^{-1} = \mathbf{X}\hat{\mathbf{P}}$ (since $\hat{\mathbf{P}}'\hat{\mathbf{P}} = I$). This is true for each individual principal component $a = 1.2. \ldots$ as well, and this gives rise to a special algorithm for estimating eigenvalues and eigenvectors.

PCA can be very useful for interpretation of a single set of variables (see e.g. Rao, 1965; Wold et al., 1984; Martens M., 1985; Kvalheim, 1987)—applicable to spectroscopy, chromatography, image analysis data and any other multivariate instrumentation or combination of instruments. It is then important to decide how many principal components $a = 1, 2, \ldots, A$ are necessary for capturing the essential information in $\mathbf{X}$. For graphical inspection one usually looks at plots of the scores and loadings of the principal components corresponding to the largest eigenvalues— the number of eigenvectors to include can also be decided by e.g. cross-validation (Wold, 1978; Eastment and Krzanowski, 1982) or by certain types of significance tests (Anderson, 1958; Malinowski and Howery, 1980).

In PCR, where the scores from $\mathbf{X}$ are used as regressors for $\mathbf{Y}$, the factor selection is usually done the same sequential manner. Alternatively, a stepwise selection procedure based on stepwise MLR of $\mathbf{Y}$ on $\hat{\mathbf{T}} = \{\hat{\mathbf{t}}_a, a = 1, 2, \ldots . A\}$ can also be used. This will be discussed in section 3.4.6. The choice of model complexity A may again be based on graphical model inspection and predictive validation (Chapter 4).

3.4.2 ALGORITHMS FOR EIGENVECTOR DECOMPOSITION

A number of different numerical algorithms lead to the same resulting PCA solution see, for instance, Golub and Van Loan (1983). In many algorithms it is normal to compute all the non-zero eigenvalues and their eigenvectors simultaneously, either from the data centered table $\mathbf{X}$ directly, or via the $\mathbf{X}'\mathbf{X}$ or $\mathbf{X}\mathbf{X}'$ cross-product matrices.

For calibration it is often enough to extract the first few principal components corresponding to the largest eigenvalues. It is practical to do this for one factor at a time starting from the one with largest eigenvalue, until we see that the data contain no more valid information. Also, working directly on the data table $\mathbf{X}$ is less alienating than going via more abstract cross-product tables.

A simple algorithm for computing the largest eigenvalues with presumably the most important eigenvectors is given in Frame 3.3 (located at the end of section 3.4.5, page 111). This NIPALS algorithm (Wold, 1966) employs the fact that the principal components are orthogonal both in scores and loadings, to extract one single factor at a time, $a = 1$, $a = 2, \ldots$.

For each factor it employs an iterative method to obtain the loading vector $\hat{\mathbf{p}}_a$ and the score vector $\hat{\mathbf{t}}_a$ from the residual power-matrix obtained after estimation of the previous $a - 1$ factors. That preliminary residual matrix can be termed $\hat{\mathbf{E}}$, but in order to facilitate the description of the PCA algorithm we shall here call it $\mathbf{X}_{a-1}$. It is defined by $\mathbf{X}_{a-1} = \mathbf{X} - \mathbf{1}\bar{\mathbf{x}}' - \hat{\mathbf{t}}_1\hat{\mathbf{p}}_1' - \ldots - \hat{\mathbf{t}}_{a-1}\hat{\mathbf{p}}_{a-1}'$.

Starting e.g. with some guessed scores, the iteration goes as follows: The loading estimate is improved by regression of $\mathbf{X}_{a-1}$ on a previous score estimate, and the score estimate is improved by regression of $\mathbf{X}_{a-1}$ on the improved loading estimate, and so on until convergence.

3.4.3 PRINCIPAL COMPONENT REGRESSION EQUATION

Principal component regression of J different Y-variables on K X-variables is equivalent to J separate principal component regressions on the same K X-variables—one for each Y-variable. Thus we here only give attention to the case of one single Y-variable, y.

The principal component regression is obtained by regressing y on the $\hat{t}$'s obtained from the PCA of X as indicated in Figure 3.6c. The regression coefficients $\hat{b}$ for each y can according to equation (3.43) be written

$$\hat{b} = \hat{P}\hat{q} \qquad (3.48)$$

where X-loadings $\hat{P} = \{\hat{p}_{ka}, k = 1, 2, \ldots, K \text{ and } a = 1, 2, \ldots, A\}$ represent the PCA loadings of the A factors employed, and Y-loadings $\hat{q} = (\hat{q}_1, \ldots \hat{q}_A)'$ are found the usual way by least squares regression of y on $\hat{T}$ from the model y=Tq+f. Since the scores in $\hat{T}$ are uncorrelated, this solution is equal to

$$\hat{q} = (\text{diag}(1/\hat{\tau}_a))^1 \hat{T}' y \qquad (3.49)$$

Inserting this in (3.48) and replacing $\hat{T}$ by $X\hat{P}$, the PCR estimate of b can be written as

$$\hat{b} = \hat{P}(\text{diag}(1/\hat{\tau}_a))\hat{P}' X' y \qquad (3.50)$$

which is frequently used as definition of the PCR (Gunst and Mason, 1979). When the number of factors A equals K, the PCR gives the same $\hat{b}$ as the MLR. But the X-variables are often intercorrelated and somewhat noisy, and then the optimal A is less than K: In such cases MLR would imply division by eigenvalues $\hat{\tau}_a$ close to zero, which makes the MLR estimate of b unstable. In contrast PCR attains a stabilized estimation of b by dropping such unreliable eigenvalues.

3.4.4 PCA AND PCR ILLUSTRATED BY SIMULATED DATA

In Table 3.2 calibration data for three X-variables and one Y-variable are given for 5 objects. The data are artificial, but illustrates a simple spectroscopic calibration problem.

We assume that the X-data represent a linearly responding spectrophotometer: We want to determine an analyte y from spectrophotometric absorbances X.

Let us assume that wavelength channel x_1 is known to be best for determining the analyte, so a predictor of the type $\hat{y} = x_1\hat{b}_1$ is expected. Channels x_2 and x_3 are only included in the modelling for good measure.

Unknown to us, the samples also contained an unidentified chemical constituent with overlapping X-spectrum. An awkward experimental design had unfortunately been chosen: The concentration of this interferent happened to vary with the analyte concentration y in a curved pattern in the calibration set.

Figure 3.7a shows a plot of the input data for the three X-variables vs. y, with

102

Table 3.2 Artificial calibration data

Obj. no	x_1	x_2	x_3	y
1	11.5	10	8	1
2	7	4.5	3	2
3	5.5	2	1	3
4	8	3.5	3	4
5	13.5	8	8	5

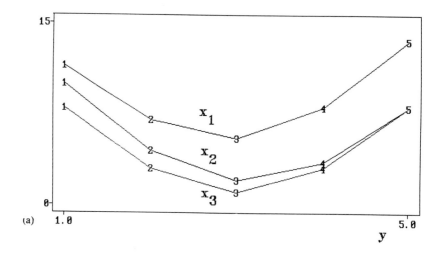

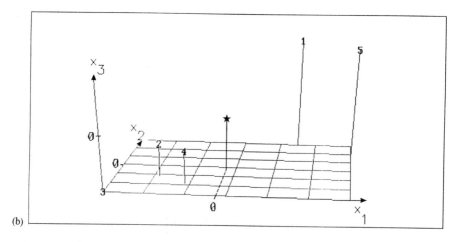

Figure 3.7 a) Instrument response at the three X-variables plotted against analyte concentration y in the 5 calibration objects. b) The five objects plotted for the three centered X-variables, X_0, i.e. after having subtracted $\bar{x}$

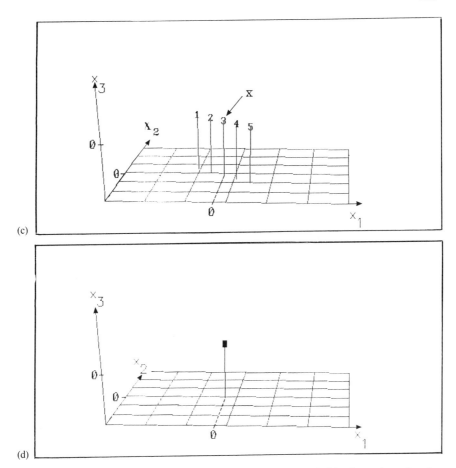

(c)

(d)

Figure 3.7 (*cont.*) c) The five objects plotted the same way as in b), after subtracting the first principal component from $\mathbf{X}_0$, i.e. $\mathbf{X}_1$. d) The five objects plotted the same way as in c), after subtracting the first principal component from $\mathbf{X}_1$, i.e. $\mathbf{X}_2$, which here represents the final residuals, $\hat{\mathbf{E}}$

the level of y written as numbers 1–5. Due to an interference effect, univariate calibration is difficult in this case. None of the X-variables can individually yield a reliable straight calibration line for y. (Note that the curvature in the figure is a consequence of interference present in an awkward experimental design, and not due to nonlinear instrument response!)

Is it then possible to predict y from linear combinations of the three X-variables? Table 3.3 gives the detailed PCR analysis, which is also illustrated in Figures 3.7b–i. The averages $\bar{\mathbf{x}}'$ and $\bar{y}$ are first subtracted from the input data $\mathbf{X}$ and $\mathbf{y}$, yielding the centered data here denoted $\mathbf{X}_0$ (Figure 3.7b) and $\mathbf{y}_0$ (i.e. 'residuals after 0 factors').

The first principal component's scores $\hat{\mathbf{t}}_1$ and loadings $\hat{\mathbf{p}}_1$ are computed from $\mathbf{X}_0$, and $\hat{\mathbf{q}}_1$ is then obtained by regression of $\mathbf{y}_0$ on $\hat{\mathbf{t}}_1$.

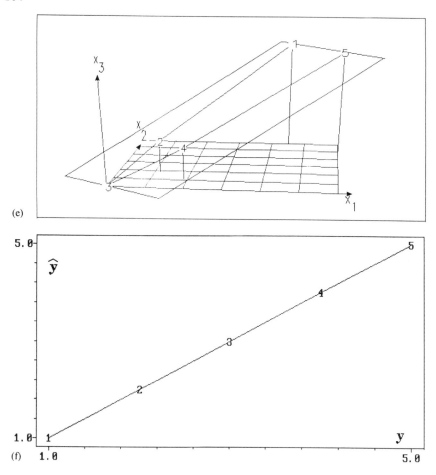

(e)

(f)

Figure 3.7 (*cont.*)   e) Plot of input data repeated to show how the five objects fall in a plane in the original X-space. f) Observed y vs the predictions from the linear 2-factor PCR model: $\hat{y} = \hat{b}_0 + x_1\hat{b}_1 + x_2\hat{b}_2 + x_3\hat{b}_3$, for the five calibration objects

The effect of factor 1 is then subtracted from $\mathbf{X}_0$ and $\mathbf{y}_0$, yielding residuals after 1 factor, $\mathbf{X}_1$ (Figure 3.7c) and $\mathbf{y}_1$. The process is repeated, yielding the second principal component with scores $\hat{\mathbf{t}}_2$ and loadings $\hat{\mathbf{p}}_2$, and subsequently $\hat{q}_2$ by regressing $\mathbf{y}_1$ on $\hat{\mathbf{t}}_2$. The resulting $\hat{\mathbf{b}}$-vector for the two-factor model becomes $(1. \ -1. \ 0)'$ with $b_0 = -0.5$. (As the reader might have guessed, this was in fact how the Y-data were constructed: $y \equiv -0.5 + x_1 - x_2$.)

After similarly subtracting $\hat{\mathbf{t}}_2\hat{\mathbf{p}}_2'$ from $\mathbf{X}_1$ and $\hat{\mathbf{t}}_2\hat{q}_2$ from $\mathbf{y}_1$, the final residuals $\hat{\mathbf{E}} = \mathbf{X}_2$ (Figure 3.7d) and $\hat{\mathbf{f}} = \mathbf{y}_2$ are zero in this artificial, error-free example. (For practical data the final X- and y-residuals will usually reflect measurement noise, unmodelled curvatures etc. unless too many PCA factors have been included in the model.)

The good description of the X-data by two PCA factors is due to the fact that

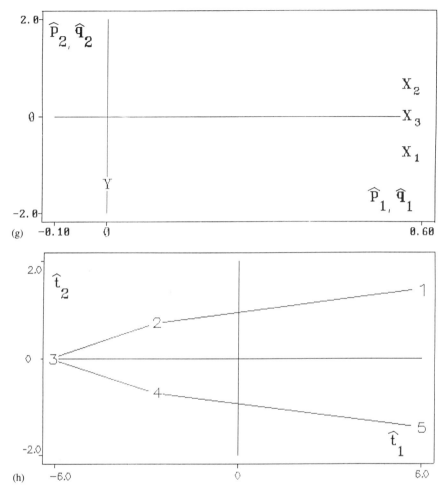

Figure 3.7 (*cont.*)   g) Principal component loading plot. The two principal component axes are defined by their loadings $\widehat{p}_{ka}$ (symbolized by X_1, X_2, X_3) and $\widehat{q}_a$ (symbolized by Y) for factors $a=1,2$. h) Principal component score plot. The plane in Figure 3.7e is modelled by the two first principal components of **X**. The coordinates of the 5 objects along these principal component axes are their scores $\widehat{t}_1$ and $\widehat{t}_2$

the data points here lie in a single *plane* in the X-space (Figure 3.7e). This can also be seen from the fact that there are only two positive eigenvalues in the centered 3×3 matrix $\mathbf{X}_0'\mathbf{X}_0$, (the eigenvalues are 123.6, 50 and 0). Hence the X-data can be adequately described by the plane of the first two principal components.

In this example, the first PCA factor extracted from **X** had little relevance for y (small $\widehat{q}_1$), while the second one had more. Figure 3.7f shows how y is linearly modelled by the two-factor PCR model. Thus, in spite of the lack of simple X–Y linear relationship in the input data (Figure 3.7a), a good, linear predictor is obtained after having modelled the X–X relationship first.

106

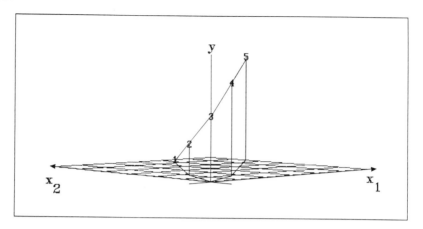

Figure 3.7 (*cont.*) i) Checking the model interpretation back into raw data for mental validation: Input variable *y* plotted against the two *X*-variables found most relevant in the PCR analysis, here rotated to an informative angle

Table 3.3 PCR calibration results

	x_1	x_2	x_3	y	
Average, $\bar{\mathbf{x}}'$, $\bar{y}$	9.1	5.6	4.6	3	
Object no		$\mathbf{X}_0$		y_0	$\hat{\mathbf{t}}_1$
1	2.4	4.4	3.4	−2	5.9
2	−2.1	−1.1	−1.6	−1	−2.8
3	−3.6	−3.6	−3.6	0	−6.2
4	−1.1	−2.1	−1.6	1	−2.8
5	4.4	2.4	3.4	2	5.9
Factor a=1:					
$\hat{\mathbf{p}}_1'$, $\hat{q}_1$	0.58	0.58	0.58	0	
$\hat{\mathbf{b}}'$ and $\hat{b}_0$:	0	0	0	3	
Object no		$\mathbf{X}_1$		y_1	$\hat{\mathbf{t}}_2$
1	−1	1	0	−2	1.4
2	−0.5	0.5	0	−1	0.7
3	0	0	0	0	0
4	0.5	−0.5	0	1	−0.7
5	1	−1	0	2	−1.4
Factor a=2					
$\hat{\mathbf{p}}_2'$, $\hat{q}_2$	−0.71	0.71	0	−1.41	
$\hat{\mathbf{b}}'$ and $\hat{b}_0$:	1.0	−1.0	0	−0.5	
Object no		$\mathbf{X}_2$		y_2	
1	0	0	0	0	
2	0	0	0	0	
3	0	0	0	0	
4	0	0	0	0	
5	0	0	0	0	

The bilinear modelling can also provide further insight. Figure 3.7g shows the loadings of y and the X-variables, and 3.7h the corresponding object scores $\hat{t}_2$ vs. $\hat{t}_1$, spanning the plane drawn in Figure 3.7e. These plots provide relevant 'windows' into the higher-dimensional data space, summarizing the relevant information.

The score plot reveals the main pattern of the objects in the data set, and the loading plot shows the main relationships between the variables, as estimated from the data set. Remember that if an object had fallen in the origin in the score plot $(\hat{t}_1 = \hat{t}_2 = 0)$, then we would expect it to have data x_i similar to $\bar{x}$. If an X- or Y-variable had fallen in the origin of the loading plot $(\hat{p}_1 = \hat{p}_2 = 0$ or $\hat{q}_1 = \hat{q}_2 = 0)$, that would mean a variable with no variability in the calibration set, or a variable whose variability had not been picked up by the X-dimensional bilinear approximation.

The loading plot here shows that all three X-variables correlate the same way with factor 1, but this factor displays no correlation to y. This was not at all the relationship originally expected from the X-instrument! The score plot shows this factor mainly to span the difference between objects 1 and 5 on one hand, and object 3 on the other. So in this factor objects 1 and 5 would tend to have higher-than-average levels of x_1, x_2 and x_3, while objects 3, 2 and 4 would tend to have lower-than-average levels of these variables.

Factor 2 reveals an additional relationship: Decreasing levels of Y corresponds to decreasing levels of x_1 and increasing levels of x_2, and vice versa, with x_3 being unaffected by this relationship. This corresponded more to the relationship originally expected. The score plot shows this factor to span mainly the difference between objects 1 and 5. So, in this factor objects 1 and 2 would tend to have higer-than-average levels of x_2 and lower-than-average levels of y and x_1, and vice versa; objects 4 and 5 would tend to have higer-than-average levels of y and x_1 and lower-than-average levels of x_2.

In this artifical data set only factor $a = 2$ is relevant for y. But normally the factors should not be interpreted individually, because it is the combination of the factors that counts. Here the two factors together indicate e.g. that object 5 has the highest level of x_1 and object 1 the highest level of x_2.

Having yielded $A = 2$ valid factors, it seems that the X-data have been affected by two different phenomena. What could they be? Let us illustrate bilinear modelling as a hypothesis generating tool:

The model plots indicate that these X-data, in addition to the expected y-effect, may have been affected by an unexpected baseline variability (all X-variables vary the same way with factor 1). In addition, there is the problem of explaining why the absorbance at channel x_2 tends to increase when the analyte concentration decreases (factor 2). This could be due to a second unknown constituent with predominant absorbance at channel x_2. If the sum of the analyte's and this second constituent's concentrations were always 100%, there would be a negative correlation between y and absorbance x_2, although the effect of two constituents would appear in the same principal component. But of course, these hypotheses would need to be verified in new experiments.

As a result of all this, the final predictor became more complex than originally expected, but it still gave good predictions.

From this illustration, notice how the bilinear model did two things: It provided a good predictor $\hat{y} = \hat{b}_0 + \mathbf{x}\hat{\mathbf{b}}$, *and* it provided new insight into what was going on in the X-data. With so few variables and objects such compressed interpretation 'windows' may not be necessary. But when the number of variables and objects increases, we cannot possibly inspect all the input data directly. The bilinear model plots can then provide valuable insight that serves to validate prior knowledge or correct it by revealing unexpected systematic phenomena in the data.

After having obtained an apparently reasonable data model, it is advisable to check back to the raw input data to ensure that we have not misunderstood the multivariate model. In the present example this could be done as in Figure 3.7i, which shows how y in fact is modelled as a linear function of x_1 and x_2.

How would other calibration methods perform on these data? Since the last eigenvalue of the centered $\mathbf{X'X}$ was 0, the ordinary MLR in this case would have been impossible to use.

Applying SMLR instead (see section 3.7) would here eliminate x_3, resulting in a predictor $\hat{y} = b_0 + \mathbf{x'}\hat{\mathbf{b}}$ similar to that of PCR, $\hat{y} = -0.5 + x_1 - x_2$. But the interpretation of the SMLR model would be confusing, because without loadings and score plots SMLR provides little insight into the phenomena actually affecting the X-data (here exemplified by the unexpected apparent baseline-shift). If there had been even more X-variables and/or more design problems, the most interpretable X-variables might be among those excluded from the SMLR solution. And with more noisy X-data the elimination of X-variables could be a wasting of valuable information.

PCR, as performed here, does not distinguish between X-phenomena relevant for y and X-phenomena irrelevant for y. In this example the first principal component was uncorrelated with y, which means that it could have been given lower emphasis if we were primarily interested in Y-relevant phenomena in X. In the next section we will discuss PLS regression and show how that method selects this second eigenvector the first regression factor, indicating the ability of PLSR to give more parsimonious solutions than PCR.

The above artificial illustration can alternatively be thought of as a demonstration of how nonlinear instrument responses can sometimes be handled by multivariate bilinear modelling. In that case one might think of the X-data as being caused only by one single constituent, y, but in an extremely nonlinear way. This concept will be covered in more detail in Chapter 7.

3.4.5 PCR ILLUSTRATED BY REAL DATA

A real example of PCR applied in calibration is given in Figure 3.8, which concerns the determination of fat content in hamburger meat by the relatively simple near-infrared reflectance. Diffuse reflectance R was measured at 19 fixed wavelength channels between 1445 and 2400 nm, i.e. in a region where all the other major components (water, protein, starch) also absorb light, and where light

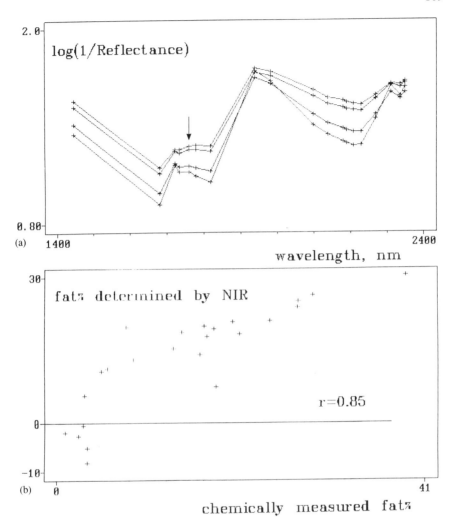

Figure 3.8 Principal component regression used in calibrating for y = fat percentage in hamburger meat from X = high-speed near infrared reflectance measurements at K = 19 wavelength channels. a) Optical density log(1/Reflectance) spectra vs. wavelength for some typical objects. The arrow marks the X-channel corresponding to a wavelength typical for lipids, 1759 nm. b) Prediction based on univariate calibration for 24 test samples: Abscissa: y, the chemically determined fat percentage. Ordinate: ŷ, fat percentage predicted from the optical density at wavelength channel $x_{1759\,nm}$

scattering variations due to homogenization differences are expected to give major interferences.

The X-data were obtained with a Technicon InfraAlyzer 400 and transformed to optical density X = log(1/Reflectance) in 37 calibration objects for which the fat percentage (y) had been determined by a traditional extraction and weighing procedure. Similarly, NIR data and fat percentage data were obtained from 24 independent test objects. Figure 3.8a shows the X-spectra for some of the calibration

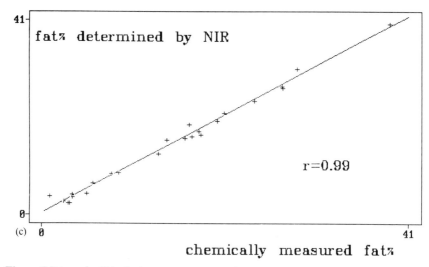

Figure 3.8 (*cont.*)   Principal component regression used in calibrating for y = fat percentage in hamburger meat from X = high-speed near infrared reflectance measurements at K = 19 wavelength channels. c) Prediction based on multivariate PCR calibration for 24 test samples: Abscissa: y, the chemically determined fat percentage. Ordinate: $\hat{y}$, fat percentage predicted from a linear combination of all 19 NIR channels based on 5 principal components

objects as a function of wavelength. The X-data were highly collinear with a condition index (or number) approximately equal to 6500.

The PCR analysis of the centered calibration data indicated that the A = 5 first principal components should be used in the calibration model (by leverage-correction, see section 4.3.2.4). The fat percentage was then predicted from the 19 NIR-variables in the 24 test objects and this gave an average root-mean-squared error of prediction (RMSEP, see section 4.3.1) equal to 1.1 percent fat. In contrast, full MLR on the same data (A = 19) gave RMSEP equal to 1.87.

Figures 3.8b and c compare the prediction results for the test objects, using univariate and multivariate calibration: Figure 3.8b illustrates predictions based on only one single X-variable (one of the wavelength channels most selective for fat, 1759 nm). Figure 3.8c shows the corresponding PCR prediction based on all 19 log(1/Reflectance) values. The PCR gave reasonably good results with five principal components, while the log(1/Reflectance) at 1759 nm alone could not give precise predictions of fat percentage.

This univariate calibration could possibly have been improved by proper pre-linearization of the X-data. But it would never reach the level of selectivity of the PCR solution, because many interferences affect these NIR reflectance measurements—variations in water/protein ratio, variations in sample preparation, temperature etc. Note that even the nonlinear instrument response at 1759 nm (Figure 3.8a) was compensated for by the other wavelengths in the PCR (see section 7.3 for detailed treatment of nonlinearities).

Graphical inspection of the loadings and scores (not shown here) revealed, among other things, a strong influence of light scattering variations in the first principal component (see section 7.4).

It is important to emphasize that PCR, in contrast to MLR, can also be applicable and give good results when there are more X-variables than objects, as long as the predictive information is in the first eigenvectors. In this situation the PCR can be performed with the NIPALS algorithm which works in the same way irrespective of the objects/variables ratio.

Frame 3.3

The NIPALS algorithm for PCA

The algorithm extracts one factor at a time. Each factor is obtained iteratively by repeated regressions of X on scores $\hat{t}$ to obtain improved $\hat{p}$, and of X on these $\hat{p}$ to obtain improved $\hat{t}$. The algorithm proceeds as follows:

Pre-scale the X-variables to ensure comparable noise-levels. Then center the X-variables, e.g. by subtracting the calibration means $\bar{x}'$, forming X_0. Then for factors $a = 1, 2, \ldots, A$ compute $\hat{t}_a$ and $\hat{p}_a$ from X_{a-1}:

Start:

Select start values, e.g. $\hat{t}_a$ = the column in X_{a-1} that has the highest remaining sum of squares.
Repeat points i) to v) until convergence.

i) Improve estimate of loading vector $\hat{p}_a$ for this factor by projecting the matrix X_{a-1} on $\hat{t}_a$, i.e.

$$\hat{p}_a' = (\hat{t}_a'\hat{t}_a)^{-1}\hat{t}_a' X_{a-1}$$

ii) Scale length of $\hat{p}_a$ to 1.0 to avoid scaling ambiguity:

$$\hat{p}_a = \hat{p}_a(\hat{p}_a'\hat{p}_a)^{-0.5}$$

iii) Improve estimate of score $\hat{t}_a$ for this factor by projecting the matrix X_{a-1} on $\hat{p}_a$:

$$\hat{t}_a = X_{a-1}\hat{p}_a(\hat{p}_a'\hat{p}_a)^{-1}$$

(*continued on next page*)

112

Frame 3.3 *continued*

iv) Improve estimate of the eigenvalue $\hat{\tau}_a$:

$$\hat{\tau}_a = \hat{\mathbf{t}}_a{'}\hat{\mathbf{t}}_a$$

v) Check convergence: If $\hat{\tau}_a$ minus $\hat{\tau}_a$ in the previous iteration is smaller than a certain small pre-specified constant, e.g. 0.0001 times $\hat{\tau}_a$, the method has converged for this factor. If not, go to step i).
Subtract the effect of this factor:

$$\mathbf{X}_a = \mathbf{X}_{a-1} - \hat{\mathbf{t}}_a\hat{\mathbf{p}}_a{'}$$

and go to *Start* for the next factor.

————————————— Statistical extensions —————————————

3.4.6 SELECTION OF PRINCIPAL COMPONENTS AND PROPERTIES OF PCR IN PREDICTION

PCR has been compared to other regression methods both from a theoretical and a practical point of view (see e.g. Berk, 1984; Næs et al., 1986; Gunst and Mason, 1979; Gunst and Mason, 1980; Joliffe, 1986; Cowe and McNicol, 1985; Devaux et al., 1987; Cowe et al., 1985).

In addition to the conceptual advantages discussed above, the PCR method has the advantage that it is rather well understood from a statistical point of view (Gunst and Mason, 1979; Mandel, 1982; Joliffe, 1986). Several papers exist describing its merits in estimation of the regression coefficients $\hat{\mathbf{b}}$ as well as in the prediction of future y's in the linear regression equation

$$y_i = b_0 + \sum_{k=1}^{K} x_{ik}b_k + f_i \tag{3.51}$$

In the present treatment we shall consider the predictive performance for one single future object i. Alternative formulae integrating the quantities above over future objects $i = 1, 2, \ldots$ can be found in Gunst and Mason (1979).

The main reasons for the theoretical simplicity of PCR are the facts that the scores $\hat{\mathbf{T}}$ are orthogonal in the calibration set, and that we can condition on $\hat{T}$ in the same way as on X for LS regression. In other words, the X-data are taken as given, irrespective their quality, because the same quality of X-data are expected also in the future. This conditioning implies that we can also condition on the principal component structure of X, $\hat{T}$. With such conditioning we can study uncertainty distributions of PCR regression coefficients and predictors in the same way as for MLR.

The following distributional deductions require the residual f_i to be random, independent and identically distributed, in accordance with usual statistical thinking. So for instance, the level of measurement noise in the data should not depend on the actual data themselves, and the obtained residual f_i should not reflect systematic model errors. If in practice the measurement noise in y_i increases with y_i itself (e.g. proportional error), or the residuals f_i contain a major contribution from unmodelled X–Y curvature, the theoretical results in principle do not apply. However, they may still shed light on how PCR compares to MLR.

Given the chosen conditioning and distributional assumptions, the uncertainty variance of $\hat{q}_a$ then equals $\sigma_f^2/\hat{\tau}_a$ (because $\hat{q}_a$ is found by regression of y on $\hat{t}_a$). Here σ_f^2 is the uncertainty variance of f. Since $\hat{y}$ is predicted from $\Sigma \hat{t}_{ia}\hat{q}_a$, this calibration variance in $\hat{q}_a$ of course later gives uncertainty in the subsequent prediction $\hat{y}_i$ in a future object i. The size of this potential error in $\hat{y}_i$ depends also on $\hat{t}_{ia}$, i.e. on how far its vector x_i' is from the calibration center, $\bar{x}'$, in component directions $a = 1, 2, \ldots, A, \ldots$.

Under model (3.51) with the usual regression assumptions (see section 2.1.2.3) the bias, variance and MSE of the PCR predictor $\hat{y}_i$, based on A principal components, can for object i be written as a function of its scores $\hat{t}_{ia}$:

$$\text{bias} = -\sum_{a=A+1}^{K} (\hat{t}_{ia}/\sqrt{\hat{\tau}_a})\theta_a \qquad (3.52)$$

$$\text{Variance} = \sigma_f^2 \sum_{a=1}^{A} \hat{t}_{ia}^2/\hat{\tau}_a + \sigma_f^2/I \qquad (3.53)$$

$$\text{MSE} = \text{Variance} + \text{bias}^2 + \sigma_f^2 \qquad (3.54)$$

where θ_a is equal to $q_a\sqrt{\hat{\tau}_a}$. It should be mentioned that $\theta_a, a = 1, \ldots, K$ correspond to the regression coefficients of y on $\hat{U}$ in the singular value decomposition of X defined in section 3.4.1.

The noise variance σ_f^2 is added in (3.54) because we consider a prediction MSE as $E(y - \hat{y})^2$ instead of $E(E(\hat{y}) - \hat{y})^2$. The σ_f^2/I in (3.53) corresponds to the contribution from $\bar{y}$ to the final $\hat{y}_i$.

Notice that the above formulae are conditional on the given X-data in the calibration set and on the given X-data for the new object (implicitly through its scores $\hat{t}_{ia}$). This is perhaps not the most natural thing to do in all cases (see further discussion in Chapter 4). But the formulae are still useful for discussing the selection of principal components to use in regression.

For comparison, the corresponding quantities for predictions $\hat{y}_i$ based on calibrations using the full LS regression (MLR, i.e. PCR with A=K) can be written as

$$\text{bias} = 0 \tag{3.55}$$

$$\text{Variance} = \sigma_f^2 \sum_{a=1}^{K} \hat{t}_{ia}^2 / \hat{\tau}_a + \sigma_f^2 / I \tag{3.56}$$

$$\text{MSE} = \text{Variance} + \sigma_f^2 \tag{3.57}$$

The comparison shows that PCR has a bias which is not present in MLR . When the X-data are collinear (some eigenvalues being very small), the uncertainty variance of PCR can, however, be substantially smaller than that of MLR, which implies that the MSE of PCR can be much smaller than MSE of the MLR predictor. In other words, by allowing the predictor to have a bias, the MSE can be substantially decreased (see also the discussion in Chapter 4 about overfitting). The difference between the MSE's if model (3.51) applies can be written

$$\text{MSE(MLR)} - \text{MSE(PCR)} =$$

$$\sigma_f^2 \left[\sum_{a=A+1}^{K} \hat{t}_{ia}^2 / \hat{\tau}_a - \left(\sum_{a=A+1}^{K} \theta_a \hat{t}_{ia} / \sqrt{\hat{\tau}_a} \sigma_f \right)^2 \right] \tag{3.58}$$

It follows from this that the size of $\hat{t}_{ia}$ compared to the corresponding eigenvalue $\hat{\tau}_a$, and the size of θ_a, are the important quantities for the comparison. The quantity $\hat{t}_{ia} / \sqrt{\hat{\tau}_a}$ is a measure of how *representative* the calibration objects were for the new object i. If the ratio is high, then object i is outside the range of the calibration set for that factor a. The *relevance* parameter θ_a is the regression coefficient of y on $\hat{t}_a / \sqrt{\hat{\tau}_a} = \hat{u}_a$, i.e. the ability of factor a to decrease the residual sum of squares in $\mathbf{y}$ in the calibration set.

If then for instance θ_a is small (because the corresponding principal component reflects some X-phenomenon irrelevant for y), while $\hat{t}_{ia} / \sqrt{\hat{\tau}_a}$ is large for a new object i (because this phenomenon was not adequately spanned by the calibration set), then deleting the correponding principal component gives an improved prediction ability for this new object: A full MLR which includes this factor would have implied overfitting.

If the $\hat{t}_{ia}/\sqrt{\hat{\tau}_a}$ is small or moderate, deleting this principal component will only have a minor effect on the performance of the predictor for that object.

If, on the other hand, θ_a is large (indicating that this factor does correlate with **y**), deletion of this eigenvector from the PCR model gives an increase in MSE (underfitting). Note also that if in addition $\hat{t}_{ia}/\sqrt{\hat{\tau}_a}$ is large (due to incomplete calibration set, or abnormal new object i), the prediction ability will probably be bad for the new object, whether we delete the eigenvector or not.

In applications with *collinear* X-variables, eigenvector scores with small eigenvalue will have a tendency to be relatively more influenced by X-noise than eigenvector scores with large eigenvalues. This implies that in many such applications the last factors will tend to have low correlation with y and consequently have a small θ-value. Exceptions exist, but in e.g. NIR analysis where spectral data are highly collinear, this has been clearly demostrated in practice (Næs and Martens, 1988; Næs and Isaksson, 1988).

Eigenvectors corresponding to small eigenvalues generally have higher sensitivity to random data noise than eigenvalues with large eigenvalues. This can be seen clearly for X-data from scanning spectophotometers: The principal components then display increasingly noisy eigenvector spectra $\hat{\mathbf{p}}_a'$ with decreasing eigenvalues. This is in fact so evident that graphical inspection of the loading $\hat{\mathbf{p}}_a'$ is a good way of determining the optimal number of bilinear factors, A! This phenomenon will be illustrated later in the book.

Therefore, the representativity of the later eigenvectors is not as good as that of the first ones with highest eigenvalues—the ratio $\hat{t}_{ia}/\sqrt{\hat{\tau}_a}$ will have a tendency to increase with decreasing $\hat{\tau}_a$. Exceptions also exist here, but this phenomenon is again clearly demonstrated for NIR analysis (Næs and Martens, 1988; Næs and Isaksson, 1988).

All these facts are good arguments for selecting eigenvectors from the top, beginning with the highest eigenvalue etc. Our experience with this strategy is good, but it must be stressed that it is a strategy particularly suitable in cases where θ's correponding to small $\hat{\tau}$ are small (factors unimportant in **X** are unimportant for modelling **y**). For instance, how to determine A, i.e. stop incorporating new eigenvectors from **X** into the predictor, can be decided by cross-validation techniques for **y**, as dicussed in Chapter 4, by cross-validation on **X** itself as described in Wold (1978), or the quick ('and dirty') method of leverage correction (Martens and Næs (1987)), and by visual inspection of loading spectra etc. We refer to Næs and Martens (1988) for a more fundamental discussion of this issue.

Other selection strategies based on various types of significance criteria can be useful, provided that the lack of fit $\hat{f}_i$ is dominated by random, independent, identically distributed errors (as opposed e.g. to random proportional errors or non-random model errors, which are sometimes seen in practice).

The simplest of these alternatives is an ordinary t-test, but more sophisticated tools like C_p, R^2 etc. also exist (see section 3.7.4.1). Common to many of these statistical criteria is that they need a computation of s_f for the full model ($A = K$), and this requires more objects than X-variables, which is often not the case. For an overview we refer to Joliffe's book (1986) on principal component analysis. A combination of the selection from the top and by significance criteria could also be envisioned, for instance selection of the most significant components among the eigenvectors with largest eigenvalues (see Næs and Isaksson, 1988).

Till now the performance of PCR as a predictor has been treated. Since multivariate calibration also requires some degree of interpretation, it is useful to have approximate significance tests or confidence intervals on its model parameters. These can be useful for deciding if certain X-variables are useless, or if certain X-variables or certain objects appear to be equal. Such formulae will not be covered here, but are presented in e.g. Mardia et al. (1980).

Mansfield et al. (1977) considered combinations of variable selection and principal component analysis. In the present book variable selection is considered a part of the data pre-processing required prior to the final bilinear calibration regression (section 7.1). Complete elimination of an X-variable then corresponds to giving it an a priori scaling weight of 0 for the final PCR, chosen e.g. from preliminary PCR modelling of the same data.

——————————— End of statistical extensions ———————————

3.5 PARTIAL LEAST SQUARES REGRESSION (PLSR)

SUMMARY   The philosophy behind partial least squares (PLS) regression is summarized together with a brief historical perspective. The PLS regression is presented as a member of the bilinear class of methods. Different algorithms are given and properties of the method are discussed with special emphasis on the relation to PCR. PLS regression for a single Y-variable is demonstrated for artificial data, for spectroscopic determination of litmus, and for determination of protein by NIR indirect analysis of whole wheat grain. Extensions of PLS regression, modelling several Y-variables simultaneously, are presented and illustrated for NIR analysis of solvent mixtures.

3.5.1 COMPRESS X TO ITS MOST RELEVANT FACTORS, $\widehat{\mathbf{T}}$

Bilinear modelling is a powerful, flexible approach to multivariate calibration. It can yield informative, reliable predictors $\widehat{\mathbf{Y}} = f(\mathbf{X})$ by projecting the many variables

$$\mathbf{X} = (\mathbf{x}_1, \mathbf{x}_2, \mathbf{x}_3, \ldots, \mathbf{x}_K)$$

onto a few variables

$$\widehat{\mathbf{T}} = (\widehat{\mathbf{t}}_1, \widehat{\mathbf{t}}_2, \ldots, \widehat{\mathbf{t}}_A),$$

i.e.

$$\hat{\mathbf{T}} = \mathbf{X}\hat{\mathbf{V}} \qquad (3.59)$$

and using these compressed variables $\hat{\mathbf{T}}$ as regressors for $\mathbf{y}$ (Frame 3.2).

Thereby the common structures in X-variables are compressed into a stabilized, more easily interpretable model, leaving out much of the noise etc. as residuals. The important problem is, however, how to define the most relevant factors $\hat{\mathbf{T}}$ both from an interpretation and prediction point of view.

In section 3.4 we studied one such bilinear method, namely PCR, which defines the scores $\hat{\mathbf{T}}$ as projections on the dominant phenomena in $\mathbf{X}$. It was argued that with collinear data, the method could give substantial improvement over ordinary MLR. The choice of eigenvectors for prediction was also discussed, and arguments for selection from the top (those with largest eigenvalues) were given.

But it was also pointed out that improvements in PCR can sometimes be obtained by leaving out some major eigenvectors, since they correspond to phenomena in $\mathbf{X}$ of no relevance for modelling $\mathbf{Y}$. In section 3.4.4 an artificial data example illustrated this; the first principal component from $\mathbf{X}$ had no relevance for $\mathbf{Y}$. Usually, however, this is difficult in practice.

In the present section we consider another bilinear method, the partial least squares (PLS) regression (here denoted PLSR to correspond to e.g. PCR). As earlier illustrated in Figure 3.5, it differs from PCR by using the Y-variables actively during the bilinear decomposition of $\mathbf{X}$. By balancing the X- and Y-information the method reduces the impact of large, but irrelevant X-variations in the calibration modelling.

The PLSR can be understood from various perspectives—a regression method, a way to compute generalized matrix inverses, a method for systems analysis and pattern recognition as well as learning algorithm. In this book we shall present it as a regression method with a particularly relevant data compression.

Thereby PLSR can yield somewhat simpler calibration models than PCR. For high-precision data like NIR spectroscopy this aids the graphical model interpretation, because more of the Y-relevant information in $\mathbf{X}$ is displayed in the first few factors. But the optimal number of biliner factors is often similar for the two methods (see e.g. Næs et al., 1986; Martens and Næs, 1987). For low-precision data PLSR can sometimes give useful results where PCR fails.

It will also be shown that the basic PLSR algorithm allows great flexibility and useful method extensions that cannot be attained in PCR.

However, the simultaneous use of X and Y-information does give PLSR some drawbacks compared to PCR. One of them concerns orthogonality: Either the PLSR solutions has to be expressed so that it needs two sets of 'loading' vectors in $\mathbf{X}$ (termed $\hat{\mathbf{W}}$ and $\hat{\mathbf{P}}$), or it is expressed so that the scores $\hat{t}_a, a = 1, 2, \ldots$ are intercorrelated, which requires simultaneous regression of $\mathbf{Y}$ on all the scores $\hat{\mathbf{T}}$. Thus, the PLSR is slightly more complex than PCR.

Another drawback is that for noisy Y-data the PLSR has a stronger tendency to overfit than PCR; hence the topic of validation (Chapter 4) is of particular importance.

The PLS principle

'Partial least squares, PLS' is a loose term for a family of philosophically and technically related multivariate modelling methods derived from Herman Wold's basic concepts of iterative fitting of bilinear models in several blocks of variables. These concepts arose around 1975 (see e.g. H. Wold, 1981, 1983) as a practical solution to concrete data-analytic problems in econometrics and social sciences: When multivariate statistical modelling tools based on the maximum likelihood principle were applied to real data with many collinear variables and a modest number of observations, grave identification and convergence problems were often encountered. Some of these problems were overcome by the more empirical PLS approach, based on a series of local least-squares fits instead of maximum likelihood (Jøreskog and Wold, 1981).

The PLS principle can be applied to a variety of modelling problems. The most common implementation in econometrics has been one-factor path-modelling of multi-block relationships. These relationships can be estimated as either 'formative models' ('mode B PLS': latent variables = f(measured variables)), or 'reflective models' ('mode A PLS': measured variables = f(latent variables)). Combinations of the two ('mode C PLS') also exist.

The PLSR calibration considered here is a so-called mode A PLS regression method. In its basic form it models the relation between two blocks of variables, X and Y. Contrary to most econometrics work on PLS, its chemometric applications imply that more than one factor is extracted, in order to model the different analyte and interference effects in the X-data.

In our experience, PLSR stimulates users to an open-minded approach to data analysis in general and to calibration of analytical instruments in particular. Several applications have shown that it solves the multivariate prediction problem for collinear data with satisfactory predictive ability. The resulting model often has good interpretation properties as well, due to its dimensional parsimony seen in many comparisons with e.g. PCR.

It is generally dangerous to develop causal explanations for results that do not have predictive validity; the results could be nonsense! Conversely, it is dangerous to apply a statistical predictor if the user does not understand what it means. The apparent predictive success could be based on some grave misunderstanding in the laboratory.

PLSR is designed to follow the declaration: 'No predictor without interpretation, no interpretation without predictive ability.' Interpretation requires simplicity, e.g. low-dimensional models. Predictive ability also requires such simplicity in order to avoid over-fitting (see Chapter 4). Hence, the intention of PLSR regression in calibration is to optimize parsimony: Produce bilinear calibration models with as few dimensions as possible and in such a way that these dimensions are as relevant as possible.

By approximating complicated multivariate input data by a few bilinear PLSR factors, the user can plot simplified 'maps' of the main relevant information in the data. This allows the person who needs the information, and who knows the data and their context, to bring important background knowledge and intuition into the interpretation interactively—information that would be far too complex to be represented as explicit numerical information in the statistical modelling.

The chemometric version of PLSR with extraction of A factors was originally developed as a two-block algorithm, consisting of a sequence of simple, partial models fitted by least-squares (Wold et al., 1983a). It was motivated intuitively by practical experience (e.g. 'all variables may have errors') and geometrical considerations (a sequence of orthogonal projections) rather than from a statistical optimization perspective. This empirical and algorithmically oriented approach was satisfactory to many chemometricians (see e.g Lindberg et al., 1983; Martens et al., 1987; Martens and Jensen, 1983).

But statisticians and others found this difficult to understand or follow. A slightly different and more compact presentation of the method, as described e.g. in Høskuldsson (1988) and Frank (1987), is found conceptually simpler by some people. Here we use this latter presentation as a definition of the PLSR method. The original approach is used in describing two equivalent PLSR algorithms (due to S.Wold and H. Martens respectively) in Frames 3.4, 3.5 and 3.6.

3.5.2 PLSR FOR ONE Y-VARIABLE

The PLS regression method in its basic form applies for one single Y-variable and is non-iterative. It can be expressed in various equivalent ways and also be modified to accomodate two or more Y-variables simultaneously (section 3.5.4) and various other additional requirements, such as smoothing nonlinear or dynamic relationships, elimination of irrelevant variables etc.

If the PLSR scores $\widehat{\mathbf{T}}$ are desired to be orthogonal, then $\widehat{\mathbf{V}}$ for PLSR is difficult to interpret compared to $\widehat{\mathbf{V}}$ for PCR. We therefore first present the PLSR method in a simpler, slightly rotated way and return to the PLS formula for $\widehat{\mathbf{V}}$ afterwards (see equation (3.60)). This approach needs an additional set of 'loadings' called loading weights $\widehat{\mathbf{W}}$.

Let A_{max}; be the maximum number of PLSR factor to be computed; this number should be higher than the number of phenomena expected to be seen in $\mathbf{X}$, in order to permit unexpected phenomena to be modelled.

First the input variables $\mathbf{X}$ are enhanced by pretreatment scaled to comparable noise levels (section 7.1) to ensure that no variable over-shadows others. The X- and y-variables are then centered, yielding $\mathbf{X}_0$ and $\mathbf{y}_0$ (see section 3.3). Then steps 1 to 6 are performed for each factor $a = 1, 2, \ldots, A_{max}$:

1) Find loading weight vector $\widehat{\mathbf{w}}_a$ by maximizing the covariance between the linear combination $\mathbf{X}_{a-1}\widehat{\mathbf{w}}_a$ and $\mathbf{y}$ under the constraint that $\widehat{\mathbf{w}}_a'\widehat{\mathbf{w}}_a = 1$. This corresponds to finding the unit vector $\widehat{\mathbf{w}}_a$ that maximizes $\widehat{\mathbf{w}}_a'\mathbf{X}_{a-1}'\mathbf{y}_{a-1}$ i.e. the scaled covariance between $\mathbf{X}_{a-1}$ and $\mathbf{y}_{a-1}$.
2) Find factor scores, $\widehat{\mathbf{t}}_a$ as the projection of $\mathbf{X}_{a-1}$ on $\widehat{\mathbf{w}}_a$, i.e. $\widehat{\mathbf{t}}_a = \mathbf{X}_{a-1}\widehat{\mathbf{w}}_a$
3) Regress $\mathbf{X}_{a-1}$ on $\widehat{\mathbf{t}}_a$ to find the loadings $\widehat{\mathbf{p}}_a'$, i.e. $\widehat{\mathbf{p}}_a = \mathbf{X}_{a-1}'\widehat{\mathbf{t}}_a / \widehat{\mathbf{t}}_a'\widehat{\mathbf{t}}_a$.
4) Regress $\mathbf{y}_{a-1}$ on $\widehat{\mathbf{t}}_a$ to find $\widehat{q}_a$, i.e. $\widehat{q}_a = \mathbf{y}_{a-1}'\widehat{\mathbf{t}}_a / \widehat{\mathbf{t}}_a'\widehat{\mathbf{t}}_a$

5) Subtract $\hat{\mathbf{t}}_a \hat{\mathbf{p}}_a{}'$ from $\mathbf{X}_{a-1}$ and call the new matrix $\mathbf{X}_a$
   Subtract $\hat{\mathbf{t}}_a \hat{q}_a$ from $\mathbf{y}_{a-1}$ and call the new matrix $\mathbf{y}_a$.
6) Compute various outlier tests and validation statistics on $\mathbf{X}_a$ and $\mathbf{y}_a$ (see Chapters 4 and 5).

Finally, determine the optimal number of factors, A, from the validation statistics of factors $a = 1, 2, \ldots, A_{\max}$ (see section 4.5). If outliers were detected during the modelling (Chapter 5), and these outliers are considered erroneous, then the model estimation could be repeated with the errors corrected or deleted.

When $K = A$, the PLSR predictor is equal to the ordinary MLR predictor, if this exists. (There is also another special case where a one-factor PLSR and MLR give the same predictor, namely when the X-variables are orthonormal, i.e. orthogonal and scaled to length one, so that $\mathbf{X'X=I}$. But that is rarely encountered in multivariate calibration.)

The loading weights $\widehat{\mathbf{W}} = \{\hat{\mathbf{w}}_a\}$ thus obtained are orthogonal, and so are the scores $\widehat{\mathbf{T}} = \{\hat{\mathbf{t}}_a\}$. The estimated loadings $\widehat{\mathbf{P}}$ are, however, generally nonorthogonal for PLSR, although they usually resemble $\widehat{\mathbf{W}}$. It should also be mentioned that the subtraction from $\mathbf{y}_{a-1}$ in 5) is unnecessary for the model estimation (see. e.g. Manne, 1987), but it simplifies the computation of the validation statistics.

With the above definition of $\widehat{\mathbf{W}}$, the $\widehat{\mathbf{V}}$ matrix for PLSR can be written as (see e.g. Helland, 1988)

$$\widehat{\mathbf{V}} = \widehat{\mathbf{W}}(\widehat{\mathbf{P}}'\widehat{\mathbf{W}})^{-1} \qquad (3.60)$$

As we see, this matrix is somewhat difficult to interpret and therefore the other, equivalent definition of the PLSR can be chosen:

The original and computationally simplest algorithm for the PLSR method was developed by Svante Wold, as given in e.g. Wold et al. (1983). It is presented in Frame 3.4. Due to the orthogonality of the scores $\widehat{\mathbf{T}}$, it is named the orthogonalized PLSR.

An alternative PLSR algorithm was developed by Harald Martens. It is described in e.g. Martens and Næs (1987). Compared to the orthogonalized algorithm it yields the same prediction formula, but only one set of 'loadings' is required, which is identical to the loading weights $\widehat{\mathbf{W}}$ in the orthogonalized PLSR algorithm. However, the scores $\widehat{\mathbf{T}}$ in this algorithm are rotated compared to the scores in the method definition (and in Wold's algorithm). But for simplicity the same symbol $\widehat{\mathbf{T}}$ is still used.

This PLSR algorithm has fewer parameters, but at the expense of more complicated estimation of $\mathbf{q}$, the loadings of $\mathbf{y}$. The reason is that these present scores $\widehat{\mathbf{T}}$ are no longer orthogonal ($\widehat{\mathbf{T}}'\widehat{\mathbf{T}}$ is not diagonal). Hence the regression of $\mathbf{y}$ on $\widehat{\mathbf{T}}$ cannot be performed for each factor independently, but requires simultaneous LS regression of $\mathbf{y}$ on all the obtained vectors $\widehat{\mathbf{T}} = (\hat{\mathbf{t}}_1, \hat{\mathbf{t}}_2, \ldots, \hat{\mathbf{t}}_a)$ simultaneously in order to reestimate $\mathbf{q} = (q_1, q_2, \ldots, q_a)'$. Therefore this is termed the non-orthogonalized PLSR algorithm (Frame 3.5).

As we see, no $\widehat{\mathbf{P}}$ matrix is required in this non- orthogonalized PLSR algorithm.

A basic difference between the two PLSR algorithms is thus the subtraction of factors in $\mathbf{X}$ ($-\hat{\mathbf{t}}_a\hat{\mathbf{p}}_a{}'$ vs $-\hat{\mathbf{t}}_a\hat{\mathbf{w}}_a{}'$).

Note also that for the non-orthogonalized PLSR, the $\hat{\mathbf{V}}$- matrix is identical to $\widehat{\mathbf{W}}$.

If so desired, a secondary loading matrix $\hat{\mathbf{P}}$ can of course be computed even here (as in the general framework), by MLR of $\mathbf{X}$ on $\hat{\mathbf{T}}$ in analogy to the projection of $\mathbf{y}$ on $\hat{\mathbf{T}}$ to obtain $\mathbf{q}$.

Frame 3.4

Orthogonalized PLSR algorithm for one Y-variable PLS1

Calibration:

C 1   The scaled input variables $\mathbf{X}$ and $\mathbf{y}$ are first centered, e.g.

$$\mathbf{X}_0 = \mathbf{X} - \mathbf{1}\bar{\mathbf{x}}' \text{ and } \mathbf{y}_0 = \mathbf{y} - \mathbf{1}\bar{y}.$$

Choose $A_{\max}$ to be higher than the number of phenomena expected in $\mathbf{X}$.

For each factor $a = 1, \ldots, A_{\max}$ perform steps C 2.1– C 2.5:

C 2.1 Use the variability remaining in $\mathbf{y}$ to find the loading weights $\mathbf{w}_a$, using LS and the local 'model'

$$\mathbf{X}_{a-1} = \mathbf{y}_{a-1}\mathbf{w}_a{}' + \mathbf{E}$$

and scale the vector to length 1. The solution is
$$\hat{\mathbf{w}}_a = c\,\mathbf{X}'_{a-1}\mathbf{y}_{a-1}$$

where c is the scaling factor that makes the length of the final $\hat{\mathbf{w}}_a$ equal to 1, i.e.

$$c = (\mathbf{y}'_{a-1}\mathbf{X}_{a-1}\mathbf{X}'_{a-1}\mathbf{y}_{a-1})^{-0.5}$$

C 2.2 Estimate the scores $\hat{\mathbf{t}}_a$ using the local 'model'

$$\mathbf{X}_{a-1} = \mathbf{t}_a\hat{\mathbf{w}}_a{}' + \mathbf{E}$$

The LS solution is (since $\hat{\mathbf{w}}_a{}'\hat{\mathbf{w}}_a = 1$)

$$\hat{\mathbf{t}}_a = \mathbf{X}_{a-1}\hat{\mathbf{w}}_a$$

C 2.3 Estimate the spectral loadings $\mathbf{p}_a$ using the local 'model'

$$\mathbf{X}_{a-1} = \hat{\mathbf{t}}_a\mathbf{p}_a{}' + \mathbf{E}$$

which gives the LS solution

$$\hat{\mathbf{p}}_a = \mathbf{X}'_{a-1}\hat{\mathbf{t}}_a / \hat{\mathbf{t}}_a{}'\hat{\mathbf{t}}_a \qquad \text{(continued on next page)}$$

Frame 3.4 *continued*

C 2.4 Estimate the chemical loading q_a using the local 'model'

$$\mathbf{y}_{a-1} = \widehat{\mathbf{t}}_a q_a + \mathbf{f}$$

which gives the solution

$$\widehat{q}_a = \mathbf{y'}_{a-1}\widehat{\mathbf{t}}_a / \widehat{\mathbf{t}}_a' \widehat{\mathbf{t}}_a$$

C 2.5 Create new **X** and **y** residuals by subtracting the estimated effect of this factor:

$$\widehat{\mathbf{E}} = \mathbf{X}_{a-1} - \widehat{\mathbf{t}}_a \widehat{\mathbf{p}}_a{}'$$

$$\widehat{\mathbf{f}} = \mathbf{y}_{a-1} - \widehat{\mathbf{t}}_a \widehat{\mathbf{q}}_a$$

Compute various summary statistics on these residuals after a factors, summarizing $\widehat{e}_{ik}$ over objects i and variables k, and summarizing $\widehat{f}_i$ over i objects (see Chapters 4 and 5).
Replace the former $\mathbf{X}_{a-1}$ and $\mathbf{y}_{a-1}$ by the new residuals $\widehat{\mathbf{E}}$ and $\widehat{\mathbf{f}}$ and increase a by 1, i.e. set

$$\mathbf{X}_a = \widehat{\mathbf{E}}$$

$$\mathbf{y}_a = \widehat{\mathbf{f}}$$

$$a = a + 1$$

C 3 Determine A, the number of valid PLS factors to retain in the calibration model.

C 4 Compute $\widehat{b}_0$ and $\widehat{\mathbf{b}}$ for A PLS factors, to be used in the predictor $\widehat{\mathbf{y}} = \mathbf{1}\widehat{b}_0 + \mathbf{X}\widehat{\mathbf{b}}$ (optional, see P4 below)

$$\widehat{\mathbf{b}} = \widehat{\mathbf{W}}(\widehat{\mathbf{P}}'\widehat{\mathbf{W}})^{-1}\widehat{\mathbf{q}}$$

$$\widehat{b}_0 = \bar{y} - \bar{\mathbf{x}}'\widehat{\mathbf{b}}$$

Prediction:

Full prediction

For each new prediction object $i = 1, 2, \ldots$ perform steps P1 to P3, or alternatively, step P4.

(*continued on next page*)

Frame 3.4 *continued*

P 1 Scale input data $\mathbf{x}_i$ like for the calibration variables. Then compute

$$\mathbf{x}_{i,0}{}' = \mathbf{x}_i{}' - \bar{\mathbf{x}}{}'$$

where $\bar{x}$ is the center for the calibration objects.
For each factor $a = 1, \ldots, A$ perform steps P 2.1–P 2.2.

P 2.1 Find $\hat{t}_{i,a}$ according to the formula in C 2.2 i.e.

$$\hat{t}_{i,a} = \mathbf{x}'_{i,a-1}\hat{\mathbf{w}}_a$$

P 2.2 Compute new residual $\mathbf{x}_{i,a} = \mathbf{x}_{i,a-1} - \hat{t}_{ia}\hat{\mathbf{p}}_a{}'$

If $a <A$, increase a by 1 and go to P 2.1. If $a = A$, go to P 3.

P 3 Predict y_i by $\hat{y}_i = \bar{y} + \sum_{a=1}^{A} \hat{t}_{i,a}\hat{q}_a$

Compute outlier statistics on $\mathbf{x}_{i,A}$ and $\hat{\mathbf{t}}_i$ (Chapters 4 and 5).

Short prediction

P 4 Alternatively to steps P1–P3, find $\hat{y}$ by using $\hat{b}_0$ and $\hat{\mathbf{b}}$ in C 4, i.e.

$$\hat{y}_i = \hat{b}_0 + \mathbf{x}_i{}'\hat{\mathbf{b}}$$

Frame 3.5

Non-orthogonalized PLSR algorithm for one Y-variable

Calibration:

C 1 The scaled input variables $\mathbf{X}$ and $\mathbf{y}$ are first centered, e.g.

$$\mathbf{X}_0 = \mathbf{X} - \mathbf{1}\bar{\mathbf{x}}' \text{ and } \mathbf{y}_0 = \mathbf{y} - \mathbf{1}\bar{y}.$$

Choose $A_{\max}$ to be higher than the number of phenomena expected in X.

For each factor $a = 1, \ldots, A_{\max}$ perform steps C 2.1–C 2.4:

(*continued on next page*)

Frame 3.5 *continued*

C 2.1 Find X-loadings $\mathbf{w}_a$ using LS and the local 'model'

$$\mathbf{X}_{a-1} = \mathbf{y}_{a-1}\mathbf{w}_a' + \mathbf{E}$$

and scale the vector to length 1. The solution is

$$\widehat{\mathbf{w}}_a = c\,\mathbf{X}'_{a-1}\mathbf{y}_{a-1}$$

where c is the scaling factor to ensure length 1, i.e.

$$c = (\mathbf{y}'_{a-1}\mathbf{X}_{a-1}\mathbf{X}'_{a-1}\mathbf{y}_{a-1})^{-0.5}$$

C 2.2 Compute scores $\widehat{\mathbf{t}}_a$ and the local 'model'

$$\mathbf{X}_{a-1} = \mathbf{t}_a\widehat{\mathbf{w}}_a' + \mathbf{E}$$

The LS solution is

$$\widehat{\mathbf{t}}_a = \mathbf{X}_{a-1}\widehat{\mathbf{w}}_a$$

C 2.3 Compute the chemical loadings $q_1, \cdot, q_a$ for all the a factors obtained till now, using the local 'MLR model'

$$y_0 = \widehat{t}_1 q_1 + \ldots + \widehat{t}_a q_a + f$$

The LS solution is equal to

$$\widehat{\mathbf{q}} = (\widehat{\mathbf{T}}'\widehat{\mathbf{T}})^{-1}\widehat{\mathbf{T}}'\mathbf{y}_0$$

C 2.4 Create new $\mathbf{X}$ and $\mathbf{y}$ residuals:

$$\widehat{\mathbf{E}} = \mathbf{X}_{a-1} - \widehat{\mathbf{t}}_a\widehat{\mathbf{w}}_a' \quad \text{(as opposed to Frame 3.4)}$$

and

$$\widehat{\mathbf{f}} = \mathbf{y}_0 - \widehat{\mathbf{t}}_1\widehat{q}_1 - \ldots - \widehat{\mathbf{t}}_a\widehat{q}_a$$

Compute various summary statistics on these residuals after a factors, summarizing $\widehat{e}_{ik}$ over objects i and variables k, and summarizing $\widehat{f}_i$ over i objects (Chapters 4 and 5).
Replace the former $\mathbf{X}_{a-1}$ and $\mathbf{y}_{a-1}$ by the residuals $\widehat{\mathbf{E}}$ and $\widehat{\mathbf{f}}$ and increase a by 1, i.e. set

$$\mathbf{X}_a = \widehat{\mathbf{E}}$$

$$\mathbf{y}_a = \widehat{\mathbf{f}}$$

$$a = a + 1$$

C 3 Determine A, the number of valid PLS factors to retain in the calibration model.

(*continued on next page*)

Frame 3.5 *continued*

C 4   Compute $\widehat{b}_0$ and $\widehat{\mathbf{b}}$ for A PLS factors, to be used in the predictor
$\widehat{\mathbf{y}} = \mathbf{1}\widehat{b}_0 + \mathbf{X}\widehat{\mathbf{b}}$ (optional, see P4 below)

$$\widehat{\mathbf{b}} = \widehat{\mathbf{W}}\widehat{\mathbf{q}}$$

$$\widehat{b}_0 = \bar{y} - \bar{\mathbf{x}}'\widehat{\mathbf{b}}$$

Prediction:

Full prediction

For each new prediction object $i = 1, 2, \ldots$ perform steps P1 to P3 or alternatively, step P4.

P 1   Scale input data $\mathbf{x}_i$ like for the calibration variables. Then compute

$$\mathbf{x}_{i,0}' = \mathbf{x}_i' - \bar{\mathbf{x}}'$$

where $\bar{\mathbf{x}}$ is the model center for the calibration objects.

For each factor $a = 1, \ldots, A$ perform steps P 2.1–P 2.2.

P 2.1   Find $\widehat{t}_{i,a}$ according to the formula in C 2.2 i.e.

$$\widehat{t}_{i,a} = \mathbf{x}_{i,a-1}'\widehat{\mathbf{w}}_a$$

P 2.2   Compute new residual $\mathbf{x}_{i,a} = \mathbf{x}_{i,a-1} - \widehat{t}_{ia}\widehat{\mathbf{w}}_a'$

If $a < A$, increase a by 1 and go to P 2.1. If not, go to P3.
(Since $\widehat{\mathbf{W}}$ plays the role of $\widehat{\mathbf{V}}$ in this algorithm, steps P 2.1 and P 2.2 for the different factors can here be replaced by the simultaneous expressions:

$$\widehat{\mathbf{t}}_i' = \mathbf{x}_{i,0}'\widehat{\mathbf{W}}$$

$$\mathbf{x}_{i,A}' = \mathbf{x}_{i,0}' - \widehat{\mathbf{t}}_i'\widehat{\mathbf{W}}')$$

P 3   Predict y_i by

$$\widehat{y}_i = \bar{y} + \sum_{a=1}^{A} \widehat{t}_{i,a}\widehat{q}_a$$

Compute outlier statistics on $\mathbf{x}_{i,A}$ and $\widehat{\mathbf{t}}_i$ (Chapters 4 and 5).

Short prediction

P 4   Alternatively to steps P1–P3, find $\widehat{y}_i$ using $\widehat{b}_0$ and $\widehat{\mathbf{b}}$ in C 4, i.e.

$$\widehat{y}_i = \widehat{b}_0 + \mathbf{x}_i'\widehat{\mathbf{b}}$$

3.5.3 PLS REGRESSION ILLUSTRATED

3.5.3.1 Simulated data

In section 3.4.4 we studied a small data set constructed to illustrate solution of a selectivity problem for a linear spectrophotometer instrument. Figure 3.7a showed that none of the three available X-variables gave adequate univariate linear calibration for the analyte y.

However, the X-data of the five artificial objects were lying on a single plane in the three-dimensional X-space (absorbance at three wavelengths). Starting from top, two principal components had to be computed before adequate calibration of y was obtained (Figure 3.7b,c and d). Table 3.3 described the PCR estimation process and the resulting parameters.

Let us now perform the corresponding computations for PLSR, following the orthogonalized algorithm given in Frame 3.4. Table 3.4 gives the detailed results, and these are graphically illustrated in Figure 3.9.

This shows that a satisfactory modelling of y was attained already after one PLSR factor. Figure 3.9a shows the PLSR factor scores $\hat{t}_1$ vs. $\hat{t}_2$, and Figure 3.9b their loadings $\hat{p}_2$ vs. $\hat{p}_1$. As for PCR, if an object falls near the origin in the score plot, it resembles an average X-pattern. A variable falling near the origin in the loading plot would have zero variability explained by the bilinear model.

In the first PLS factor a tendency for higher-than-average levels of y and x_1 and lower-than-average levels of x_2 are seen to affect objects 5 and 4, and oppositely for objects 1 and 2. Factor two shows a tendency for higher-than-average levels of all three X-variables for objects 1 and 5, and correspondingly lower-than-average levels in objects 3, 2 and 4.

Notice that the PLSR solution is rotated compared to that of PCR (Figure 3.7g–h), so that the first PLSR factor gives better description of y. Using this one-factor PLSR solution, a good linear fit of $\mathbf{y}$ to $\mathbf{X}$ was attained (Figure 3.9c), even though Table 3.4 shows that $\mathbf{X}$ itself was not fully described by this one-factor solution. The 1-factor PLSR predictor $\hat{y} = -0.5 + x_1 - x_2$ is here comparable to that of the 2-factor PCR solution (Table 3.3). The 2-factor PLSR predictor is the same, since $\hat{t}_2$ in these noise-free data is totally irrelevant for y($\mathbf{q}_2 = 0$).

The PLSR is also a hypothesis generating tool: The residuals $\hat{\mathbf{E}} = \mathbf{X}_1$ after 1 factor show the same unmodelled structure at all three X-variables. The second PLS factor summarizes this effect well. It resembles the interference effect of some unexpected baseline shift in the X-data. In the 5 available objects this second phenomenon shows no linear correlation to y, but the score plot reveals a systematically curved relationship—this could be due to an awkward choice of samples—or to some inherent physical relation.

Figure 3.9 PLSR analysis of the input data in Table 3.2 previously illustrated in Figure 3.7. a) The coordinates of the 5 objects along the first two PLS factors' scores $\hat{t}_2$ vs $\hat{t}_1$. b) PLS loadings for the two first factors, $\hat{p}_2$ vs $\hat{p}_1$. The symbols X_1, X_2, X_3 represent the X-variables and the letter Y represents the Y-variable. c) Predicted $\hat{y}$ versus y for the resulting one-factor PLS calibration model

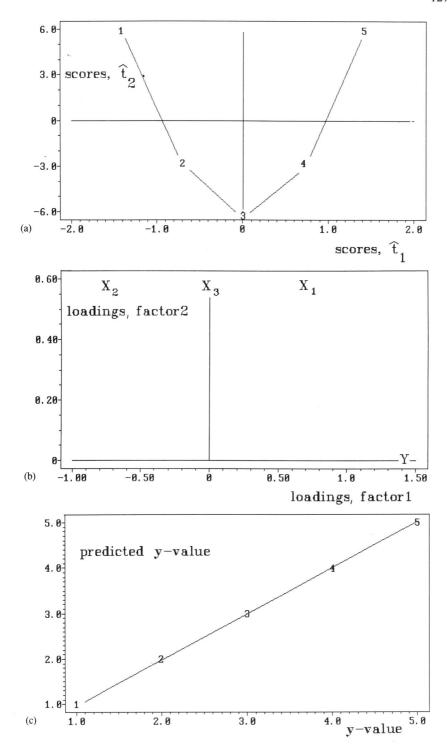

The negative $\widehat{b}_2$ coefficient for 'absorbance' at x_2 could indicate the presence of some unexpected and unidentified constituent whose concentration always descreases when the analyte concentration increases (e.g. because their sum were 100%).

These interpretation hypotheses of course would need more experimental data for verification. The interpretations in this case are basically the same as those derived from the PCR modelling.

This artificial example illustrates several things. First of all it shows that multivariate calibration can remove selectivity problems (here the univariate calibration was useless). Secondly it indicates that bilinear methods can handle collinear X-variables (here the last eigenvalue of the centered $\mathbf{X'X}$ is zero) and data with

Table 3.4 PLSR calibration

	x_1	x_2	x_3	y	
Average, $\bar{\mathbf{x}}'y$	9.1	5.6	4.6	3	
Object no.		$\mathbf{X_0}$		y_0	$\widehat{t}_1$
1	2.4	4.4	3.4	-2	-1.4
2	-2.1	-1.1	-1.6	-1	-0.7
3	-3.6	-3.6	-3.6	0	0.0
4	-1.1	-2.1	-1.6	1	0.7
5	4.4	2.4	3.4	2	1.4
Factor $a = 1$					
$\widehat{\mathbf{w}}_1'$	0.71	-0.71	0		
$\widehat{\mathbf{p}}_1'$ and $\widehat{q}_1$	0.71	-0.71	0	1.4	
$\widehat{\mathbf{b}}'$ and $\widehat{b}_0$	1.0	-1.0	0	-0.5	
Object no.		$\mathbf{X_1}$		y_1	$\widehat{t}_2$
1	3.4	3.4	3.4	0.0	5.9
2	-1.6	-1.6	-1.6	0.0	-2.8
3	-3.6	-3.6	-3.6	0.0	-6.2
4	-1.6	-1.6	-1.6	0.0	-2.8
5	3.4	3.4	3.4	0.0	5.9
Factor $a = 2$					
$\widehat{\mathbf{w}}_2'$	0.58	0.58	0.58		
$\widehat{\mathbf{p}}_2'$ and $\widehat{q}_2$	0.58	0.58	0.58	0.0	
$\widehat{\mathbf{b}}'$ and $\widehat{b}_0$	1.0	-1.0	0	-0.5	
Object no.		$\mathbf{X_2}$		y_2	
1	0	0	0	0	
2	0	0	0	0	
3	0	0	0	0	
4	0	0	0	0	
5	0	0	0	0	

curved relationships. And finally, it illustrates how PLSR regression can give simpler calibration models than PCR by enhancing X-phenomena that are correlated to Y, at the expense of X-phenomena uncorrelated to **Y** (one PLSR factor required, vs two principal components).

(It may be noted that in this dataset the second X-phenomenon was actually orthogonal to **y**, apart from round-off errors etc. That is usually not the case for real data.)

3.5.3.2 A simple example: Analyzing a dye in solution

Figure 3.10a shows the optical density between 400 and 800 nm of a dye, the

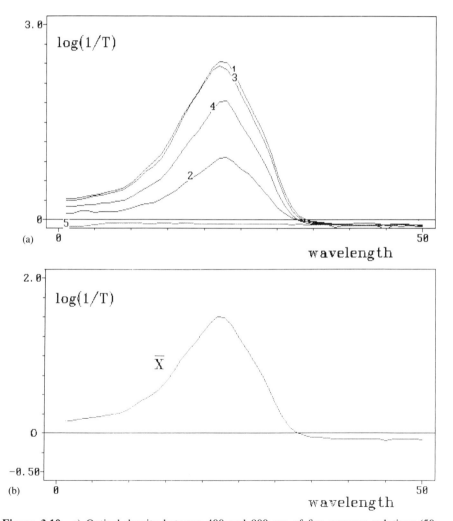

Figure 3.10   a) Optical density between 400 and 800 nm of five aqueous solutions (50 X-variables). b) Average spectrum $\bar{x}$ of the five spectra in a)

130

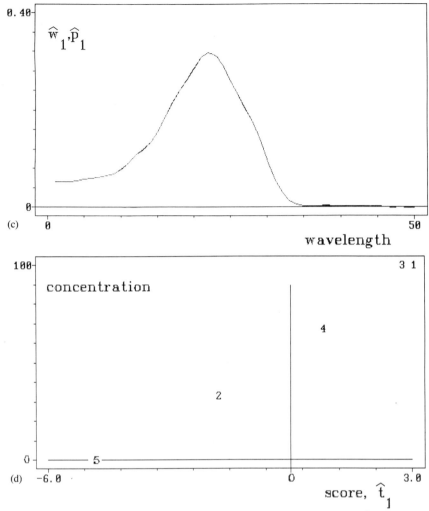

(c) 0

wavelength

(d) −6.0

score, $\widehat{t}_1$

Figure 3.10 (*cont.*) c) The loading weights $\widehat{\mathbf{w}}_1$ and loadings $\widehat{\mathbf{p}}_1$. d) y versus $\widehat{t}_1$ for the five calibration objects

traditional pH-indicator litmus. This dye actually consists of a group of compounds, but for the present purpose these are assumed to occur in fixed proportions, so that the dye behaves as one single compound. More detail on this is given in Chapter 8. The spectra represent five aqueous solutions at various dye concentrations at pH 10 (i.e. the solutions are blue-coloured).

These spectra were obtained by transmission T, using a Guided Wave 200–40 spectrophotometer: White light was passed through a several meters long optical fiber into the water-solutions (1 cm path length) and back through another optical fiber to a monochromator and detector. The transflection data, read at 50 wavelengths (every 16 nm) were linearized by O.D: = $\log(1/T)$ and used as X-variables.

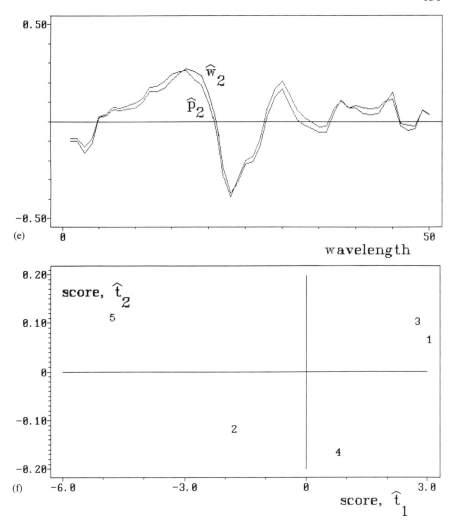

Figure 3.10 (*cont.*)   e) The loading weights $\hat{\mathbf{w}}_2$ and loadings $\hat{\mathbf{p}}_2$. f) The two first factor scores of the calibration objects

The litmus concentration (in percent of a reference concentration of 1.25 mg/ml) was used as y.

The average spectrum $\bar{\mathbf{x}}'$ is given in Figure 3.10b, and shows the typical O.D. spectrum of blue litmus. The first PLS factor's loading weight vector $\hat{\mathbf{w}}_1'$ and loading vector $\hat{\mathbf{p}}_1'$ (the two are here indistinguishable) are given in Figure 3.10c. As we see, this difference spectrum between high and low levels of factor $\hat{t}_1$ is very similar to the average spectrum in shape. Most of the spectral variation in **X** was accounted for by this first PLS factor $\hat{t}_1$, and this factor gave a sufficiently good fit to y (Figure 3.10d).

However, internal cross validation (section 4.3.2.2) indicated that a minor second PLS factor improved the modelling slightly. Figure 3.10e shows its spectral features,

132

$\hat{\mathbf{w}}_2$ and $\hat{\mathbf{p}}_2$, which in this case are slightly different from each other (indicating that the X–Y relationship and the X–X relationship call for slightly different X-modelling).

Figure 3.10f shows the 2-factor score plot for the calibration objects, scaled for maximum resolution. It indicates a curvature with increasing analyte concentration. Upon closer inspection, a small curvature could also be seen in Figure 3.10d. Upon comparing the loadings for the first and second factors, it seems that factor 2 could well be the result of a minor curvature in the instrument response at the highest O.D: Factor 2 has negative loadings at the analyte peak maximum, and positive loadings on both sides of the peak.

Whether or not to include the second factor in the calibration is debatable. With only 5 calibration objects available, the estimation of this factor is rather uncertain. (Generally, it is good calibration practice to have at least 4 objects for each bilinear factor.) Its loading is sufficiently uneven to indicate that it also reflects significant amounts of random measurement noise. Its contribution to the modelling of $\mathbf{X}$ was very minor; while $\hat{\mathbf{t}}_1'\hat{\mathbf{t}}_1$ was 44.2, $\hat{\mathbf{t}}_2'\hat{\mathbf{t}}_2$ was only 0.069, so the second factor therefore corresponds to a minor spectral effect. Figure 3.10g compares the corresponding predictor function $\hat{\mathbf{b}}$ for the one-factor and two-factor solutions. It shows that the one-factor solution is smooth (like $\hat{\mathbf{w}}_1$), while the two-factor predictor solution is strongly dominated by the small, but amplified second factor and therefore also less smooth.

Figure 3.10h shows the analyte concentration vs. the predicted analyte concentration for the 5 calibration objects plus three new 'unknown' test objects (nos. 6, 7 and 8), using the 2-factor PLS solution. A good predictive ability was attained for 'unknown' objects 6 and 7, but for 'unknown' object 8 the analyte prediction was rather inaccurate (about 75% instead of 97%). However, this 'unknown' object was automatically detected as an abnormal object (outlier,

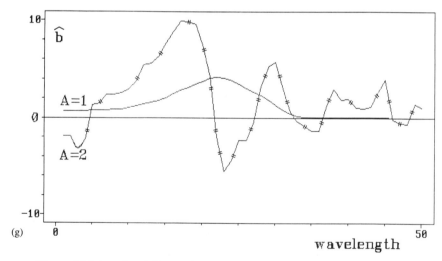

Figure 3.10 (*cont.*) g) Regression coefficients based on one and two factors

see Chapter 5) based on its X-residuals, both with the 1-factor and the 2-factor PLS solutions.

Figure 3.10i shows the residual spectra after fitting of the one-factor PLS model. The residuals for most of the objects are close to zero, as expected, but object 8 shows a strong residual spectrum. By comparison with other available spectra, this residual $\hat{e}_8$ closely resembles that of the analyte litmus in its red-coloured state (acid pH). Closer examination of object no. 8 revealed that it indeed had an abnormally low pH.

In the present example only one phenomenon was expected in the spectral data. So a univariate calibration could also have been used, but this would not have revealed the unexpected unknown object condition. By 'modelling' the full X-spectrum for each individual unknown object by a low-dimensional PLS model,

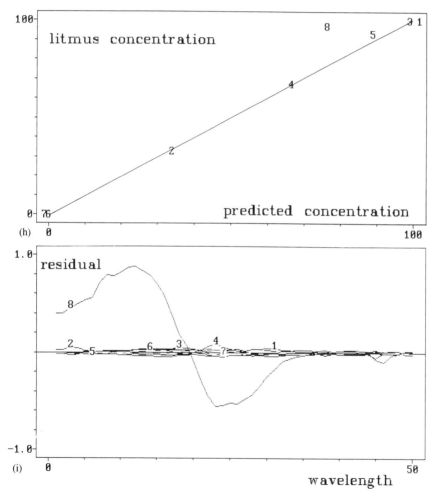

Figure 3.10 (*cont.*) h) y versus $\hat{y}$ for the five calibration objects plus 3 prediction objects.
i) The residual spectra of the objects relative to the one-factor PLS model

the abnormality was automatically detected for further identification. This is covered in more detail in Chapter 5.

The problem of determining the total concentration of litmus in the presence of various interferences was briefly mentioned in Chapters 1.1 and 1.2. Chapter 8 uses this analytical problem as an illustration of a full calibration experiment.

3.5.3.3 Spectroscopic analysis of whole wheat kernels

The next example of PLSR concerns NIR determination of protein in wheat. This resembles the high-sensitivity chromatographic dioxin example in Figure 1.3 in the sense that the analyte Y to be determined is actually a weighted sum of many different molecular species, summarized for relevance and simplicity. In Figure 1.3 the analyte was ng TCDD equivalents per sample, which summarizes the toxicity of a number of different halogenated dibenzo-p-dioxins and dibensofurans. In the present example the analyte is weight percent 'crude protein', which summarizes a number of different protein species, including enzymes, storage proteins, structural proteins etc. A third, analoguous example of calibrating for chemically complex, but relevant analytes can be mentioned: The determination of botanical components (rather than chemical constituents) in wheat flour milling streams by autofluorescence exitation/emmission UV/vis. spectroscopy (Jensen et al., 1982; Pedersen and Martens, 1989), and sensory properties from NIR data (Martens and Martens, 1986b).

The protein content of wheat is important for its nutritive value and baking quality. Traditionally, protein has been quantified by 'wet chemistry' determination of the nitrogen content of the product (the Kjeldahl method). Under the simplifying assumption that proteins are the only major nitrogen (N)-containing constituents in the objects, and that the various wheat proteins on the average contain 16 percent N, the traditional 'wet chemical' calibration consisted of the transformation

$$\text{protein } \% = \text{N } \%/0.16 = \text{N } \% * 6.25$$

However, the Kjeldahl N analysis is time-consuming, laboursome and uses noxious chemicals, and it is rather imprecise (typical error standard deviation: 0.2% protein at 10% protein, i.e. coefficient of variation = 0.02). In addition, it is not the ideal measurement of protein, since the nitrogen content of protein varies with the growth conditions etc. This is so because the ratio between the individual molecular protein species (storage proteins, enzymes etc, each with a different N-content) varies.

So, based on the pioneering work of Karl Norris and others (see e.g. Norris et al., 1976; Norris, 1983 and references in them), multi-wavelength NIR reflectance analysis of ground flour has since the mid 1970s increasingly replaced the Kjeldahl analysis, just like NIR reflectance of ground meat has replaced traditional fat, water and protein determination of meat (section 3.4.5). But since the Kjeldahl-N measurement is so well established, it still has the status of reference method for wheat protein, in spite of its questionable relevance.

Figure 3.11a shows the absorbance spectra of the major constituents in wheat,

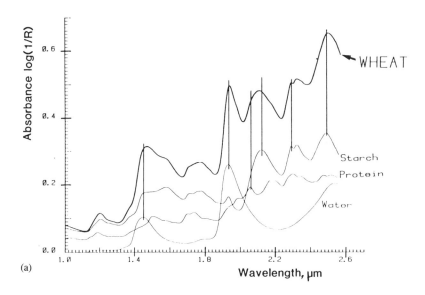

Figure 3.11 a) Absorbance spectra of the major chemical constituents in wheat: starch, protein and water, in the NIR wavelength range (courtesy Karl H. Norris, USDA)

protein, starch and water in the wavelength region 1000–2600 nm, analyzed in their pure state. The peaks represent over-harmonics and combination bands from the IR absorbance spectra of organic molecules. The figure shows that protein does display spectral information distinguishable from the other constituents, but no single wavelength can be expected to be selective. However,in analogy to the determination of fat in hamburger meat (Figure 3.8), multivariate calibration solves the selectivity problems.

Thus, in NIR spectroscopy in the wavelength range 1400–2500 nm the only sample preparation needed is a homogenization, like grinding wheat kernels to flour prior to the reflectance analysis. Several recent books give more detail about the NIR method, e.g. Osborne and Fearn (1986) and Williams and Norris (1987).

The present example shows how more recent NIR transmission instruments can be used for totally non-destructive analysis of whole wheat kernels, thereby simplifying the protein determination even further. In this application a lower NIR wavelength range is employed. In this range the absorbance is sufficiently low to allow penetration through the intact wheat kernels. Thus, for spectroscopic determination of major wheat constituents like water and protein content, there is no need for any sample preparation at all.

The NIR measurements were done in a Tecator Infratec instrument. Relative light intensities, transmitted through samples of wheat kernels were measured at many wavelengths. A subset of $K = 100$ wavelengths were chosen, and the obtained transmittance T linearized by the conventional O.D. transformation $X = \log(1/T)$. In the present illustration, a rather wide calibration was sought, with applicability for various types of wheat.

136

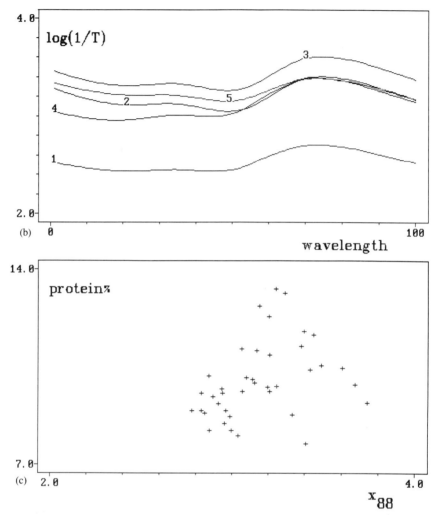

(b)

(c)

Figure 3.11 (*cont.*) b) Input spectra **X** for high-protein and low-protein wheat batches at a subset of 100 wavelength channels in a lower NIR range. c) Univariate prediction: Analyte content (protein percentage) vs. O.D. at the 'best' NIR wavelength (*X* channel 88) for the 39 wheat kernel batches in the test set

In the example, the NIR spectrum $x_{ik}, k = 1, 2, \ldots, 100$ was obtained for $i = 1, 2, \ldots, 89$ wheat batches. Their protein content (y_i), expressed as weight percentage, was measured by conventional Kjeldahl nitrogen analysis by averaging 3 replicates to a final precision of about .1 percent protein. Fifty of the batches were selected as calibration objects, the remaining 39 were employed as an independent test set. The two sets had quite similar statistical distributions of the analyte (calibration set: Range: 7.1 − 12.9, $\bar{y} = 9.9$. Test-set: Range: 7.8 − 13.3, $\bar{y} = 10.0\%$).

Figure 3.11b shows the input spectra **X** for five typical wheat kernel batches.

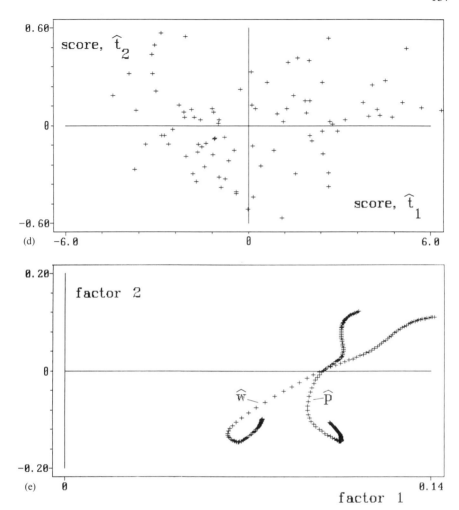

Figure 3.11 (*cont.*)   d) PLS score plot, showing the position of the 50 calibration objects on the first two factors, $\widehat{t}_2$ vs $\widehat{t}_1$. e) PLS loading weights $\widehat{w}_a$ and loadings $\widehat{p}_a$ for factor $a = 2$ vs $a = 1$

Some major and minor smooth peaks are visible, in addition to different general levels and trends. But a clear, unique pattern representing the analyte variations is not evident.

Thus, not surprisingly, univariate calibration gave bad fit to the data for the calibration set, as well as for the test set (Figure 3.11c).

The 100 X-variables and the analyte variable y for the 50 calibration objects were submitted to PLSR. To ensure sufficient modelling, a total of $A_{max} = 18$ PLS factors were computed. The last six factors will not be presented here, because they had no predictive validity.

Figure 3.11d shows the distribution of the 50 calibration objects along the two

138

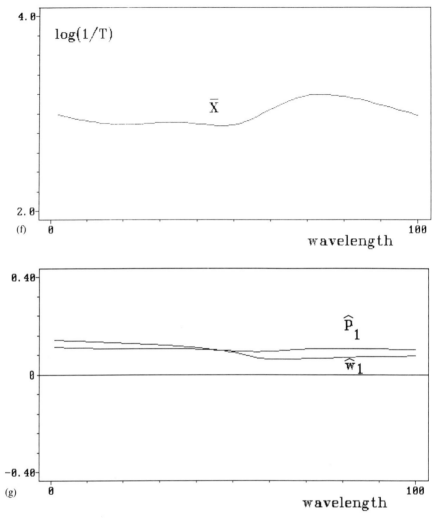

Figure 3.11 (*cont.*)   f) model center $\bar{\mathbf{x}}$ plotted vs wavelength channel no. g) loading weight $\widehat{\mathbf{w}}_a$ and loading spectrum $\widehat{\mathbf{p}}_a$ for factor $a = 1$

first factors $\widehat{t}_2$ vs. $\widehat{t}_1$ corresponding to modelling 99.8% of the total variation in the NIR calibration data. Figure 3.11e displays the PLS loading weights $\widehat{\mathbf{w}}_2$ vs $\widehat{\mathbf{w}}_1$ and the corresponding PLS loadings $\widehat{\mathbf{p}}_2$ vs. $\widehat{\mathbf{p}}_1$.

Figure 3.11f through 3.11r shows how the PLSR model developed, in terms of its average spectrum $\bar{\mathbf{x}}$ and loading spectra $\widehat{\mathbf{p}}_a$, for factors $a = 1, 2, \ldots, 12$. In figures 3.11g and 3.11h even $\widehat{\mathbf{w}}_a$ has been included, in order to explain the pattern seen in Figure 3.11e.

From purely chemical considerations it is unexpected to find that $\widehat{\mathbf{w}}_1{}'$, the vector of covariances of the NIR variables with protein, is positive at every wavelength. When the protein content goes up, the starch and water content is bound to go

139

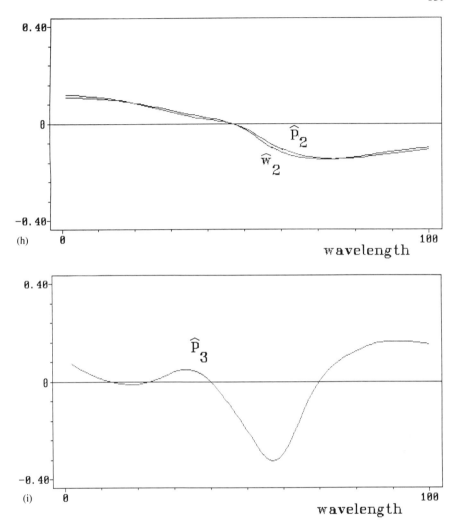

Figure 3.11 (*cont.*) h) loading weight $\widehat{\mathbf{w}}_a$ and loading spectrum $\widehat{\mathbf{p}}_a$ for factor $a = 2$. i) loading spectrum $\widehat{\mathbf{p}}_a$ for factor $a = 3$

down, since their sum is 100% in the grain. Hence a negative correlation to protein should be expected at the typical starch and water wavelengths. Instead it seems that the protein percentage y correlates with some physical phenomenon in **X** (possibly light scattering or kernel volume filling) that affects the whole O.D. spectrum.

This unidentified spectral phenomenon is not particularly important for modelling y, but it dominates the X-variations: The first PLS factor accounted for about 98.7% of the variation in $\mathbf{X}_0$. On the basis of this estimated 'intensity' level $\widehat{\mathbf{t}}_1$, the corresponding spectral fingerprint of this physical phenomenon, $\widehat{\mathbf{p}}_1$, is also estimated. Compared to $\widehat{\mathbf{w}}_1$, a general slope has been reduced, so that the loading

140

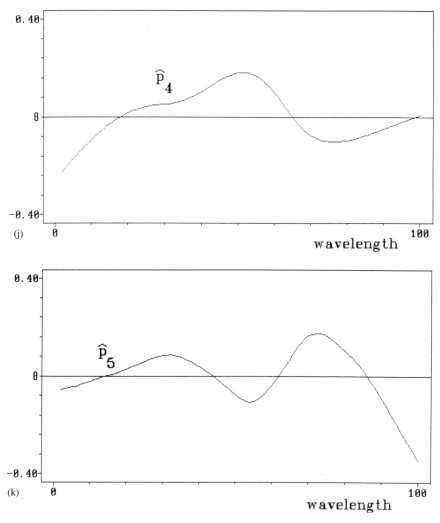

Figure 3.11 (*cont.*)   j) and k): loading spectrum $\widehat{\mathbf{p}}_a$ for factors $a = 4$ and 5

$\widehat{p}_1$ is more similar to the loading expected for a major baseline shift in the spectral data (i.e. a constant loading level of $\sqrt{1/K} = 0.10$.)

The two sets of X-loadings are much more similar for factor 2 (Figure 3.11h) and for the subsequent factors; therefore only $\widehat{p}_a$ is given for factors 3–12 (Figures 3.11i–r).

A number of spectral peaks can be seen in the loading spectra, but in contrast to the litmus example above, a causal intepretation of each individual factor is rather difficult. In addition to responding to the effect of protein variations in the NIR spectrum, the factors probably reflect both chemical interferences (from starch and water absorbances etc.) as well as changes in the analyte itself (various

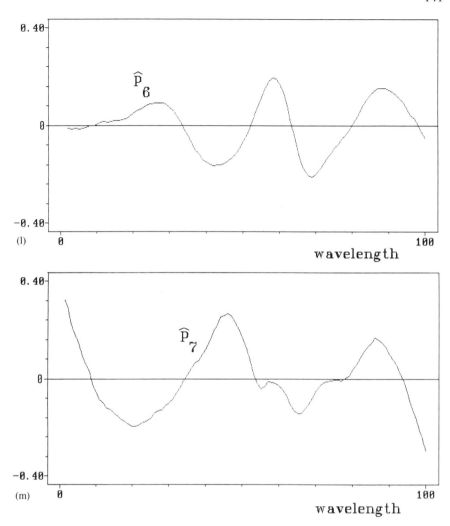

Figure 3.11 (*cont.*) l) and m): loading spectrum $\widehat{\mathbf{p}}_a$ for factors $a = 6$ and 7

protein compositions), physical interferences (differences in light scattering and kernel shape, temperature shifts in the spectra and in the monochromator etc.) and curvature effects.

Notice how the first factors seem to reflect systematic, broad spectral features, while the last spectra bring in more random noise! This corresponds to the fact that the first factors describe most of the X-variations, while the last factors only represent minor residuals, amplified by the small values of $\widetilde{\mathbf{t}}_a'\widehat{\mathbf{t}}_a$.

How many of these factors should be used in the resulting PLS calibration model? The loading spectra graphically indicate that random spectral noise increasingly affect the factors from $a = 10$ and onwards. A more explicit statistical validation

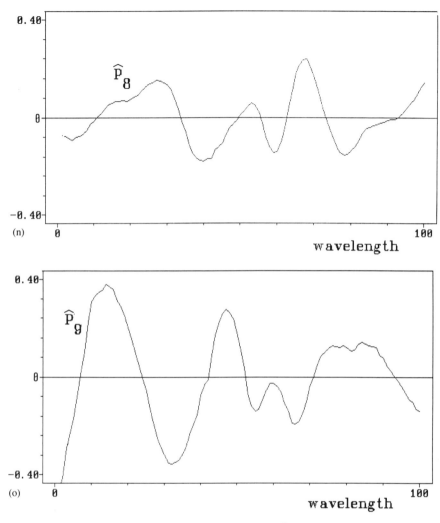

Figure 3.11 (*cont.*)   n) and o): loading spectrum $\hat{\mathbf{p}}_a$ for factors a = 8 and 9

(internal cross validation, see section 4.3.2.2) indicated that the ten or eleven first factors had predictive validity, with the nine first factors accounting for most of the predictive ability, as shown in Figure 3.11s.

Figure 3.11t shows how the corresponding predictor $\hat{\mathbf{b}}$ develops with increasing number of PLSR factors. It shows that, compared to the 'optimal' model complexity (10 factors), underfitting (here: using only 6 factors) includes too little spectral detail, while overfitting (here: 12 factors) includes too much detail in the predictor $\hat{\mathbf{b}}$. The former leaves important interference phenomena unmodelled. The latter draws low-level calibration noise into $\hat{\mathbf{b}}$, which gives systematic prediction errors. In addition this amplification of minor X-phenomena increases the absolute value

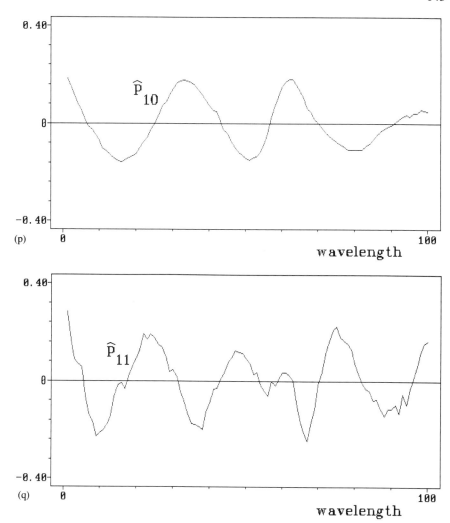

Figure 3.11 (*cont.*)   p) and q): loading spectrum $\widehat{\mathbf{p}}_a$ for factors $a = 10$ and 11

of $\widehat{\mathbf{b}}$, which leads to unnecessary amplification of measurement noise in future X-spectra (see e.g. sections 3.4.6 and 4.5).

It was thus decided to use the $A = 10$ factor PLSR solution. Finally, this obtained PLS calibration model was used for predicting protein from NIR in the independent test set. The pattern of prediction ability was similar to that obtained from the calibration set (Figure 3.11s). This is as expected, since the both calibration and test sets were relatively large and representative. A certain deviation is seen at factor 3: Apparently this factor spans a certain type of variability in the calibration set that is not important in the predition set. This shows that the calibration modelling might possibly have been optimized further, e.g by data pre-processing or removal of certain calibration objects—or that the test set lacked one type of variability.

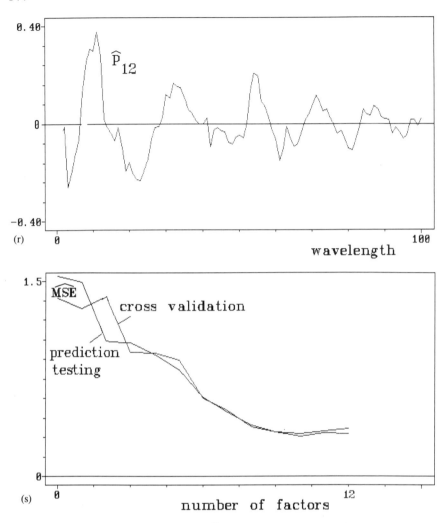

Figure 3.11 (*cont.*) r): loading spectrum $\widehat{\mathbf{p}}_a$ for factor a = 12. s) Validation of PLSR for protein in whole wheat: Average deviation (RMSEP, see Chapter 4) between reference measurement y_i and $\widehat{y}_i$ predicted from NIR data $\mathbf{x}_i'$ vs. the number of PLS factors a = 0,1,2,...,12. The two curves represent internal cross-validation within the set of 50 calibration objects, and external validation for the set of 39 test objects, respectively

Figure 3.11u compares the chemically measured analyte concentrations and the corresponding predictions obtained from the NIR data, using the PLS calibration model with 10 factors in the test set. The solid line gives the 'true' prediction line $(\widehat{y}_i = y_i)$. A correlation coefficient of 0.97 is observed between the traditional and the NIR protein determinations in these test objects.

In this example the crude protein content y in whole wheat kernels was predicted from a combination of 10 factors $\widehat{\mathbf{t}}_a, a = 1, 2, \ldots, 10$—each of which was a linear

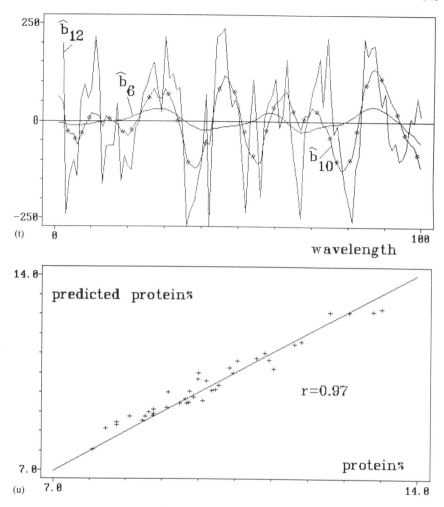

Figure 3.11 (*cont.*) t) Predictor $\widehat{\mathbf{b}}$, corresponding to the combination of PLSR factors that predicts **y** from **X**, given for 6, 10 (optimal) and 12 factors. u) Protein percentage in wheat kernels, measured by the traditional wet chemistry and the multi-channel NIR method for objects $i = 1,2,\ldots,39$ in the test set. Abscissa: y_i = Kjeldahl $- N \times$ 6.25. Ordinate: $\widehat{y}_i$ found by the 10-factor PLSR solution. Average apparent prediction error (RMSEP) is 0.3%, of which the prediction bias is -0.08%; $r = 0.97$

combination of the 100 measured NIR variables **X**. This has shown how a well established, but complicated, noxious and slow traditional analysis *y* was replaced by a high-speed, non-destructive analysis **X**. An estimated average prediction error (RMSEP, see Chapter 4) of about 0.3 percent protein was thus obtained in the present example. When instead calibrated for a more narrow population (hard winter wheat between 10 and 12 percent protein), this type of instrument has yielded average prediction errors (RMSEP) well below 0.2 percent protein—i.e.

the NIR predictions can be more precise than the reference method being replaced.

Since all organic compounds, as well as water, display absorbance patterns in the NIR range, this type of spectroscopy has great potentials even outside the agricultural field where it was developed.

In the next section a more industrially oriented application of NIR spectroscopy with multivariate calibration will be illustrated: Remote quantitative analysis of organic solvent mixtures.

3.5.4 PLS CALIBRATION FOR SEVERAL Y-VARIABLES

3.5.4.1 The PLS2 algorithm

Sometimes several constituents $y_1, y_2, \ldots, y_J$ are to be predicted from the same input data $\mathbf{X}$. How can this be done efficiently?

One aspect of this is to find a prediction procedure that does not require J different repetitions of almost the same modelling. Another aspect concerns how to estimate the J predictor formulae: Should each Y-variable be modelled separately, or should they all be modelled simultaneously?

In PCR this is not a topic at all, since $\mathbf{Y}$ does not affect the modelling of $\mathbf{X}$. During calibration, the ordinary least squares regression of the y-variables on $\hat{\mathbf{T}}$ in PCR gives the same result, whether it is performed for each y-variable individually or for all J y-variables simultaneously.

In the standard PLS calibration this is somewhat more complicated, since Y now affects the data compression modelling of $\mathbf{X}$, i.e. influences $\hat{\mathbf{V}}$. The different Y-variables calibrated for separately in PLS will therefore give somewhat different modelling of $\mathbf{X}$ and hence different regression factors ($\hat{\mathbf{T}}$).

It is easy to modify the orthogonalized PLSR algorithm to accomodate a jointly optimized calibration for several Y-variables. The only difference needed in the method formulation given in section 3.5.2 is that instead of maximizing the covariance between $\hat{\mathbf{t}}_a = \mathbf{X}_{a-1}\hat{\mathbf{w}}_a$ and a single y-variable, we maximize its covariance between $\hat{\mathbf{t}}_a = \mathbf{X}_{a-1}\hat{\mathbf{w}}_a$ and $\hat{\mathbf{u}}_a = \mathbf{Y}_{a-1}\hat{\mathbf{q}}_a(\hat{\mathbf{q}}_a{}'\hat{\mathbf{q}}_a)^{-1}$, which is a linear combination of the Y-variables.

Even in this case we have the orthogonality properties that $\hat{\mathbf{T}}'\hat{\mathbf{T}}$ is a diagonal matrix and $\hat{\mathbf{W}}'\hat{\mathbf{W}} = \mathbf{I}$. Thus $\hat{\mathbf{V}}$ can be written as in (3.60).

A simultaneous PLSR algorithm for several Y-variables, developed from the orthogonalized PLSR in Frame 3.4, is given in Frame 3.6.

As we see, this so-called 'PLS2' algorithm is iterative for each factor, because both the linear combinations $\hat{\mathbf{t}}_a$ and $\hat{\mathbf{u}}_a$ are now to be optimized with respect to each other. Additional numerical aspects can be found in Manne (1987) and Høskuldson (1988).

3.5.4.2 When should PLS2 regression be used?

All calibration problems start with an explorative phase, where over-view and detection of gross errors are the main purposes. If several Y-variables exist, it is advisable to use one single PLS2 analysis instead of J different PLS1 runs at this stage. Various extra variables, for instance design variables (1 =yes, 0 =no), indicating various experimental conditions etc, can be included in **X** or **Y** at this stage, to improve interpretability. Standardizing the variables is then often helpful.

The PLS2 analysis can also be useful during calibration, if the Y-variables are known to be strongly intercorrelated with each other. The PLS2 method uses their intercorrelation structure to stabilize the determination of $\widehat{\mathbf{W}}$ against random noise in the individual Y-variables:

Assume that we want to calibrate for one important analyte, but that the available calibration data for the analyte, variable y_1, are very imprecise. Assume also that we know a priori that the true analyte level is strongly correlated to other constituents or effects of less interest but which are easier to measure precisely (variables $y_j.j = 2,3,\ldots,J$). Then a joint PLS2 calibration for y= $(y_j.j = 1,2,\ldots,J)$ may give better understanding of the calibration data and better predictions for y_1 than PCR or a separate PLSR calibration for y_1 alone. In the PLS2 modelling the effect of random noise in calibration data y_1 will be reduced.

However, two things are worth noticing: First of all, the final Y-loadings $\widehat{q}_{ja}$ are estimated for each Y-variable separately in PLS2; so such a stabilizing improvement can only be expected in the determination of $\widehat{w}_{ka}$. Therefore the statistical benefit of PLS2 is limited.

Secondly non-linearities in the $X-Y$ relationships may be a greater problem than the random noise in the **Y** or in **X** data for many calibration situations. Forward calibration methods like PLS regression can give good approximation of many types of non-linearities as demonstrated in section 3.5.3.1 (see also Chapter 7). However, if different Y-variables have different types of curvature in their relationships to the X-variables, then the PLS2 solution will have to find a suboptimal approximation compromise between the different Y-variables. Otherwise too many PLSR factors may have to be estimated, leading to over-fitting problems if the number of calibration objects is low and/or the data are noisy. This was studied in more detail by Pedersen and Martens (1988). They used various types of PLSR in calibrating for botanical components pericarp, aleruron and endosperm in wheat milling streams from simplified multichannel autofluorescence spectroscopy.

Thus, PLS2 regression is very useful for preliminary overview in explorative data analysis. It can even work well for a simultaneous final calibration. But when non-linearity, rather than random noise, is the major problem, then it may be advantageous to use separate PLS1 modelling for each y-variable for the final calibration results. For subsequent prediction it is then practical to use only one of these PLS1 models for outlier testing etc, and predict the Y-variables via their respective predictors, $\widehat{y}_{ij} = \widehat{b}_{0j} + \sum_{k=1}^{K} \widehat{b}_{kj}X_{ik}$.

3.5.4.3 Example: Joint calibration for three analytes

The present example concerns analysis of mixtures of organic solvents—in this case mixtures of three similar alcohols, methanol, ethanol and n-propanol. Like all other organic solvents (except CCl_4) they display absorbance spectra in the near-infrared wavelength range. To allow measurements in noxious and inaccessible environments it was decided to separate the physical measurements from the sensitive instrumentation by optical fibers.

The NIR data were obtained by transmisssion T in a Guided Wave 200–40 fiber optics spectrophotometer with a germanium detector and a 600 L/mm grating: White light passing via a glass fiber of several meters length, through a 10 mm optical path length in the liquid and back via another fiber to the scanning monochromator and detector. Transmission (T) was measured at every 5 nm from 1100 to 1600 nm. For response linearization these were converted to optical densities (O.D) by taking $\log(1/T)$ at the 101 wavelength channels. The O.D. spectra were then shifted to the same average baseline in the range between channel 41 (1305nm) and 45 (1325nm); this baseline shift preprocessing was done for each spectrum to compensate for offset shift apparently due to flexing of the optical fiber cable. These readings represent the 101 X-variables in the present data set. The weight fraction of the three alcohols methanol, ethanol and n-propanol represent the three Y-variables (Bjorsvik and Martens, 1989).

Figure 3.12a shows the NIR O.D. spectra of the three pure alcohols, each given in two replicates. It shows that none of the individual wavelengths are selective for any one alcohol, although distinct differences between their NIR spectra can be seen, for instance around channel 20 and 70. The two replicates of each alcohol are virtually indistinguishable.

For these transparent solvent mixtures it might in principle have been possible to obtain calibrations by direct unmixing (section 3.6.1), but previous experience had, however, shown that the NIR spectra might be affected by nonlinear phenomena. Therefore it was decided to perform indirect calibration, based on actual mixtures. A calibration set of 15 mixtures and a test set of 11 mixtures of the three alcohols were accurately prepared and their NIR spectra measured.

Figure 3.12b shows the experimental design of the calibration set in terms of two of the analytes (the sum of the three alcohols' weight fractions is 1). The design of the test set was very similar.

The 101 X-variables and 3 Y-variables in the 15 calibration objects were modelled with the PLS2 regression without further preprocessing. With only two independent physical phenomena (three constituents, with sum 1), one would expect two PLS factors. Internal cross validation (see Chapter 4), as well as validation by test set (Figures 3.12c and d) indicated that 2 PLS factors indeed modelled most of the variation, both in the X- and the Y-variables. But 3 factors was required in order to attain fully adequate predictive ability.

Figures 3.12e, f, g and h give the average spectrum and the first three factor loadings. It shows that the loading weights $\hat{\mathbf{w}}_a$ and the loadings $\hat{\mathbf{p}}_a$ were very similar for these data.

Figure 3.12i, j and k show the two first factors plotted against each other, with respect to loadings for the X-variables and Y-variables and with respect to object scores.

In the score plot the triangular shape of the mixture system (see e.g. Martens, 1979) is evident, with the three pure alcohols at the corners, well in correspondence

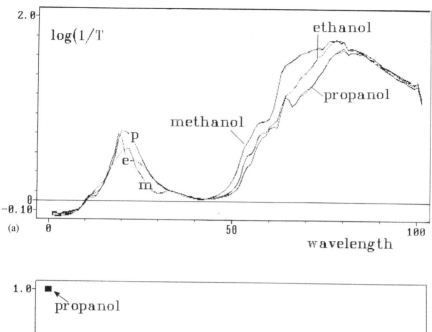

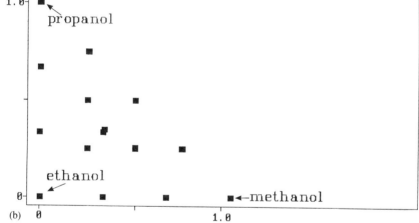

Figure 3.12 a) O.D. spectra between 1100 and 1600 nm (NIR wavelength channels $k=1,2,\ldots,100$) for the three analytes methanol (m), ethanol (e) and n-propanol (p), each given in two replicates. b) Design of the set of 15 ternary calibration mixtures given for propanol vs. methanol. The prediction objects cover approximately the same region. Note that the origin in the design corresponds to pure ethanol

with the general pattern in the Y-loadings. Since the triangle in these plots comes with a reasonably horizontal lower edge, the first factor (Figure 3.12f) can be interpreted as mainly a difference between propanol (lower right corner) and methanol (lower left corner). The second factor (Figure 3.12g) is then mainly the difference spectrum between ethanol (upper corner) and a combination of propanol and methanol.

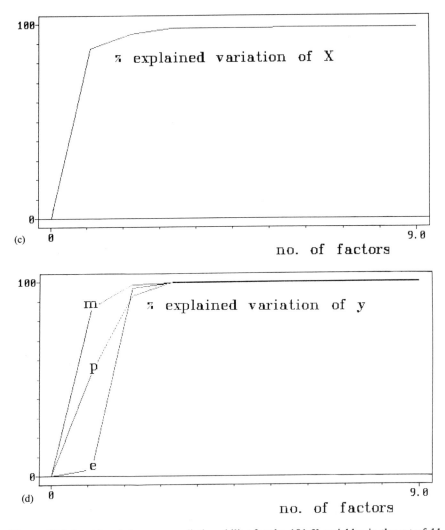

Figure 3.12 (*cont.*)   c) Average predictive ability for the 101 *X*-variables in the set of 11 test objects, given in percent of their total initial variation s_x^2 (see 2.28), as a function of the number of PLSR factors used, $a = 0,2,\ldots,9$. d) Prediction ability of the three individual *Y*-variables in percent of the total variation in the set of 11 test objects, as a function of the number of PLSR factors used

Thus, objects in the center of the score plot have *X*- and *Y*-data similar to the averages ($\bar{x}$ and $\bar{y}$). Objects to the right of the center have higher-than-average levels of y_3 and e.g. *X*-channel nos. 19, 20 and 24, and lower-than-average levels of y_1 and e.g. *X*-channels nos. 72, 66 and 63, and vice versa for objects to the left of the center. Likewise, objects above the center have higher-than-average levels of y_2 and e.g. *X*-channels no. 72 and 19, and lower-than-average levels of y_1, y_3 and e.g. *X*-channel 64 and 22, etc.

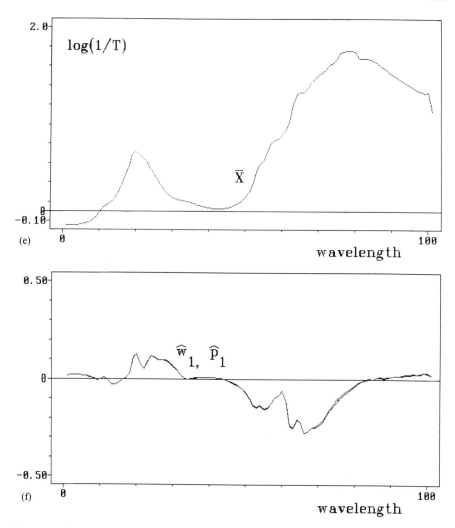

Figure 3.12 (*cont.*)   e) Model center $\bar{\mathbf{x}}$ ($\bar{\mathbf{y}}'$ = (0.34, 0.31, 0.34) for methanol, ethanol and propanol, respectively). f) $\widehat{\mathbf{w}}_1$ and $\widehat{\mathbf{p}}_1$; $\widehat{\mathbf{q}}_1$ = (−.57, 0.11, 0.45)′, $\widehat{\mathbf{t}}_1{}'\widehat{\mathbf{t}}_1$ = 3.4

Objects to the right and below the origin in Figure 3.12k thus would be expected to have particularly high levels of y_3 and x_{21} to x_{24} and particularly low levels of the X-variables in the opposite direction, x_{65} to x_{72}. Of course, checking back to the pure alcohols' spectra in Figure 3.12a confirms that propanol has particularly distinct absorbance in these wavelength ranges.

Thus the geometric interpretation of the various two-way 'maps' of variables and objects can give a lot of insight. However, care should be taken to avoid erroneous ways of comparing these 'maps'. Generally speaking, the directions in the three maps, relative to the origin, are comparable (if the plots have been suitably scaled). The absolute positions of the X- and Y-variables (loadings $\widehat{\mathbf{P}}$ and $\widehat{\mathbf{Q}}$) are

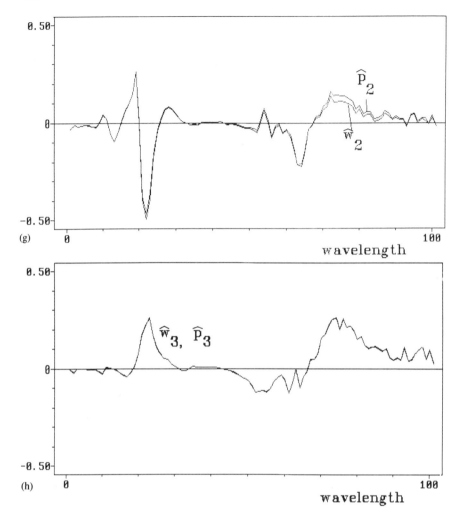

Figure 3.12 (*cont.*)   g) $\widehat{w}_2$ and $\widehat{p}_2$; $\widehat{q}_2 = (-.72, 2.00, -1.28)'$, $\widehat{t}_2{}'\widehat{t}_2 = 0.3$. h) $\widehat{w}_3$ and $\widehat{p}_3$; $\widehat{q}_3 = (-.26, -.57, 0.83)'$, $\widehat{t}_3{}'\widehat{t}_3 = 0.1$

comparable, but are sensitive to the a priori scaling weights of the variables (which here were all equal to 1). Note that the absolute positions of the variables in the loading plots are not comparable to the absolute positions of the objects in the score plots!

However, the more or less horizontal line for mixtures of methanol and propanol in Figure 3.12k is curved. The nonlinearity is even more evident in Figure 3.12l, where factor $\widehat{t}_3$ is seen plotted against factors $\widehat{t}_2$ and $\widehat{t}_1$ in a certain perspective. It thus seems that the third factor primarily spans a nonlinearity in the NIR relationship. This is in keeping with the fact the loading spectrum of factor 3 resembles the general shape of the analytes' spectra and the average spectrum $\overline{\mathbf{x}}$; it

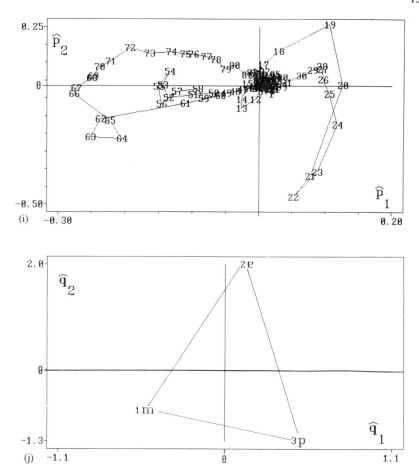

Figure 3.12 (*cont.*) i) Loadings of X-variables: $\hat{\mathbf{p}}_2$ vs $\hat{\mathbf{p}}_1$. j) Loadings of Y-variables: $\hat{\mathbf{q}}_2$ vs $\hat{\mathbf{q}}_1$

appears that the nonlinearity is strongest at the wavelength regions of highest O.D.

This slight non-additivity between the two most different solvents was expected, from earlier experience with other organic solvents using the same optical fiber probe. But like in the case of single-y PLS regression, the 'self-modelling' ability of PLS2 regression found certain combinations of wavelengths to model and automatically correct for this minor non-linearity. It is therefore examplifies a phenomenon to be taken care of, but without necessarily being a serious problem. This effect was seen very clearly in autoflourescence spectroscopy (Pedersen and Martens, 1989), and will also be demonstrated in more detail in Chapter 7.

Since the three individual analytes' original spectra overlap strongly (Figure 3.12a), not all parts of their spectra carry useful information for quantifying them in mixtures. The resulting predictor coefficients $\hat{\mathbf{B}}$, based on 3 PLSR factors, are given in Figure 3.12m, n and o for the three alcohols.

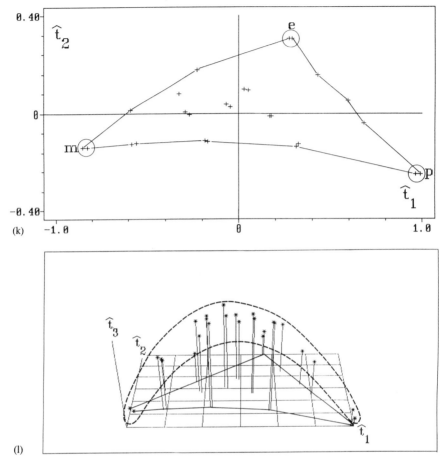

Figure 3.12 (*cont.*) k) Scores of the objects: $\hat{t}_2$ vs $\hat{t}_1$. l) Scores of the objects in three dimensions: $\hat{t}_1$ vs $\hat{t}_2$ vs $\hat{t}_3$. The design is outlined by the triangle at the foot-points in the plane $\hat{t}_2$ vs $\hat{t}_1$, and the strong curvature in dimension t_3 is outlined by the dotted circumference

How well did the obtained PLS calibration model predict the analytes in the independent set of alcohol mixtures? Using only the 2 first PLS factors in the predictor, the average (RMSEP, Chapter 4) prediction error in the test set was 3.4, 5.5 and 8.1 (expressed as weight fraction ×100) for methanol, ethanol and propanol, respectively. This decreased to 1.7, 2.0 and 1.2 when the full 3-factor predictor was used.

Thus the nonlinearity factor appared to be particularly important for the prediction of the propanol. Figure 3.12p shows this in more detail: The 2-factor model had good predictive ability, but gave too high predictions at the highest propanol levels, and also at the lowest propanol levels, particularly for those objects containing methonol. Increasing from 2 to 3 factors in the joint PLS2 model removed this nonlinear effect.

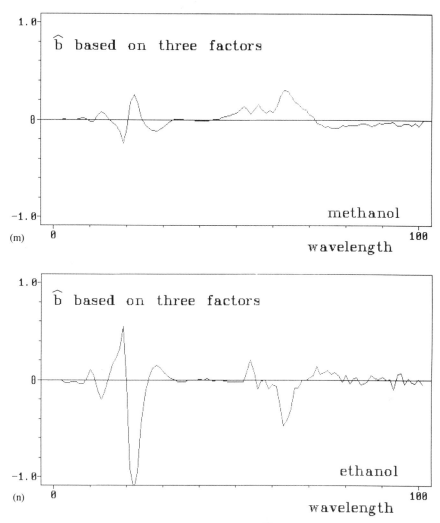

Figure 3.12 (*cont.*) m) Regression coefficients $\hat{b}$ of methanol (3 PLS factors). n) Regression coefficients $\hat{b}$ of ethanol (3 PLS factors)

This NIR example has illustrated the simultaneous calibration for three constituents. The PLS2 algorithm gave a complete, yet manageable overview of all the main relationships in the data.

For completion, we can now instead calibrate for each of the three constituents separately by the PLS1 algorithm, each time treating the other two constituents and the nonlinearity effect as unknown interferences. For each of the analytes this led to calibration models with clearly decreased prediction errors after 2 factors, but only slightly decreased errors after 3 factors, which gave the optimal calibration solutions.

156

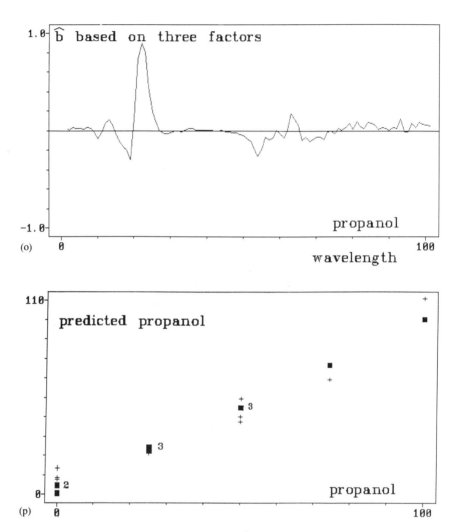

Figure 3.12 (*cont.*)   o) Regression coefficients $\widehat{b}$ of propanol (3 PLS factors). p) Predicted vs. known weight fraction of propanol, plotted for the 26 methanol/ethanol/propanol mixtures (15 calibration and 11 test objects), using different number of PLSR factors in the prediction. Crosses: 2 PLSR factors. Squares: 3 PLSR factors. (The numbers 2 and 3 indicate the number of overlapping observations)

Frame 3.6

Simultaneous PLSR calibration for several Y-variables ('PLS2 regression')

If we replace vectors **y**, **f**, and **q** in Frame 3.4 by matrices $Y(\dim I * J)$, $F(\dim I * J)$ and $Q(\dim J * A)$, the calibration in PLS2 is almost the same as for the orthogonalized PLS1. The exceptions are that y_{a-1} in C2.1 is replaced by a temporary Y-score for this factor, $\hat{u}_a$ and that two extra steps are needed between C 2.4 and C 2.5:

C 2.1 Use the temporary Y-factor $\hat{u}_a$ that summarizes the remaining variability in **Y**, to find the loading-weights $\hat{w}_a$ by LS, using the local 'model'

$$X_{a-1} = \hat{u}_a w_a{}' + E$$

and scale the vector to length 1. The LS solution is

$$\hat{w}_a = c\, X'_{a-1}\hat{u}_a$$

where c is the scaling factor that makes the length of the final $\hat{w}_a$ equal to 1, i.e.

$$c = (\hat{u}_a{}'X_{a-1}X'_{a-1}\hat{u}_a)^{-.5}$$

The first time this step is encountered, $\hat{u}_a$ has been given some start values, e.g. the column in Y_{a-1} with the largest sum of squares.

The following two extra stages are then needed between C 2.4 and C 2.5:

C 2.4b Test whether convergence has occurred, by e.g. checking that the elements $\hat{t}_{ia}$ have no longer changed meaningfully since the last iteration.

C 2.4c If convergence is not reached, then estimate temporary factor scores u_a using the 'model'

$$Y_{a-1} = u_a \hat{q}_a{}' + F$$

giving the LS solution

$$\hat{u}_a = Y_{a-1}\hat{q}_a(\hat{q}_a{}'\hat{q}_a)^{-1}$$

and go to C 2.1.

(*continued on next page*)

Frame 3.6 *continued*

If convergence has been reached, then go to step 2.5 in Frame 3.4. The expression for $\widehat{\mathbf{B}}$ is the same in this PLS2 algorithm as in the PLS1 algorithm, i.e.

$$\widehat{\mathbf{B}} = \widehat{\mathbf{W}}(\widehat{\mathbf{P}}'\widehat{\mathbf{W}})^{-1}\widehat{\mathbf{Q}}'$$

and

$$\mathbf{b}_0' = \overline{\mathbf{y}}' - \overline{\mathbf{x}}'\widehat{\mathbf{B}}.$$

——————————————— Statistical extensions ———————————————

3.5.4.4 Other modifications of the PLSR

The partial least squares regression algorithm can be modified to accomodate special needs or additional information available. Provided certain rules concerning orthogonality are followed, a number of different modifications can successfully be made. The rules are as follows:

1. The scores $\widehat{\mathbf{t}}_a, a = 1, 2, \ldots$ must be orthogonal to one another if simple treatment of individual factors $\widehat{\mathbf{t}}_a$ is to be attained. This can be attained by projecting $\mathbf{X}_{a-1}$ on $\widehat{\mathbf{t}}_a$, resulting in the loading $\widehat{\mathbf{p}}_a$ and new residuals $\mathbf{X}_a$, as in Frame 3.4. Otherwise, for factors $a = 1, 2, \ldots$, the present and all the previously estimated factors $\widehat{\mathbf{T}} = \{\widehat{\mathbf{t}}_b, b = 1, 2, \ldots, a\}$ should be used as simultaneous regressors for $\mathbf{Y}$, like in Frame 3.5.
2. For simplicity, the loading weights $\widehat{\mathbf{w}}_a$ should be orthogonal to one another. In the normal algorithm this comes automatically, but if they are modified in some way, they will be non-orthogonal. The orthogonalization can be done by projecting the preliminary modified estimate of the loading weights ($\mathbf{X}'_{a-1}\mathbf{y}_{a-1}$ in step C 2.1 in Frames 3.4 and 3.5 or $\mathbf{X}'_{a-1}\widehat{\mathbf{u}}_a$ in step C 2.1 in Frame 3.6) onto the $a - 1$ previous factors' loading weights (projection on $\widehat{\mathbf{W}}$, where $\widehat{\mathbf{W}} = \{\widehat{\mathbf{w}}_b, b = 1, 2, \ldots, a - 1\}$). The residuals after this projection are then scaled to length 1 and taken as $\widehat{\mathbf{w}}_a$.

Here are some of the extensions available (Martens 1987, Martens, 1989, QUIET, 1991):

Missing values

The problem of missing values is less common in calibration than in explorative data analysis—but also more critical. Basically there is no way to compensate for lack of information. But if there is sufficient redundancy between variables and between objects in $\mathbf{X}$, a few randomly or evenly distributed redundant elements x_{ik} with data missing do not affect the modelling in a significant way—if properly handled in the estimation algorithms.

Similarly, missing elements y_{ik} can be acceptable if **Y** has sufficient redundancy for the elements involved.

But if more or less unique variables or objects contain missing elements, this cannot be compensated for. Likewise, if there are many missing values or the missing elements come in systematic patterns, there is reason to be particularly careful, both with respect to model interpretation and validation results.

A simple way to accomodate such missing elements arises when we express every LS estimation in the algorithms as a proper univariate regression. As an approximation, the missing values can then simply be ignored. For instance, $\hat{t}_a$ in step C 2.2 in Frame 3.4 can be written as the result of a regression of $\mathbf{X}_{a-1}$ on $\hat{\mathbf{w}}_a$, yielding $\hat{t}_a = \mathbf{X}_{a-1}\hat{\mathbf{w}}_a / \hat{\mathbf{w}}_a'\hat{\mathbf{w}}_a$.

When a missing element x_{ik} is encountered, it is simply skipped in the numerator $\mathbf{X}_{a-1}\hat{\mathbf{w}}_a$, and the denominator $(\hat{\mathbf{w}}_a'\hat{\mathbf{w}}_a)$ is decreased by the subtraction of the corresponding element $\hat{\mathbf{w}}_{ka}^2$. The same technique is used in the estimation of p_{ka} and q_{ja}, and in PLS2 also for u_{ia}. For w_{ka} the denominator correction it is unnecessary, since $\hat{\mathbf{w}}_a$ is scaled to length 1 afterwards in any way.

Strictly speaking, this simplified method violates the orthogonality requirements, particularly if the outlier elements are not sufficiently redundant. It may then be necessary to replace the simple univariate regressions in the PLS algorithms, be it over objects or over variables, by multivariate regressions on the present and all previous factors, in analogy to step C 2.3 in Frame 3.5.

Modification of the variables' loading weights $\hat{w}_{ka}$

Modifying the parameters of the X-variables in the PLSR allows us to optimize the modelling statistically, and it allows us to select informative subsets of variables.

This can be done in two ways: Iteratively adjusting the a priori scaling weights and repeating the standard PLSR, or modifying the preliminary PLS loading weights (see, for instance, section 2.1.2.3) in an augmented PLSR algorithm.

The a priori scaling weights that adjust the relative impact of the different X-variables (or the different Y-variables, if so desired), are normally based on prior knowledge about the general noise level of the different variables (section 7.1). However, they can be modified one or more times, based on the validation results from preliminary A-dimensional PLSR models. An X-variable x_k with high residual variance (large elements $\hat{e}_{ik}, i = 1, 2, \ldots, I$) and/or low unique Y-relevance (small elements $|\hat{b}_{kj}|, j = 1, 2, \ldots, J$) can for instance be given lower weights or eliminated altogether.

Other criteria, involving price and user-friendliness etc. for the different

X-variables can also be incorporated into such weight expressions, e.g. in order to select which X-variables to reject and which to use in later analyses.

Such an iteration between a priori weighting—PLSR modelling—model validation—re-weighting—new PLSR modelling—new validation etc. can be automated to form a flexible self-modelling scheme, but must be used with care.

Modification of the impact of various variables can also be attained inside the PLS algorithms. For instance, the preliminary estimates of $\mathbf{w}_a$ in step C 2.1, $\widehat{\mathbf{w}}_a = \mathbf{X}'_{a-1}\mathbf{y}_{a-1}(\mathbf{y}'_{a-1}\mathbf{y}_{a-1})^{-1}$ (Frames 3.4 and 3.5) or $\widehat{\mathbf{w}}_a = \mathbf{X}'_{a-1}\mathbf{u}_a(\mathbf{u}_a'\mathbf{u}_a)^{-1}$ (Frame 3.6) can be modified in various ways before being re-orthogonalized and scaled to length 1 (Martens, 1985, Frank, 1987). The expressions for estimating the loadings $\widehat{\mathbf{p}}_a$ may not have to be modified.

One such modification is truncation: The preliminary elements in $\widehat{\mathbf{w}}_a$ can be checked to see if they are meaningfully different from zero. Preliminary elements $\widehat{w}_{ka}$ that are judged to be small and uncertain can simply be replaced e.g. by zeroes (Martens, 1985 and Frank, 1987).

Assume that from prior knowledge or from preliminary PLSR modelling we have estimated the uncertainty standard deviation of each variable x_k, $k = 1, 2, \ldots, K$; this is termed $s(\widehat{e}_k)$. The truncation judgement can consist of a comparison of $\widehat{w}_{ka}$ to an approximate estimate of its uncertainty, e.g. $s(\widehat{w}_{ka\ \text{preliminary}}) \approx s(\widehat{e}_k)(\mathbf{y}'_{a-1}\mathbf{y}_{a-1})^{-0.5}$.

This rough analogue to a t-test is based on the simplifying assumptions that $\mathbf{y}_{a-1}$ is fixed and orthogonal for different a. Additional uncertainty in $\widehat{w}_{ak}$ may come from noise in $\mathbf{y}$ itself.

These considerations also indicate how the uncertainty of the preliminary $\widehat{\mathbf{w}}_a$ in PLSR increases for successive factors as $\mathbf{y}_{a-1}$ decreases with $a = 1, 2, \ldots$. Of course this effect is further amplified when the final $\widehat{\mathbf{w}}_a$ has been scaled to length 1, since the scale factor $c = (\mathbf{y}'_{a-1}\mathbf{X}_{a-1}\mathbf{X}'_{a-1}\mathbf{y}_{a-1})^{-0.5}$ will increase as the covariance $\mathbf{y}'_{a-1}\mathbf{X}_{a-1}$ decreases. A similar effect is seen for $\widehat{\mathbf{p}}_a$. This is why for scanning spectrophotometry the later PLSR factors' loading spectra invariably show increasing noise levels (cf. Figure 3.11f–r).

Another PLSR modification is smoothing. If the X-variables are expected to represent only smoothly varying phenomena, the preliminary loading weights $\widehat{\mathbf{w}}_a$ can be smoothed (section 7.5). This can to some degree reduce the senstivity of $\widehat{\mathbf{w}}_a$ to the random noise described above. It can therefore be useful e.g. in scanning spectrometers with narrow wavelength span between neighbouring variables, or in statistical counting histograms, e.g. image summaries at pixel gray tone levels $k = 1, 2, \ldots, K$ (Martens et al., 1983a).

A further modification is applicable if the K element in the vectors $\mathbf{x}_i'$ actually can be expressed as a two-way matrix $\mathbf{X}_i = (x_{m,n})_i$, e.g. coming from a second-order instrument like a chromatograph with some multichannel detector (channels $m = 1, 2, \ldots, M$ for for each retention time $n = 1, 2, \ldots, N$). Each preliminary loading vector can be expressed as $\widehat{\mathbf{W}}_a = (\widehat{w}_{m,n})_a$ and can then be modelled further in terms of a bilinear structure plus residual, and thereby be further refined. Other higher-order modelling methods also exist for such data (Wold et al., 1987; Esbensen et al., 1988); see also section 3.7.3.

A third modification is to change the preliminary loading weight estimate on the basis of external data. Assume for instance that we have additional data vectors $\mathbf{Z} = (z_{km}, k = 1, 2, \ldots, K; \; m = 1, 2, \ldots, M)$ that contain information about the X-variables (e.g. spectra of suspected analytes or interferents). One possibility is then to fit the preliminary $\widehat{\mathbf{w}}_a$ to $\mathbf{Z}$ ($\widehat{\mathbf{w}}_a = \mathbf{Zb} + \mathbf{e}$, using e.g. a separate PLS-modelling, S. Wold, pers.com. 1988). This is an alternative to using this extra Z-information in a generalized a priori weighting (section 7.1). Depending on whether $\mathbf{Z}$ represents desirable or undesirable information, one could then use the projection $\mathbf{Z}\widehat{\mathbf{b}}$ or the residual $(\widehat{\mathbf{w}}_a - \mathbf{Z}\widehat{\mathbf{b}})$ as new preliminary $\widehat{\mathbf{w}}_a$. In the former case, if $\mathbf{Z}$ consisted of sine and cosine functions, a solution similar to Fourier regression (section 3.7) would be obtained.

Another possibility is to compare the preliminary $\widehat{\mathbf{w}}_a$ to a data base and replace it by the candidate that seems most likely. This is similar to Hruschka regression (section 3.7). (The data base must be projected on the previous factor loading weights prior to the comparison and replacement, otherwise the estimation of $\mathbf{t}_a$ must be done by regression on all the selected factors spectra $\widehat{\mathbf{w}}_b$, $b = 1, 2, \ldots, a$ simultaneously; Kvalheim, 1988.)

The Y-variables can probably best be modified by iteratively changing their a priori scaling weights; modifying them inside the PLSR algorithm may give confusing results.

Modification of the objects' scores $\widehat{t}_{ia}$

The same type of PLSR modifications can be applied to the objects:

First of all the a priori weighting of the objects (section 7.1) can be changed on the basis of how well the objects fitted a preliminary PLSR modelling in $\mathbf{X}$ and $\mathbf{Y}$. The simplest of this is of course the elimination of outliers prior to re-modelling (Chapter 5), and the inclusion of two or more copies of particularly important objects in the calibration set.

PLSR algorithms themselves can be modified to accomodate various types of external information. For instance, when objects $i = 1, 2, \ldots$ represents a smooth time series, it is possible to smooth and reorthogonalize the preliminary regressor $\mathbf{y}_{a-1}$ (or $\widehat{\mathbf{u}}_a$, Frame 3.6) used for estimating $\mathbf{w}_a$ (Esbensen and Wold, 1983).

It is also possible to modify $\widehat{\mathbf{t}}_a$ in similar ways within the PLS algorithm, e.g. by including dynamic PLS forecasting (Martens, 1989 and Ricker, 1988).

Likewise, these preliminary regressors from $\mathbf{X}$ and/or $\mathbf{Y}$ can be projected to a third set of external variables, over some or all of the objects. Either the projection or the residual from that projection can be used as new regressor in subsequent PLSR algorithm steps (e.g. for estimation of $\mathbf{w}_a$).

Such modifications of the PLSR are related to the various multi-block PLS versions used in e.g. social sciences (H. Wold, 1981).

The PLSR algorithms can also be modified to accomodate nonlinear X–Y relationships. The simplest modification is to add a regression of $\mathbf{X}_{a-1}$

and $\mathbf{Y}_{a-1}$ on $\hat{\mathbf{t}}_a{}^2$ etc, i.e. a polynomial regression of $\mathbf{X}_{a-1}$, $\mathbf{Y}_{a-1}$ on $\hat{\mathbf{t}}_a$. Certain types of nonlinearities would require a nonlinear modelling even in the estimation of $\mathbf{w}_a$, (Wold, S., pers.com.) but this will not be covered here.

Robust regression (Huber, 1981) is also possible within the PLSR framework. For instance objects with extreme preliminary values of $\hat{t}_{ia}$ or $\hat{u}_{ia}$ for some factor can be weighted down in the regressions on $\hat{\mathbf{t}}_a$ and $\hat{\mathbf{u}}_a$. The scores will then be non-orthogonal.

If so desired, it is possible to skip non-informative objects in the algorithm, for instance objects with particularly high relative uncertainty in $\hat{t}_{ia}$, as approximated by $\hat{t}_{ia}/s(\hat{e})$ (where $s(\hat{e})$ is the general noise standard deviation of $\mathbf{X}$; Chapter 5). This can alternatively be based on the signal to noise ratio of $\hat{u}_{ia}$, as approximated by $\hat{u}_{ia}(\hat{\mathbf{q}}'_a\hat{\mathbf{q}}_a)^{0.5}/\hat{\sigma}f$ (where $\hat{\sigma}_f$ is the general noise level in $\mathbf{Y}$, Chapter 4).

In general it may be noted that the PLSR algorithms can be reformulated using weighted least squares instead of ordinary least squares at each of the regression stages. Even generalized least squares can be used, if there are certain covariance structures that ought to be weighted down or up in the model estimation. When combined with non-orthogonalized formulations (employing for each factor also the previous $a - 1$ factors in the regressions), the PLSR can thus be expressed as a very flexible modelling tool.

However, conceptual simplicity is important in order that users can feel confident with a modelling tool like PLSR. The simplest PLSR algorithm versions (Frames 3.4, 3.5 and 3.6) are therefore recommended whenever possible. More advanced a priori knowledge about interference structures etc. may then be applied in the pre-processing stage, e.g. generalized a priori weighting (section 7.1), response linearizations (section 7.3) or multiplicative signal correction (section 7.4).

It may also be noted that even in PCA some types of external information about variables and objects can be incorporated, based on the NIPALS algorithm.

Updating PLSR calibration models

Related to this is the topic of updating former bilinear calibration models as new (X,Y)-data arrive. Of course this can be done pragmatically by adding the new objects to the old calibration set, and recalibrating on the full extended calibration set. If the population in question (e.g. a chemical process) changes over time, or the instrument changes over time, the oldest measurements can be given lower a priori weights or ignored. Alternatively, the most recent measurements can be given particularly high weights (e.g. simply by including two or more copies of each of the lastest objects in the calibration set).

However, if it is important to save time and storage space without losing too much information, it would be desirable to represent the main information in a calibration set of many objects $i = 1, 2, \ldots, I$ by a few artificial 'objects'. This can be done in several ways. The simplest is to use a PCR or PLSR calibration model to construct a small set of (x,y) vectors that when used again as a calibration set gives the same PCR or PLSR calibration model as the original calibration set of I real objects.

The space spanned by each bilinear factor $a = 1, 2, \ldots, A$ can be described by the two objects (ignoring possible a priori weights):

$$(\bar{\mathbf{x}}' + \sqrt{(\hat{\mathbf{t}}_a{}'\hat{\mathbf{t}}_a/I)}\hat{\mathbf{p}}_a{}', \bar{\mathbf{y}}' + \sqrt{(\hat{\mathbf{t}}_a{}'\hat{\mathbf{t}}_a/I)}\hat{\mathbf{q}}_a{}')'$$

$$(\bar{\mathbf{x}}' - \sqrt{(\hat{\mathbf{t}}_a{}'\hat{\mathbf{t}}_a/I)}\hat{\mathbf{p}}_a{}', \bar{\mathbf{y}}' - \sqrt{(\hat{\mathbf{t}}_a{}'\hat{\mathbf{t}}_a/I)}\hat{\mathbf{q}}_a{}')'$$

where I is the number of calibration objects. This set of $2A$ artificial objects for $a = 1, 2, \ldots, A$ has the same mean $(\bar{\mathbf{x}}, \bar{\mathbf{y}})$ as the original data, and spans the same relative variability types as the original data.

This small dataset of $2A$ 'objects' can later be joined with $(\mathbf{X},\mathbf{Y})$ data for new objects, and an updated bilinear calibration model can be re-estimated. It may then be useful to ascribe higher a priori weights to the $2A$ artifical objects than to the new objects (e.g. weights $\sqrt{(I/2)}$ and 1, respectively), in order to ensure a reasonable balance between old and new information.

This is also a way to map together information from several multivariate bilinear models, e.g. multidimensional scaling (MDS, M. Martens et al., 1988; Popper et al., 1988).

It is possible to represent the calibration space by even fewer objects: The $1 + A$ vectors $(\bar{\mathbf{x}}', \bar{\mathbf{y}},)$ and $\sqrt{(\hat{\mathbf{t}}_a{}'\hat{\mathbf{t}}_a/I)}(\hat{\mathbf{p}}'_a, \hat{\mathbf{q}}_a{}')', a = 1, 2, \ldots, A$ span the calibration information relative to the origin. To merge this with new $(\mathbf{X},\mathbf{Y})$ information requires a slight modification of the PLS algorithm in order to separate the estimation of model center and factors (H. Martens, 1987, QUIET, 1991). That will not be pursued here.

3.5.5 PROPERTIES OF PLSR

PLSR has been tested in numerous applications and from these results one can certainly conclude that the method has given useful results both from a prediction and interpretation point of view (see e.g. Wold et al., 1983; Lindberg et al., 1983; Otto and Wegsheider, 1985; Martens and Næs, 1987; Esbensen and Martens, 1987; Næs et al., 1986; Martens and Martens, 1986a; Haaland, 1988; Haaland and Thomas, 1988a,b; and references in these.

On the theoretical side, however, the statistical and mathematical properties of PLSR as a regression method are not yet fully known. In this section we will shortly summarize some of the available results and comment on their importance. Further details can be found in the references given.

In addition to the basic properties $\hat{\mathbf{T}}'\hat{\mathbf{T}} =$ diagonal matrix and $\hat{\mathbf{W}}'\hat{\mathbf{W}} = \mathbf{I}$, there are a number of additional relations that hold for the matrices in the PLS algorithm. These can be found in many of the references to PLSR given above; of special importance are Helland (1988), Geladi and Kowalski (1986), M. Martens, (1987), Høskuldsson (1988) and Manne (1987).

In Næs and Martens (1985) a definition of relevant and irrelevant factors (for one Y-variable) in the X-space is given related to a population formulation. Roughly, the relevant factors are the population eigenvectors with nonzero correlation with y and the irrelevant factors are the eigenvectors with zero correlation with y. It was shown that in the case of A relevant population factors, PLS with A factors is a consistent estimator of the best linear predictor (see e.g. Searle, 1974; and section 3.6.3). In other words, the A first PLS factors estimate the same space as that spanned by the A relevant factors. Therefore, in this sense the PLS method is as parsimonious as possible. The same concepts were considered more elegantly and in deeper detail by Helland (1988) who also gave alternative characterizations for the relevance concept and the PLSR estimator. He also pointed out directions for further research and modifications using the relevance idea and restricted ML estimation.

In Helland (1988) was also given an alternative basis for the PLSR space (one y-variable), i.e the space spanned by the loading weights $\hat{\mathbf{w}}_1, \ldots \hat{\mathbf{w}}_A$. From the nonorthogonalized formulation of PLSR as described in Frame 3.5 it is known that PLS can be found as the regression of y on the projection of X down on this space. This means that it is of fundamental importance to know what this space looks like. In fact he showed that the PLSR space is spanned by a so-called Krylow sequence (see e.g. Manne, 1987). This basis was for interpretation purposes reformulated in the following way by Næs (1987b):

$$\hat{\mathbf{z}}_1 = \sum_{k=1}^{K} \hat{\mathbf{p}}_k (\hat{\theta}_k \sqrt{\hat{\tau}_k})$$

.

.

$$\hat{\mathbf{z}}_A = \sum_{k=1}^{K} \hat{\mathbf{p}}_k (\hat{\theta}_k \hat{\tau}_k^{A-1} \sqrt{\hat{\tau}_k})$$

(3.61)

where $\hat{\mathbf{p}}_k$ is now the i'th eigenvector of $X'X$ (centered X), $\hat{\tau}_k$ is the i'th eigenvalue and $\hat{\theta}_k$ is the estimated regression coefficient obtained by regressing y on principal component $nr . i$ weighted to variance 1 (see section 3.4.6). Notice that these $\hat{\theta}$'s are comparable for the different i since they are related to variables with equal variance (Jfr. singular value decomposition in section 2.1).

This basis shows that both the size of the eigenvalue in X, $\hat{\tau}_k$, and their Y-relevance as measured by $\hat{\theta}_k$ are important in the PLS weighting of the eigenvectors $\hat{\mathbf{P}}$. This is contrary to both PCR with selection of eigenvectors from the top and the PCR with selection of eigenvectors according to predictive relevance. Furthermore we also see that the large $\hat{\tau}$'s have larger influence as A increases.

From this basis one can also easily argue that PLS is a member in a class of methods giving different emphasis on the $\hat{\tau}$ and $\hat{\theta}$. For instance $\hat{\theta}^2$ instead of $\hat{\theta}$ in the $\hat{\mathbf{z}}$ is another of infinitely many possibilities. A discussion of how to use prior information in the weighting can be found in Næs (1987).

It is also easily seen from the definition of PLS in section 3.5.2, that PLSR is a natural element in a class of methods. From this definition it follows that each PLSR factor is found as a compromise between maximal correlation to $\mathbf{y}$ and maximal exlained variance of $\mathbf{x}$ (Frank, 1987) and infinitely many such compromises exist. Ideas of considering PLS as a member of a class of methods can also be found in Lorber et al. (1987) and in Høskuldson (1988).

Based on earlier results by S.Wold et al.(1984), Manne considered in detail the relation between PLSR and the conjugate gradient method of generalized inverse. He showed that from an algorithmic point of view the method is simply an already established numerical algorithm for solution of linear equations (see Golub and Koshan, 1965 and Lanczos, 1950). Manne (1987), however, did not consider the statistical properties of stopping before $A = K$, but his results could be a sensible starting point for a further study of the properties of PLSR.

Manne (1987) also gives an excellent clarification of the relationship between the different PLS alternatives existing in the literature. (See also Helland, 1988.)

From a statistical point of view very little is known about the PLS2 method. It seems, however, that it is difficult to study theoretically due to iterations and the subtraction of factors involved.

It is important to notice the similarity between the PLS2 analysis as described here and the statistically well known canonical correlation analysis as described in Mardia et al. (1980). The main difference is that maximization of the X–Y covariance in the PLS2 modelling has replaced the maximization of correlation in the latter. Thereby the Y-relevant covariance structure in $\mathbf{X}$ helps stabilize the model estimation against noise and guards against overfitting.

Since collinearity is no problem in PLS modelling, there is no need for variable selection, neither in $\mathbf{X}$ nor in $\mathbf{Y}$: The number of X-variables $k = 1, 2, \ldots, K$ and/or Y-variables $j = 1, 2, \ldots, J$ in Mode A PLS methods like the PLS2 regression can well be higher than the number of objects $i = 1, 2, \ldots, I$. This is impossible in canonical correlation, which belongs to the Mode B PLS class (Jøreskog and Wold, 1981) and therefore requires full column rank in $\mathbf{X}$ and $\mathbf{Y}$.

In addition, there is a difference in symmetry between the two methods: In the PLSR methods the linear combinations of $\mathbf{X}$, $\hat{\mathbf{t}}_a$, are used for modelling both $\mathbf{X}$ and $\mathbf{Y}$; the Y-combinations, $\hat{\mathbf{u}}_a$, are only used as an intermediate estimation tool. This implies that we can extract more PLS factors than the number of Y-variables ($A > J$). In contrast, $\mathbf{X}$ is modelled in terms of $\hat{\mathbf{t}}$ and $\mathbf{Y}$ in terms of $\hat{\mathbf{u}}$ in the canonical correlation analysis.

As stated above, PLSR can be considered a forward regression method and therefore shares the general properties of such methods discussed in section 3.6.3. (See also Figure 4.2.)

——————— End of statistical extensions ———————

3.6   MULTIVARIATE CALIBRATION BASED ON THE LINEAR MIXTURE MODEL

SUMMARY   The multivariate mixture model Beer's law in spectroscopy is presented and discussed. Problems with use of the model are illustrated. The effect of closure among the constituents is treated in detail. Extensions of the mixture model are given together with a treatment of alternative calibration methods and a discussion of their statistical properties.

In the previous sections an approach to multivariate calibration in terms of bilinear modelling by PCR and PLS regressions was given. In the present section we shall instead discuss multivariate calibration in the framework of the more causal concept of an additive mixture model, often termed 'Beer's Law' in spectroscopy.

Two different but related applications of the mixture model will be covered:

* Direct calibration: Estimating unknown concentrations from known analyte and interferent responses

and

* Indirect calibration: First estimating unknown analyte and interferent responses from known concentrations, then determining unknown concentrations from the estimated responses.

Both cases will be illustrated by two examples: A simple, artificial example simulating chromatography, and a real spectroscopy example from process analysis (mixtures of the three solvents methanol, ethanol and propanol measured by fiberoptic NIR spectrocopy, also used in section 3.5.4.3).

Calibration methods developed in the linear mixture model can be used to calibrate for several constituents or analytes simultaneously. Therefore, we present our predictors.

$$\hat{y}_i' = x_i'\hat{B} \tag{3.62}$$

$$(\text{or} \quad \hat{y} = \hat{B}'x)$$

where $\hat{B}$ is the matrix of regression coefficients for the constituents considered.

Additive mixture model

The additive mixture model can be written:

$$x_{ik} = \sum_{j=1}^{J}(y_{ij}k_{kj}) + e_{ik} \tag{3.63}$$

or, in matrix form:

$$X = YK' + E \tag{3.64}$$

Here X represents the spectral data for mixtures $i = 1, 2, \ldots, I$ at channels $k = 1, 2, \ldots, K$:

$$X = \begin{bmatrix} x_{11} \cdot \cdot & & x_{1k} \\ & \cdot \cdot & & \cdot \\ & & \cdot & \\ & \cdot \cdot & & \cdot \\ x_{I1} \cdot \cdot & & x_{IK} \end{bmatrix}$$

K represents unit spectra of the constituents $j = 1, 2, \ldots, J$:

$$K = \begin{bmatrix} k_{11} \cdot & k_{1J} \\ \cdot & \cdot \\ \cdot & \cdot \\ k_{K1} \cdot & k_{KJ} \end{bmatrix} = (k_1 \ldots, k_J)$$

Y represents concentrations of the same J constituents:

$$Y = \begin{bmatrix} y_{11} & y_{1J} \\ \cdot & \cdot \\ \cdot & \cdot \\ y_{I1} & y_{IJ} \end{bmatrix}$$

and the matrix E represents residuals between the mixture model and the data, due to random errors in the data and inadequacies in the linear mixture model (unmodelled interferents, nonlinearities etc.).

In some cases it is more realistic to assume that the model

$$X = 1k_0' + YK' + E \tag{3.65}$$

is true where k_0 represents a background or baseline offset. For mathematical simplicity, (3.65) is often expressed in the form of equation (3.64), with $Y = (1, \text{old } Y)$ and $K = (k_0, \text{old } K)$. But in the present section we shall include the offset explicitly as in equation (3.65) whenever it is used.

In some cases the spectral values and also the concentrations are presented as deviations from their mean (or from some other value). In such cases the model (3.65) can be written in the form of (3.64), i.e. the common 'baseline offset' of the X-data vanishes.

The mixture model or its offset extension can be used for a variety of purposes as will be demonstrated in the subsequent sections. Different approaches to calibration under the assumption of linear mixture model are covered in section 3.7.

3.6.1 SELECTIVITY ENHANCEMENT BY DIRECT CALIBRATION (DIRECT UNMIXING)

In this section we assume that the unit spectra $\mathbf{K}$ of all the constituents are known from prior experiments, and that no other types of interference exist in the X-data. The ordinary least squares 'unmixing' solution to (3.64) is as described in section 2.1.2.3:

$$\hat{\mathbf{Y}} = \mathbf{X}\mathbf{K}(\mathbf{K'K})^{-1} \tag{3.66}$$

implicitly assuming that the columns of $\mathbf{K}$ are linearly independent or equivalently that the $\mathbf{K}$ matrix has full rank if $J < \mathbf{K}$. If $\mathbf{K'K}$ cannot be inverted in a meaningful way ($\mathbf{K}$ is said to be collinear, see section 2.3), it means that some of the constituents have too similar unit spectra to be distinguishable in more detail: one or more constituents' unit spectra are too similar to some linear combination of some of the other unit spectra in $\mathbf{K}$. This is a fundamental selectivity problem (Bergmann et al., 1987). The problem can sometimes be overcome by treating several similar constituents as one single group 'constituent'.

Like all linear calibration methods this direct unmixing can be expressed as:

$$\hat{\mathbf{Y}} = \mathbf{X}\hat{\mathbf{B}} \tag{3.67}$$

where $\hat{\mathbf{B}} = (\hat{\mathbf{b}}_{kj})$ is the matrix of estimated calibration coefficients for the analytes $j = 1, 2, \ldots, J$ at X-variables $k = 1, 2, \ldots, K$.

If there is an offset spectrum in the model, then equation (3.67) is modified to:

$$\hat{\mathbf{Y}} = (\mathbf{X} - \mathbf{1k_0}')\mathbf{K}(\mathbf{K'K})^{-1} \tag{3.68}$$

or

$$\hat{\mathbf{Y}} = (\mathbf{X} - \mathbf{1k_0}')\hat{\mathbf{B}} \tag{3.69}$$

In both cases the calibration coefficients are defined by

$$\hat{\mathbf{B}} = \mathbf{K}(\mathbf{K'K})^{-1} \tag{3.70}$$

If prior information is available about the noise levels of the wavelengths (error variances $\mathbf{V} = \text{diag}(\sigma_k^2, k = 1, 2, \ldots, K)$ (not the same $\mathbf{V}$ as used in bilinear modelling)), then weighted least squares (WLS) can be used instead of (3.70):

$$\hat{\mathbf{B}} = \mathbf{V}^{-1}\mathbf{K}(\mathbf{K'V}^{-1}\mathbf{K})^{-1} \tag{3.71}$$

Further extensions of the WLS into the Generalized Least Squares predictor (GLS) are treated in section 3.6.3.

Assuming e_{ik} to be random, independently distributed with expectation 0 and standard deviation σ_i for an object i, the covariance matrix of the estimated concentrations $\hat{\mathbf{y}}_i = (\hat{y}_{ij}, j = 1, 2, \ldots, J)'$ in object i from (3.69)–(3.70) is

$$\text{Cov}(\hat{\mathbf{y}}_i) = (\mathbf{K'K})^{-1}\sigma_i^2 \tag{3.72}$$

For (3.71) the equivalent expression is $(\mathbf{K}'\mathbf{V}^{-1}\mathbf{K})^{-1}$.

For an application we refer to Lea et al. (1984) and to Bergmann et al. (1987). The latter gives more theoretical detail on the definition of sensitivity and selectivity under this rather restrictive direct unmixing model ('Beer's Law' in spectroscopy).

Direct unmixing, followed by partial spectral reconstruction, is used in the QUIET package (1991) for Spectral Interference Subtraction (SIS) preprocessing.

The following subsections will serve as further illustrations of the described estimation procedure, and also to point out difficulties and problems.

3.6.1.1 The mixture model illustrated for chromatography

Assume that we want to determine the concentration of a certain analyte by chromatography. We want a calibration that gives the analyte concentration y_i from 4 peaks in the chromatogram $\mathbf{x}_i = (x_{ik}, \ k = 1, 2, 3, 4)'$ measured for unknown mixture no. i.

Figure 3.13a gives the evaluation pattern of the analyte (artificial data). Simulating a good chromatographic detector, the peak hight is considered to respond linearly with the analyte concentration and there is no baseline offset. The problem is varying levels of contamination of an interferent that eluate at exactly the

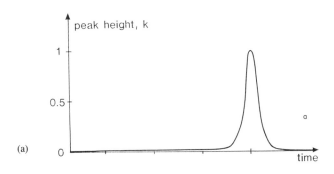

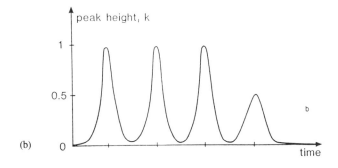

Figure 3.13 The known instrument responses used for simulating the chromatography data. a) Analyte, $\mathbf{k}_1 = (0,0,0,1)'$ b) Interferent, $k_2 = (1,1,1,.5)'$

same retention time. Therefore the peak at retention time $k = 4$ is not proportional to the analyte concentration alone.

However, in our chromatograms of this type of objects we have observed peaks at four different retention times (X-variable 1,2,3,4), the last of which corresponds to the analyte's and the interferent's retention time. In this simplified case let us assume that we know that the interferent at retention time 4 is directly proportional to each of three other interferents that are responsible for the three first peaks. The reason for this could be that the four interferents come from the same common contamination source. The chromatogram for the interferent alone is given in Figure 3.13b. These individual instrument responses of the analyte ($j = 1$) and interferent ($j = 2$) at the four instrument channels ($k = 1, 2, 3, 4$) are termed matrix **K** (not to be confused with X-variable index k).

Fully specified model

Table 3.5 shows the chromatographic responses $\mathbf{k}_1'$ and $\mathbf{k}_2'$(row vectors) for the analyte and interferent, as well as the corresponding X-data for five 'unknown' mixtures of the analyte and the interferent.

Table 3.5 Analyte and interferent signals **K**, and the error-free mixture data **X**

		X-channel no.			
		k=1	k=2	k=3	k=4
K′					
Analyte $\mathbf{k}'_1$	$j = 1$	0	0	0	1
Interferent $\mathbf{k}'_2$	$j = 2$	1	1	1	.5
	$i = 1$	0	0	0	1
X	$i = 2$	.1	.1	.1	.95
Mixture	$i = 3$	.5	.5	.5	.75
no.	$i = 4$	.9	.9	.9	.55
	$i = 5$	1	1	1	.5

For the **K**-data in the two first rows of Table 3.5 this gives the calibration coefficients $\widehat{\mathbf{B}}$ (from (3.70)) for the analyte ($j = 1$) and the interferent ($j = 2$) shown in Table 3.6.

Now we can predict the analyte concentration y_{i1} from the measurements of each unknown mixture, $\mathbf{x}_i'$, by

$$\hat{y}_{i1} = \mathbf{x}_i'\widehat{\mathbf{b}}_1. \tag{3.73}$$

(Likewise we can predict the interferent concentration, if so desired.) The results are given in column 2, Table 3.7 together with the true analyte concentrations in column 1. For these artificial, error-free data, we of course get a perfect fit and perfect prediction results.

Table 3.6 Matrix $\widehat{\mathbf{B}}$ for the joint direct calibration of the analyte and the interferent

		Constituent no.	
		$j = 1$	$j = 2$
X-channel no.	$k = 1$	−.17	.33
	$k = 2$	−.17	.33
	$k = 3$	−.17	.33
	$k = 4$	1	0

Table 3.7 Results from direct and indirect unmixing. Analyte concentration $\widehat{y}_{i1} = \mathbf{x}_i{}'\widehat{\mathbf{b}}_1$ given for the five test mixtures together with the X-residuals' standard deviation (see Chapter 5) and their total average. Numbers in paranthesis represent calibration objects, for which 'too good' results can be expected (see Chapter 4 about overfitting)

	Model no.	1	2	3	4	5	6
	y_1	True		Estimated			
					Direct		Indirect
	Noise in $\mathbf{X}$	0	0	5%	0	0	5%
	Noise in $\mathbf{K}$	0	0	0	5%	0	est.
	Model	—	J=2	J=2	J=2	J=1	J=2
	Results for $\mathbf{y}_1$:						
	$i = 1$	1	1	.95	1.05	1	1.03
Mixture	$i = 2$	.9	.9	.86	.94	.95	(.93)
no.	$i = 3$	.5	.5	.52	.52	.75	(.50)
	$i = 4$	.1	.1	.11	.09	.55	(.08)
	$i = 5$	0	0	−.03	−.02	.5	−.03
	Residual s.dev. in $\mathbf{X}$, $s(\widehat{\mathbf{e}}_i)$:						
	$i = 1$	0	0	0	0	0	0.027
Mixture	$i = 2$	0	0	.003	.004	.10	(0.020)
no.	$i = 3$	0	0	.029	.021	.50	(0.011)
	$i = 4$	0	0	.055	.039	.90	(0.039)
	$i = 5$	0	0	.049	.044	1.0	0.046
	Average	0	0	.035	.028	.64	0.0101

Effect of noise in the mixture data $\mathbf{X}$

Of course measured data are never perfect. Table 3.8 shows the same X-data after the addition of about 5% random proportional and normally distributed errors simulating the real-world measurement noise. Column 3 in Table 3.7 shows that the corresponding predictions of analyte concentrations are now less perfect but still reasonably precise. In addition, all the objects display estimated residuals of the same order of magnitude as the proportional noise in their X-data.

Table 3.8 Error-free analyte and interferent signals **K**, and mixture data **X** with 5% random proportional noise added

		X-channel no.				
		k=1	k=2	k=3	k=4	
K′						
Analyte	j = 1	0	0	0	1	No errors
Interferent	j = 2	1	1	1	.5	
	i = 1	0	0	.0	.948	With 5% errors
X	i = 2	.102	.097	.103	.911	
Mixture	i = 3	.557	.507	.508	.782	
no.	i = 4	.876	.981	.902	.570	
	i = 5	.998	1.093	1.070	.496	

Effect of errors in the model data **K**

What if there are errors in the analyte and interferent data **K**, on which the direct calibration was based? This type of model error are illustrated in the first two rows of Table 3.9.

This error of course leads to errors in the calibration coefficients $\widehat{\mathbf{B}}$ (Table 3.10) which in turn leads to systematic errors in the resulting analyte predictions, even for the mixtures (Table 3.9) for which error-free X-data are avaliable. The resulting predictions are given in column 4 in Table 3.7. In this case the noise effects have

Table 3.9 Analyte and interferent data with 5% error added to **K** from Table 3.5, and error-free mixture data **X**

		X-channel no.				
		k=1	k=2	k=3	k=4	
K′						
Analyte	j = 1	0	0	.0	.951	With 5% error
Interferent	j = 2	.959	1.045	.991	.515	
	i = 1	0	0	.0	1	
X	i = 2	.1	.1	.1	.95	Error free
Mixture	i = 3	.5	.5	.5	.75	
no.	i = 4	.9	.9	.9	.55	
	i = 5	1	1	1	.5	

Table 3.10 Matrix $\widehat{\mathbf{B}}$ for the joint direct calibration of the analyte and the interferent, based on the noisy analyte and interferent data in Table 3.9

		Constituent no.	
		j = 1	j = 2
	k = 1	−.173	.320
X-channel	k = 2	−.189	.349
no.	k = 3	−.179	.331
	k = 4	1.051	0

been permanently 'baked' into the calibration model, leading to systematically wrong predictions in spite of the perfect X-data for future mixtures.

Model mis-specification: Unmodelled interference

There is a type of model error that may be even more serious than random noise in the model data. That is model mis-specification, which can lead to systematic errors in the calibration model. Such a case arises when important constituents or other interference phenomena are mistakenly left out of the mixture model.

Assume for instance that we forgot, or did not know, that there was an interferent in the chromatographic data in the present illustration. Thus we would calibrate, based on the analyte's instrument response $\mathbf{k}_1$ alone (Figure 3.13a).

In the present simplified case this would of course lead to a calibration coefficient vector $\hat{\mathbf{b}}_1 = \mathbf{k}_1/(\mathbf{k}_1'\mathbf{k}_1) = (0,0,0,1)'$, which only has a contribution at peak no. 4. Table 3.7, column 5 shows that this leads to very bad prediction of the analyte concentration when there is interference present. For instance, mixture 5 appears to have an analyte concentration of 0.5, while in fact it has zero. But since in the present calibration we have assumed that there should be no chromatographic signal at the first three channels, the presence of the interferent is easily detected in outlier warnings using e.g. the residual standard deviations (see Chapter 5).

This so-called 'alias effect' leads to bias in the predictions and can be described as follows:

The mixture model (3.64) can be rewritten:

$$\mathbf{X} = \mathbf{Y}_1\mathbf{K}_1' + \mathbf{Y}_2\mathbf{K}_2' + \mathbf{E} \tag{3.74}$$

where $\mathbf{Y}_1$ represents the concentrations we are interested in or have data for, with instrument response $\mathbf{K}_1$ and $\mathbf{Y}_2$ represents interferents or other analytes with instrument response $\mathbf{K}_2$ overlapping that of $\mathbf{K}_1$. The interference responses $\mathbf{K}_2$ can always be decomposed into one part that is confused with $\mathbf{K}_1$ and one that is orthogonal to $\mathbf{K}_1$:

$$\mathbf{K}_2 = P(\mathbf{K}_1)\mathbf{K}_2 + (\mathbf{I} - P(\mathbf{K}_1))\mathbf{K}_2 \tag{3.75}$$

where $P(\mathbf{K}_1)$ means projection on the space spanned by $\mathbf{K}_1$ (see section 2.1.1.12). This implies that the two matrices in the sum are orthogonal to each other, i.e. each column in one of them is orthogoal to all the columns in the other. The confused part of $\mathbf{K}_2$ can be written as $P(\mathbf{K}_1)\mathbf{K}_2 = \mathbf{K}_1\mathbf{D}$ where $\mathbf{D}$ is a matrix of dimension $J_1 * J_2$ where J_1 is the number of columns in $\mathbf{K}_1$ and J_2 is the number of columns in $\mathbf{K}_2$. For simplicity we denote the last, orthogonal part of $\mathbf{K}_2$ by matrix $\mathbf{G}$. Thus we can alternatively write (3.75) as $\mathbf{K}_2 = \mathbf{K}_1\mathbf{D} + \mathbf{G}$. This implies that the mixture model can be rewritten:

$$\mathbf{X} = \mathbf{Y}_1\mathbf{K}_1' + \mathbf{Y}_2\mathbf{D}'\mathbf{K}_1' + \mathbf{Y}_2\mathbf{G}' + \mathbf{E} \tag{3.76}$$

or

$$\mathbf{X} = (\mathbf{Y}_1 + \mathbf{Y}_2\mathbf{D}')\mathbf{K}_1' + \mathbf{Y}_2\mathbf{G}' + \mathbf{E} \tag{3.77}$$

Thus, by forgetting to include the interferent responses $\mathbf{K}_2$ into the mixture model, we must expect the obtained $\widehat{\mathbf{Y}}_1$ to be approximately equal to $\mathbf{Y}_1 + \mathbf{Y}_2\mathbf{D}'$ instead of the correct value $\mathbf{Y}_1$. This alias problem always arises for incomplete modelling, except of course when the responses $\mathbf{K}_1$ and $\mathbf{K}_2$ are orthogonal ($\mathbf{K}_1'\mathbf{K}_2 = 0$ and hence $\mathbf{D} = 0$).

In the present simple case with one analyte and one inteferent $d = (0.5)$ and $g = (1, 1, 1, 0)$. This means that the obtained analyte concentration will be $y_1 + 0.5y_2$ when we forget to include the interferent in the unmixing model (Table 3.7, column 5).

Another source of model mis-specification in unmixing is that *in situ* signal contribution of some important constituents are different from the signals measured for the isolated constituents and used in the mixture model. This difference between the isolated and in situ signals will be illustrated in section 3.6.2. But let us first give a practical example of the simplicity and dangers of the direct unmixing based on some real data.

3.6.1.2 Direct unmixing of alcohols from NIR spectroscopy

The present illustration concerns mixtures of the three alcohols methanol, ethanol and n-propanol. The same data were used in the previous section to illustrate simultaneous calibration by PLS2 regression. Near-infrared transmission spectra were measured by transflection at every 5 nm in the wavelength range 1100 to 1600 nm, in total 101 wavelength channels.

The measurements were done with a Guided Wave fiber optic monochromator instrument, simply dipping the fiber-optic probe into beakers with different alcohol mixtures.

Various pure constituent objects and mixtures of the three alcohols were prepared and analyzed with respect to their O.D. spectrum log(1/Transmittance) ($= X$).

All-constituents model

Figure 3.14a–c shows the O.D. spectra $\mathbf{k}_{\text{meth.}}$, $\mathbf{k}_{\text{eth.}}$ and $\mathbf{k}_{\text{prop.}}$, for the three alcohols measured individually in the pure state. The additive mixture model was first applied using all three constituents known to be present. The model *without* an offset was assumed (equation (3.64)). Equation (3.70) was thus applied to these three spectra $\mathbf{K} = (\mathbf{k}_{\text{meth.}}, \mathbf{k}_{\text{eth.}}, \mathbf{k}_{\text{prop.}})$.

Figure 3.14d, e and f show the columns of $\widehat{\mathbf{B}} = \mathbf{K}(\mathbf{K}'\mathbf{K})^{-1}$ corresponding to the absorbance information from each of the three alcohols. Notice that these calibration coefficient spectra are quite similar to those obtained by PLS regression in the previous section.

Figure 3.14g, h and i show the corresponding 'true' vs. predicted concentrations of the three constituents for the 11 test objects. It shows that the predictions for methanol are now reasonably accurate over the whole concentration range from 0 to 100 percent. But for ethanol and propanol it is rather unsatisfactory.

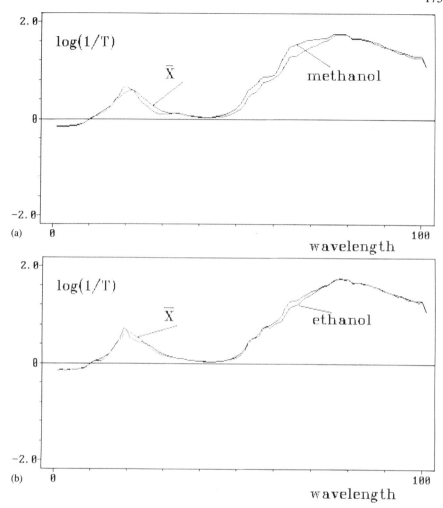

Figure 3.14 Direct unmixing of spectra of alcohol mixtures, based on the fully specified model: NIR spectrum = f(methanol, ethanol and propanol). Pure-state spectra k_j directly measured: a) methanol and $\bar{x}$ b) ethanol and $\bar{x}$

It may appear that the methanol/propanol 'interaction' to some degtree resembles an ethanol effect on the NIR spectrum.

This illustrates that the simple mixture model may be unsatisfactory when there are interactions between the constituents so that the constituents' *in situ* spectra, i.e. their contribution to the X-data in mixtures, are different from their spectra in the pure, isolated state.

Effect of gross model mis-specification: Propanol forgotten

Let us then see what happens when we even forget to include one of the constituents (propanol) in the mixture modelling. The mixture model (3.64) was now applied with $J = 2$, assuming that the mixtures only contained methanol and ethanol. The resulting calibration coefficients $\widehat{\mathbf{B}}$ were now quite different (Figure 3.15a,b). As we see from Figure 3.15c and d erroneous concentration predictions were obtained, since part of the absorbance caused by the propanol was mistakenly ascribed to

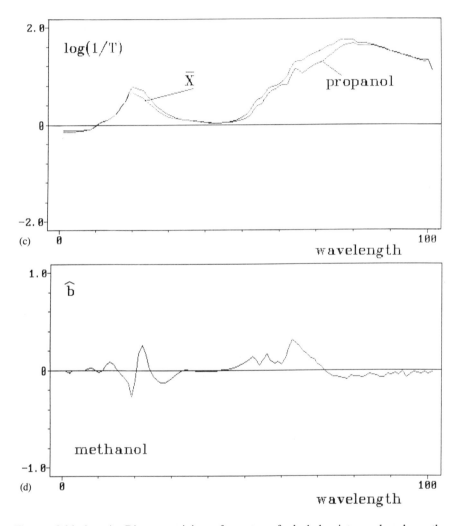

Figure 3.14 (*cont.*) Direct unmixing of spectra of alcohol mixtures, based on the fully specified model: NIR spectrum = f(methanol, ethanol and propanol). c) Pure state spectrum k_j and $\bar{\mathbf{x}}$ for propanol calibration coefficients $\widehat{\mathbf{b}}_j$. for d) methanol e) ethanol f) propanol. Predicted $(\hat{y})$ vs 'true' concentrations (y) in test-set of 11 mixtures: g) methanol

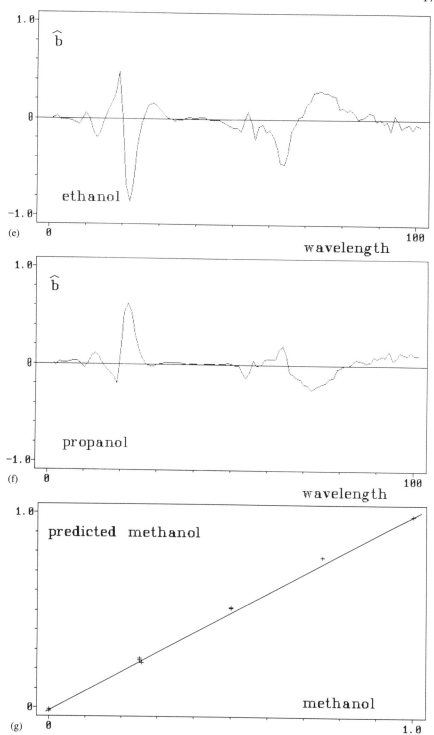

178

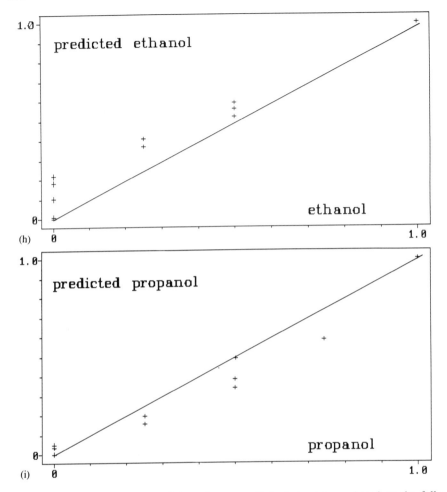

Figure 3.14 (*cont.*)   Direct unmixing of spectra of alcohol mixtures, based on the fully specified model: NIR spectrum = f(methanol, ethanol and propanol). Predicted ($\widehat{y}$) vs 'true' concentrations (y) in test-set of 11 mixtures: h) ethanol i) propanol

Figure 3.15   Direct unmixing of spectra of alcohol mixtures, based on the incomplete model: NIR spectrum = f(methanol and ethanol) Calibration coefficients $\widehat{\mathbf{b}}_j$: a) methanol b) ethanol. c) Predicted ($\widehat{y}$) vs 'true' concentrations (y) in the test-set of 11 mixtures for methanol

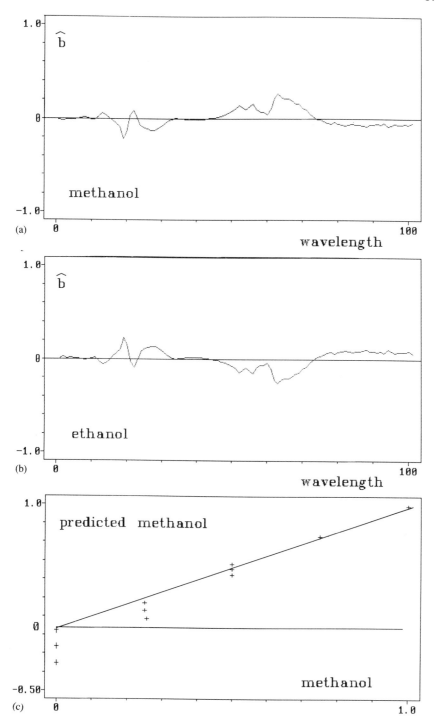

(a) methanol

(b) ethanol

(c) predicted methanol

the effect of the two other alcohols ('alias error'). In statistical language we say that we have a biased predictor. It may be noted that most of the mistakes come in the ethanol results. This was expected, since propanol resembles ethanol more than it resembles methanol in spectral response.

However, most of the propanol-containing objects were now detected as outliers, due to their high NIR residuals, $\hat{e}$. The residual spectra might then be analysed and compared to various potential constituents, whose spectra are read from a data bank and fitted to the mixture model (see Chapter 5).

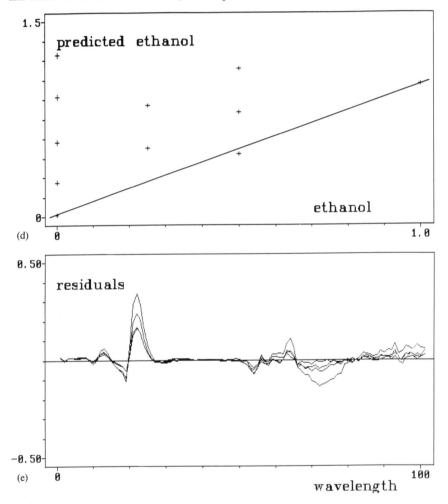

Figure 3.15 (*cont.*) Direct unmixing of spectra of alcohol mixtures, based on the incomplete model: NIR spectrum = f(methanol and ethanol) Predicted ($\hat{y}$) vs 'true' concentrations (y) in the test-set of 11 d) mixtures for ethanol e) Residual NIR spectrum $\hat{e}_i$ of three of the most extreme mixture outliers, compared to the corresponding residual of pure propanol. The latter was obtained by fitting its spectrum z_m to the same two-component mixture model, using pure-state spectra **K** of methanol and ethanol

Figure 3.15e shows that we could find propanol to be the cause of the residuals. So, after this detective work, we could now expand the unmixing to the all-constituents model described above. But we would still not be able to handle the non-additivity tendency discussed.

3.6.2 INDIRECT CALIBRATION OR UNMIXING: ESTIMATING AND USING THE IN SITU SPECTRA

The mixture model can be used in different ways. Till now we have used it for predicting concentrations **Y** from known constituent spectra **K** and mixture spectra **X**. Let us now see how we can use the mixture model to estimate the constituents' *in situ* spectra, $\widehat{\mathbf{K}}$. These can then be used for interpretation as well as for subsequent unmixing (see (3.70) and (3.71)) and extensions (section 3.6.3).

From known concentrations **Y** and mixture spectra **X** we can estimate **K** by regressing each column of **X** on **Y** over a set of calibration objects $i = 1, 2, \ldots, I$. If the mixture model is assumed to pass through the origin (equation 3.64), then a minimum of $I = J$ calibration objects is required and the solution is $\widehat{\mathbf{K}}' = \mathbf{Y}^{-1}\mathbf{X}$ (if **Y** is invertible). But in order to stabilize the estimation against the effect of measurement noise in the calibration data, we should increase the number of calibration objects, and use the least squares estimator for **K**, which is equal to:

$$\widehat{\mathbf{K}}' = (\mathbf{Y}'\mathbf{Y})^{-1}\mathbf{Y}'\mathbf{X} \tag{3.78}$$

If, however, a baseline offset is included as in the model (3.65), the estimator is equal to

$$(\mathbf{k}_0, \widehat{\mathbf{K}})' = ((\mathbf{1}, \mathbf{Y})'(\mathbf{1}, \mathbf{Y}))^{-1}(\mathbf{1}, \mathbf{Y})'\mathbf{X} \tag{3.79}$$

(Weighted least squares versions of these estimators can alternatively be used, if prior information about the noise level of the different calibration objects is available, see section 2.1.2.3.)

These estimated constituent 'spectra' or instrument responses $\widehat{\mathbf{K}}$ can be studied graphically and used statistically for further chemical identification of the constituents, if necessary. Comparison of $\widehat{\mathbf{K}}$ with known **K** can indicate the nature of possible interactions in the mixtures or nonlinearities in the instrument responses.

The matrix inversions in (3.78) and (3.79) require that the constituent concentrations in the calibration set vary 'independently' of one another. Otherwise it is impossible to determine what X-variations to ascribe to what individual constituent (collinearity). If this is a problem for a certain calibration situation, there are ways to overcome it. For instance, two or more constituents that more or less vary in proportion to one another can be grouped together as one single 'group constituent'. If the sum of the constituent concentrations is always fixed (e.g. 100%), one of the constituents can be ignored in the modelling.

As we shall see below, if the mixture model is incorrectly specified, then the obtained estimates $\widehat{\mathbf{K}}$ can be very misleading. If the X-data are affected by other

182

chemical constituents or physical effects that have been ignored in equation (3.66), for instance because their 'concentration' or level is unknown or insufficiently spanned in the calibration objects, as we shall see shortly, the resulting unmixing will be aliased just like the direct unmixing with constituents forgotten (Table 3.7, column 5; Figure 3.15). However, later sections of this chapter show how statistical extensions of the mixture model can overcome such unidentified interference effects by methods similar to the bilinear regressions described in sections 3.3–3.5.

3.6.2.1 Indirect unmixing illustrated for chromatography

Estimating in situ responses $\hat{\mathbf{K}}$

For the artificial 'chromatography' data in the beginning of the previous section, we can estimate $\mathbf{K}$ on the basis of the noisy X- data from Table 3.8 and the corresponding true analyte and interferent concentrations $\mathbf{Y}$ (column 1, Table 3.7). But note that the calibration data $\mathbf{Y}$ in practice may be far from 'true'—they may be contaminated by noise and operator mistakes. In such cases the use of bilinear method with explicit modelling of such errors may be used instead (sections 3.3–3.5). The present linear mixture models can still work as data approximation tools under such conditions, but they must then be interpreted with care, and they lose some of their theoretical properties (section 3.6.3).

The simulated chromatography data set used here was generated with no general offset affecting the data and we therefore assume here that $\mathbf{k}_0 = (0, 0, \ldots, 0)'$. Thus we can base the estimation of the constituents' responses on the mixture model that passes through the origin (3.64), which leads to (3.78). (The alternative model, (3.65) cannot be used here, because since the two constituents' concentrations sum to 1.0 in every calibration object, the inverse in (3.79) does not exist. This so-called 'closure' phenomenon is treated in more detail later in this section.)

The X- and Y-data for the calibration objects are given in Table 3.11. We shall use only the three mixtures' data, to show explicitly that the pure constituents are unnecessary in this estimation of $\mathbf{K}$.

Table 3.11 Calibration data: Mixture data X, after addition of about 5% random proportional noise, and error-free constituent concentrations y_1 and y_2 from Table 3.5

		X-channel no.				
		1	2	3	4	
X:	$i = 2$	.102	.097	.103	.911	Chromatography peaks,
Mixture	$i = 3$	.557	.507	.508	.782	with 5% noise
no.	$i = 4$	.876	.981	.902	.570	
		Constituent				
		1	2			
Y:	$i = 2$	.9	.1		Constituent concentrations,	
Mixture	$i = 3$	.5	.5		error free	
no.	$i = 4$	.1	.9			

Equation (3.78) was applied to the data in Table 3.11. The resulting estimated in situ constituent responses $\widehat{\mathbf{k}}_1$ and $\widehat{\mathbf{k}}_2$ are shown in Figure 3.16, together with the 'true' pure-constituent responses used for generating the mixture data. As we see they are quite similar to each other.

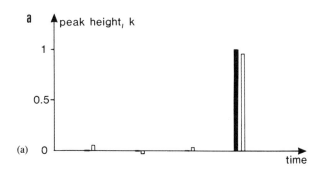

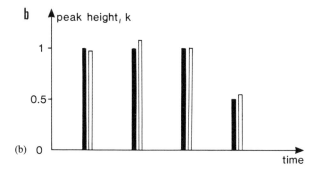

Figure 3.16 The estimated instrument responses $\widehat{\mathbf{k}}_1$ and $\widehat{\mathbf{k}}_2$, for artificial chromatography data, compared to the 'true' responses $\mathbf{k}_1$ and $\mathbf{k}_2$ (black columns). a) The analyte; $\mathbf{k}_1 = (0, 0, 0, 1)'$, $\widehat{\mathbf{k}}_1 = (0.03, -0.02, 0.005, 0.97)'$ b) The interferent; $\mathbf{k}_2 = (1, 1, 1, 0.5)'$, $\widehat{\mathbf{k}}_2 = (0.99, 1.08, 1.00, 0.54)'$

Unmixing based on the estimated in situ spectra $\widehat{\mathbf{K}}$

We can now use the obtained $\widehat{\mathbf{K}} = (\widehat{\mathbf{k}}_1, \widehat{\mathbf{k}}_2)$ in a subsequent unmixing as if it had been measured directly as in section 3.6.1. This would lead to predictors $\widehat{\mathbf{Y}} = \mathbf{X}\widehat{\mathbf{B}}$. Table 3.7, column 6 shows the resulting concentrations $\widehat{\mathbf{Y}}$ obtained when applying the calibration model to the original error-free X-data (Table 3.5). It illustrates that for this linear, well characterized system we got good estimates of the analyte concentration in all mixtures of the two constituents.

If we had ignored the second constituent y_2 in the mixture modelling, the resulting $\hat{\mathbf{k}}_1$ could have been seriously misleading: It would contain contributions from $\mathbf{k}_2$ proportional to the degree of correlation between their concentrations in the calibration set, y_1 and y_2. In the present case the two constituents are strongly intercorrelated, and the resulting analyte spectrum $\hat{\mathbf{k}}_1{}'$ is in fact $(0.43, 0.41, 0.41, 1.18)'$ instead of the correct $(0, 0, 0, 1)'$. So interpretation of $\hat{\mathbf{K}}$ has to be done with care because of possible alias effects. This is treated in more detail in the next illustration.

In summary, the artificial chromatography data have served to illustrate both direct calibration (from known $\mathbf{K}$ to $\hat{\mathbf{Y}}$ in unknown mixtures) and indirect calibration (from $\mathbf{Y}$ in known mixtures, via $\hat{\mathbf{K}}$ to $\hat{\mathbf{Y}}$ in unknown mixtures).

3.6.2.2 Indirect unmixing of alcohols from NIR specroscopy

Let us now return to the real NIR-data for the mixtures of methanol, ethanol and propanol. We shall estimate the in situ constituent spectra $\mathbf{K}$ of the alcohols, from spectra $\mathbf{X}$ in calibration mixtures with known concentrations $\mathbf{Y}$. Then we combine these estimated constituent spectra $\hat{\mathbf{K}}$ into calibration coefficients $\hat{\mathbf{B}}$, which we finally apply to new spectral data in 'unknown' mixtures, in order to predict concentrations.

Assume that our aim is to study the alcohols in mixtures, not in their pure state. Therefore we shall not use the measured pure-state spectra as $\mathbf{K}$-matrix, but use the mixtures of the alcohols to estimate $\mathbf{K}$. This can be important in many practical situations; for instance the NIR spectrum of pure liquid water is different from the in situ NIR spectrum of water in wheat grain, and from the spectrum of water vapour which also can affect NIR readings. In the present case the effect of hydrogen bonding or different refractive indices of the different alcohols may cause different instrument responses for different mixture ratioes.

Thus, from the 15 calibration objects used in section 3.5.4.3 (PLS2) we shall use only the 12 binary and ternary mixtures spanning the range 0.1–0.75 for methanol, ethanol and propanol. Ten of them are shown in Figure 3.17. The second set, consisting of other 11 mixtures of the same alcohols, will then be used for testing the predictive ability of the calibrations.

The closure problem

Let us say that a general offset effect in the NIR spectra can not be excluded a priori, and we assume the model to be

$$\mathbf{X} = \mathbf{1k}_0{}' + \mathbf{y}_{\text{meth.}}\mathbf{k}'_{\text{meth.}} + \mathbf{y}_{\text{eth.}}\mathbf{k}'_{\text{eth.}} + \mathbf{y}_{\text{prop.}}\mathbf{k}'_{\text{prop.}} + \mathbf{E} \qquad (3.80)$$

which equivalently can be written

$$\mathbf{X} = (\mathbf{1}, \mathbf{Y})(\mathbf{k}_0, \mathbf{K})' + \mathbf{E} \qquad (3.81)$$

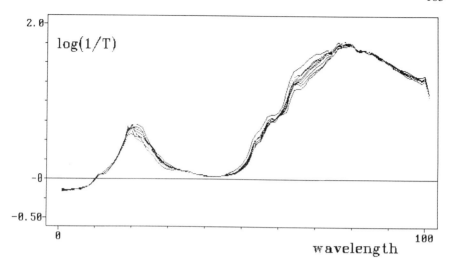

Figure 3.17 NIR spectra of mixtures of methanol, ethanol and propanol for 10 of the calibration mixtures

where $(1, \mathbf{Y}) = (1, \mathbf{y}_{meth.}, \mathbf{y}_{eth.}, \mathbf{y}_{prop.})$ and
$(\mathbf{k}_0, \mathbf{K}) = (\mathbf{k}_0, \mathbf{k}_{meth.}, \mathbf{k}_{eth.}, \mathbf{k}_{prop.})$

In the present data set, however, it is not possible to solve (3.79) directly, since the constituent concentrations y_{meth}, y_{eth} and y_{prop} sum up to 100%, i.e.

$$y_{meth.} + y_{eth.} + y_{prop.} = 1 \tag{3.82}$$

for each mixture or object. This mathematical effect is called 'closure' and such dependencies creates rank problems for $((1,\mathbf{Y})'(1,\mathbf{Y}))$ which implies that it can not be inverted. In other words, it is impossible for the estimator to differentiate between the offset vector $(\mathbf{1})$ and the sum of the constituent concentrations.

It must be emphasized that closure is not always present in multi-constituent data and for many types of methods it is no problem (see e.g. section 3.3). But closure is a prevalent phenomenon in nature, and when present, it requires conscious interpretation. In geochemistry, for instance, the concentration variation of micro-elements (e.g precious metals) in rock samples is often dominated by their indirect correlation to the single major element, Si, which often has nothing to do with the mineralization of interest.

The closure restriction in the present alcohol mixtures is seen in Figure 3.18. The three pure alcohols (not present in the calibration set) form the corners of a two-dimensional triangle in the Y-space.

As we shall see, there are several ways of overcoming the closure problem in multivariate calibration. One way is to include extra assumptions; this can give good results both for $\hat{\mathbf{K}}$ and subsequent $\hat{\mathbf{Y}}$, but only if the assumptions are valid. Another approach is to search for good predictors $\hat{\mathbf{Y}}$, reducing the ambition of obtaining

186

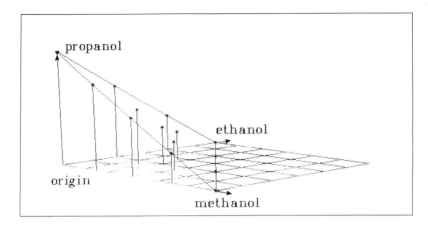

Figure 3.18 Closure in three-constituent calibration mixtures. Concentrations of methanol, ethanol and propanol in their mixtures

the 'true' in situ spectra for every individual constituent. Here we consider both approaches.

Solution 1: Forcing model through the origin In the present example it may be tempting to assume the absorbances to be proportional to concentrations all the way down to zero for all the three alcohols, as a first approximation. This means that we assume the baseline spectrum to be have zero absorbance at all wavelengths ($k_{k0} = 0$ for $k = 1, 2, \ldots, K$). The mixture model then becomes:

$$\mathbf{X} = \mathbf{YK}' + \mathbf{E} = \mathbf{y}_{\text{meth.}}\mathbf{k}_{\text{meth.}}' + \mathbf{y}_{\text{eth.}}\mathbf{k}_{\text{eth.}}' + \mathbf{y}_{\text{prop.}}\mathbf{k}_{\text{prop.}}' + \mathbf{E} \qquad (3.83)$$

With this assumption we simply ignore the offset and directly estimate the three alcohol spectra from the input spectra of the 12 calibration mixtures. The matrix $\mathbf{Y'Y}$ is now 3×3 and invertible, since the data span the full 3 dimensions of the Y-space (see Figure 3.18).

How meaningful these three estimates are depends on how realistic the extra assumption is. Ideally, if the instrument response is linear, the three pure alcohol spectra $\mathbf{k}_{\text{meth.}}$, $\mathbf{k}_{\text{eth.}}$ and $\mathbf{k}_{\text{prop.}}$ should represent the corners of a triangle in the X-space, with all the mixtures inside this flat, two-dimensional triangle. Figure 3.19 indicates this to be a reasonable representation for the three wavelengths selected.

Figure 3.20 shows the in situ spectra $\hat{\mathbf{K}}$ estimated from the alcohol mixtures, compared to the pure-state $\mathbf{K}$. The two spectra are almost indistinguishable for ethanol, and for methanol and propanol they are also quite similar. But they differ appreciably in a wavelength region corresponding to the highest O.D. peaks. This could be due to some errors in the assumption about the model passing through the origin, or to non-additivity due to e.g. optical refractive-index effects or chemical interactions. But at least it seems that the zero offset assumption allowed reasonable estimates of the constituents' in situ spectra.

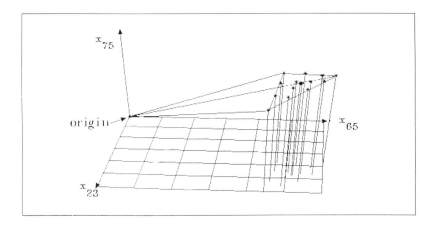

Figure 3.19 Overcoming closure by forcing the solution through the origin. Absorbances plotted at three NIR wavelengths (channels 23, 65 and 75, corresponding to 1215, 1425 and 1475 nm) for the 12 calibration mixtures. The origin and the three pure alcohols methanol, ethanol and propanol represent the corners of the tetrahedron to be modelled

Figure 3.21a–c show the resulting calibration coefficient spectra $\hat{\mathbf{b}}_j$. They show for instance that the predictors contain amplification of spectral structures around channel 20. This is particularly evident for ethanol, as expected, since the spectrum of ethanol is particularly difficult to distinguish from a combination of the methanol and propanol spectra.

Figure 3.21d–f show the corresponding concentration predictions $\hat{y}_j$ for the 11 test objects, using the obtained $\hat{\mathbf{k}}_{\text{meth.}}$, $\hat{\mathbf{k}}_{\text{eth.}}$, $\hat{\mathbf{k}}_{\text{prop.}}$. The predictions are reasonably good, except for the methanol and propanol predictions for the pure alcohols. The latter may be due to nonlinearities, or because our assumption $\mathbf{k}_0 = 0$ is not correct.

Thus it seems that forcing the causal linear model through the origin did give information about the in situ constituent spectra, and these in turn did have predictive ability. These results are in harmony with those from direct calibration where also an intercept was excluded. Further improvement in the predictive ability might even have been attained by giving lower a priori weight to certain wavelength regions.

Solution 2: Ignoring one constituent   Let us now look at an alternative and less demanding way of overcoming the closure problem.

The closure restriction allows us to write the model as

$$\mathbf{X} = \mathbf{1}\mathbf{k}_0' + \mathbf{y}_{\text{meth}}\mathbf{k}_{\text{meth}}' + \mathbf{y}_{\text{eth}}\mathbf{k}_{\text{eth}}' + (1 - \mathbf{y}_{\text{meth}} - \mathbf{y}_{\text{eth}})\mathbf{k}_{\text{prop}}' + \mathbf{E} \qquad (3.84)$$

which can be rewritten as

$$\mathbf{X} = \mathbf{1}\mathbf{k}_{0p}' + \mathbf{y}_{\text{meth.}}\mathbf{k}_{\text{mp.}}' + \mathbf{y}_{\text{eth.}}\mathbf{k}_{\text{ep.}}' + \mathbf{E} \qquad (3.85)$$

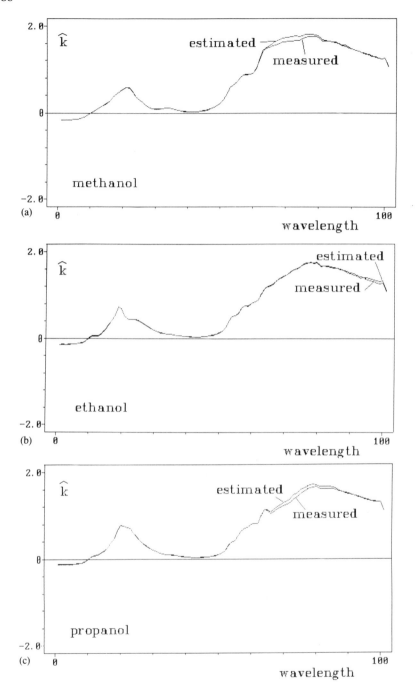

Figure 3.20 Indirect unmixing without offset, for NIR spectra of alcohols: Estimated *in situ* constituent spectra, $\hat{\mathbf{k}}_j$, obtained by forcing the model through the origin, compared to the directly measured *pure-state* spectra $\mathbf{k}_j$. a) Methanol b) Ethanol c) Propanol

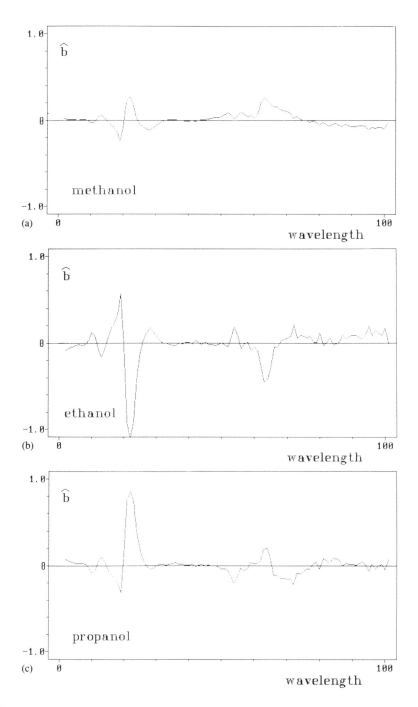

Figure 3.21 Predictions from indirect unmixing without offset for NIR determinations of alcohols: Estimated calibration coefficients $\widehat{\mathbf{b}}_j$. a) Methanol b) Ethanol c) Propanol

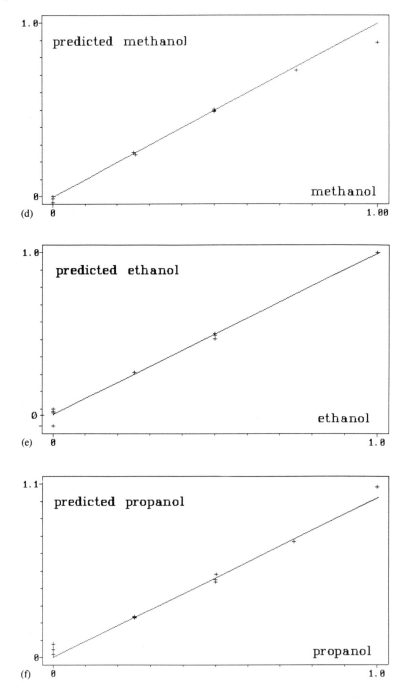

Figure 3.21 (*cont.*) Predictions from indirect unmixing without offset for NIR determinations of alcohols: Predicted vs 'true' concentrations ($\widehat{y}_{ij}$ vs y_{ij}) in test-set of $i = 1,2,\ldots,11$ mixtures: d) methanol e) ethanol f) propanol

where

$$k_{0p} = k_0 + k_{prop},$$
$$k_{mp} = k_{meth} - k_{prop} \qquad (3.86)$$
$$k_{ep} = k_{eth} - k_{prop}.$$

This implies that interpretation of the spectra in model (3.85) is now more complicated. Used in prediction, however, they can give excellent results as will be demonstrated.

Solving for k_{0p} and $\mathbf{K}_p = (k_{mp}.k_{ep}.)$ in (3.78) gave the in situ estimates shown in Figure 3.22. The two measured pure-state spectra are included for comparison. The figure shows quite dramatic differences between estimated and measured constituent spectra. The negative peaks in $\hat{k}_{mp}$ and $\hat{k}_{ep}$ correspond to the main unique absorption range of propanol. The estimated baseline spectrum k_{0p} shows a great resemblance to the average spectrum of the calibration set. This means that we have developed a calibration model that consists of an average spectrum plus two difference spectra that span the main variability of the ternary mixtures under closure (see Honigs et al., 1984, for a similar discussion).

Figure 3.23a–b shows the resulting calibration coefficients for methanol and ethanol obtained by LS using model (3.85). We see $\hat{\mathbf{B}}$ for the two constituents modelled is quite similar to that obtained in the direct unmixing (Figure 3.14), as well as to that obtained in the PLS modelling (section 3.5.4.3). This is so, even though propanol was ignored in the explicit unmixing model, and shows the effect of closure.

Figure 3.23c–e shows the corresponding predicted alcohol concentrations $\hat{\mathbf{Y}} = (\mathbf{X} - \mathbf{1}\hat{k}_0{}')\hat{\mathbf{B}}$ for 11 test objects not used in the calibration, plotted against the 'true' concentrations. Based on the closure condition, the weight fraction of propanol was simply determined by difference:

$$\hat{y}_{prop} = 1 - \hat{y}_{meth} - \hat{y}_{eth} \qquad (3.87)$$

These prediction results are similar to those obtained by forcing the calibration model through the origin.

Thus we have used two quite different ways to overcome the closure problem. The first one assumed that the model is linear and additive all the way to the origin, and the other one actually used the actual closure condition as part of the calibration model. The predictors $\hat{\mathbf{B}}$ and the resulting prediction abilities were quite similar in the present case.

Over-simplified model: Ignoring more than one constituent. Ignoring one constituent successfully overcame the closure problem in calibration, although the estimated in situ constituent spectra were rather strange due to the indirect correlations caused by the closure. What happens if we simplify the calibration model even further?

Let us now ignore two of the three alcohols, using only offset and methanol in the mixture model:

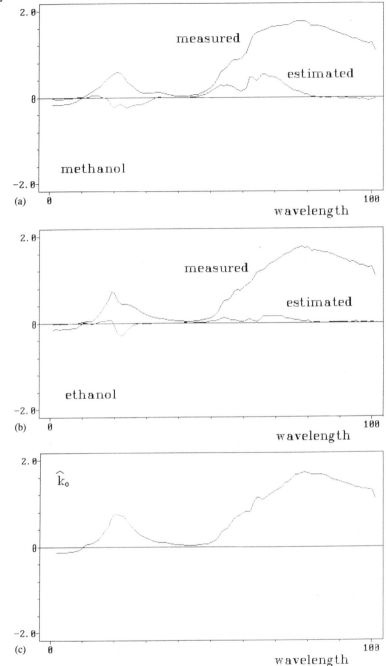

Figure 3.22 Indirect unmixing with offset, but with one constituent intentionally left out, for NIR spectra of alcohols: Estimated apparent in situ constituent spectra $\hat{\mathbf{k}}_j$, obtained from the model: NIR = f(offset+methanol+ethanol), compared to the directly measured *pure-state* spectra $\mathbf{k}_j$. a) Methanol b) Ethanol c) Offset, $\hat{\mathbf{k}}_0$

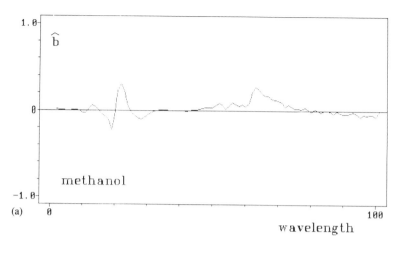

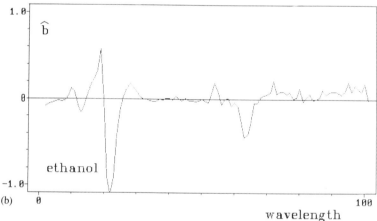

Figure 3.23   Indirect unmixing of spectra of alcohol mixtures, based on the reduced model: NIR spectrum = f(offset, methanol and ethanol) Calibration coefficients $\widehat{\mathbf{B}}$: a) methanol b) ethanol

$$\mathbf{X} = (\mathbf{1}, \mathbf{Y})(\mathbf{k}_0, \mathbf{K})' + \mathbf{E} = \mathbf{1k}_0{}' + \mathbf{y}_{meth.}\mathbf{k}_{meth.}{}' + \mathbf{E} \qquad (3.88)$$

For simplicity we use this old notation, but as for model (3.85) $\mathbf{k}_0$ and $\mathbf{k}_{meth}$ will contain information about all three constituents and might therefore be difficult to interpret. Figure 3.24 shows this estimated methanol spectrum to be very different from that of pure methanol; its negative peaks may be expected to be due to both ethanol and propanol. The offset spectrum again resembles an average spectrum.

Can these strange offset and in situ methanol spectra $\widehat{\mathbf{k}}_{meth}$ and $\widehat{\mathbf{k}}_0$ from the model be used for predicting the methanol concentrations in future mixtures? The calibration coefficient spectrum $\widehat{\mathbf{b}}_{meth.}$ is shown in Figure 3.25a. Compared to the

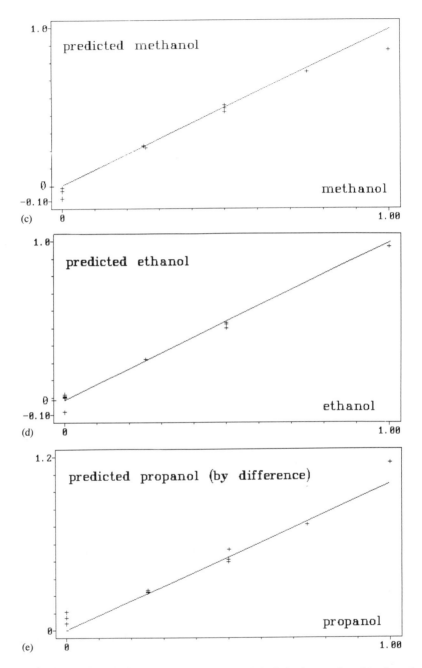

Figure 3.23 (*cont.*)   Indirect unmixing of spectra of alcohol mixtures, based on the reduced model: NIR spectrum = f(offset, methanol and ethanol). Predicted vs 'true' concentrations, ($\hat{y}_j$ vs y_j) in test-set of 11 mixtures: c) methanol d) ethanol e) propanol (determined by difference)

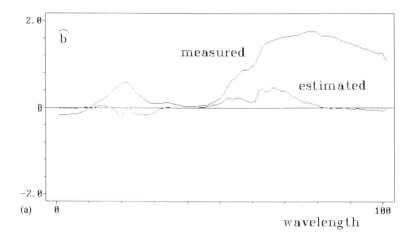

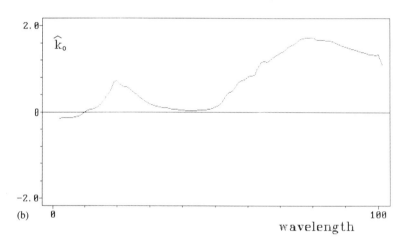

Figure 3.24 Indirect unmixing with two constituents intentionally left out, for NIR spectra of alcohols: Estimated *in situ* constituent spectra $\widehat{k}_j$, obtained from the model: NIR = f(offset+methanol), compared to the directly measured *pure-state* spectrum **k**. a) Methanol. b) Offset $\widehat{k}_0$

corresponding $\widehat{b}_{meth.}$ from the (offset+methanol+ethanol) model (Figure 3.23a), the (offset+methanol) model gives a quite different calibration coefficient spectrum for the analyte.

When this coefficient spectrum $\widehat{b}_{meth.}$ was applied to the NIR data for the 11 test mixtures, the predicted methanol concentrations were far too high, (Figure 3.25b). Obviously, ignoring two of the three constituents in the mixture modelling was too naive.

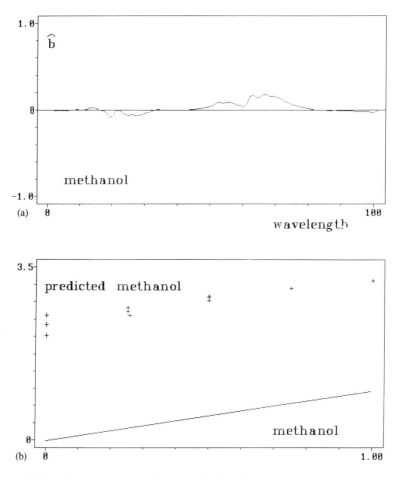

Figure 3.25 Indirect unmixing of spectra of alcohol mixtures, based on the extremely reduced model: NIR spectrum = f(offset + methanol). a) Calibration coefficient spectrum $\widehat{b}_{meth.}$. b) Predicted vs 'true' methanol concentrations in test-set of 11 mixtures. The solid line represents 100% correct predictions

Expanding from unmixing to 'unscrambling' The closure condition could only create implicit modelling of one, not two 'unknown' constituents. Thus, if we were in a situation where only the methanol concentration y_{meth} were known, while the parameters y_{eth}, y_{prop}, k_{meth}, k_{eth}, k_{prop} and k_0 were unknown, then the indirect unmixing would have been unsatisfactory. Now the problem is not the closure, but the fact that the methanol effect on the X-data was 'scrambled' with other 'unidentified' effects—interferences for which we know neither their concentration levels nor their in situ spectra. These 'unidentified' intereferences in the present case represent varying ethanol/propanol ratio, non-additivities etc.

However, there are ways of 'unscrambling' such under-specified problems. The forward bilinear calibration methods treated in sections 3.3–3.5 do this by ignoring the explicit estimation of $\widehat{\mathbf{K}}$, spanning instead the X-space by estimated loading vectors $\widehat{\mathbf{P}} = (\widehat{\mathbf{p}}_a, a = 1, 2, \ldots, A)$.

A similar approach can be used presently, extending the mixture model $\widehat{\mathbf{K}}$ with estimated loading vectors that span the unmodelled systematic variability remaining as residuals $\widehat{\mathbf{E}}$ in the calibration data $\mathbf{X}$:

$$\widehat{\mathbf{E}} = \mathbf{X} - (\mathbf{1}\widehat{\mathbf{k}}_0' + \mathbf{Y}_{Y\_known}\widehat{\mathbf{K}}'_{Y\_known}) \qquad (3.89)$$

which in this case is

$$\widehat{\mathbf{E}} = \mathbf{X} - (\mathbf{1}\widehat{\mathbf{k}}_0' + \mathbf{y}_{meth}\widehat{\mathbf{k}}'_{meth}) \qquad (3.90)$$

Figure 3.26 shows some residual spectra $\widehat{\mathbf{E}}$ from the (offset+methanol) model for 10 methanol/ethanol/propanol calibration mixtures. Not surprisingly, it shows that there is a large amount of unmodelled structure in the NIR data after only having subtracted an average and one type of chemical variability.

Let us assume the following model for the calibration data:

$$\mathbf{X} = \mathbf{1}\mathbf{k}_0' + \mathbf{Y}_{Y\_known}\mathbf{K}'_{Y\_known} + \mathbf{Y}_{Y\_unknown}\mathbf{K}'_{Y\_unknown} + \mathbf{E}^* \qquad (3.91)$$

We assume that the elements of $\mathbf{E}^*$ are random and independent of $\mathbf{Y}$, which implies that we can regard $\mathbf{Y}'_{Y\_known}\mathbf{E}^* \approx 0$ and $\mathbf{Y}'_{Y\_unknown}\mathbf{E}^* \approx 0$.

In the present case $\mathbf{Y}_{Y\_known}$ corresponds to $\mathbf{y}_{meth}$. The 'unidentified interferents' concentration matrix $\mathbf{Y}_{Y\_unknown}$ corresponds e.g. to the two columns $\mathbf{y}_{eth}$ and $\mathbf{y}_{prop}$, or to, for instance, the single column $(\mathbf{y}_{eth} - \mathbf{y}_{prop})$ (due to the restriction $\mathbf{y}_{meth} + \mathbf{y}_{eth} + \mathbf{y}_{prop} = 1$).

As in (3.75) we can always regard the unknown constituents' concentrations $\mathbf{Y}_{Y\_unknown}$ as a sum of one part that is linearly dependent of the known analyte concentrations $\mathbf{Y}_{Y\_known}$ and one part that is orthogonal to $\mathbf{Y}_{Y\_known}$:

$$\mathbf{Y}_{Y\_unknown} = \mathbf{Y}_{Y\_known}\mathbf{D} + \mathbf{G} \qquad (3.92)$$

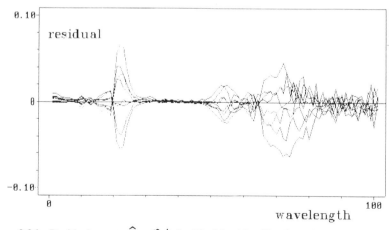

Figure 3.26 Residual spectra $\widehat{\mathbf{E}} = (\widehat{\mathbf{e}}_i')$ for 10 of the 12 calibration mixtures after unmixing based on the over-simplified model: NIR $= f$(offset + methanol) (Figure 3.25)

where $\quad \mathbf{Y'}_{Y\_known}\,\mathbf{G} \;=\; 0.$

By inserting (3.92) into (3.91) we obtain

$$\mathbf{X} = \mathbf{1k_0}' + \mathbf{Y}_{Y\_known}(\mathbf{K}_{Y\_known} + \mathbf{K}_{Y\_unknown}\mathbf{D'})' + \mathbf{GK}_{Y\_unknown}' + \mathbf{E}^* \qquad (3.93)$$

Thus, by the projection of $\mathbf{X}$ onto $(\mathbf{1}, \mathbf{Y}_{Y\_known})$ the obtained $\hat{\mathbf{K}}_{Y\_known}$ (e.g. Figure 3.24a) must be expected to be

$$\hat{\mathbf{K}}_{Y\_known} \approx (\mathbf{K}_{Y\_known} + \mathbf{K}_{Y\_unknown}\mathbf{D'}) \qquad (3.94)$$

Hence, the obtained in situ spectrum of the contituents with known concentrations will be aliased with contributions from the unidentified interferents, and the degree of this contamination alias will depend on the degree of correlation between $\mathbf{Y}_{Y\_known}$ and $\mathbf{Y}_{Y\_unknown}$.

However, if the linear model holds and the data are precise enough to give good estimation, the residual spectra $\hat{\mathbf{E}}$ after the estimation of $\mathbf{K}_{Y\_known}$ contain only the spectral effects of the unknown constituents:

$$\hat{\mathbf{E}} \approx \mathbf{GK'}_{Y\_unknown} + \mathbf{E}^* \qquad (3.95)$$

Hence, under the condition that $\mathbf{G'E}^* \approx 0$ (which is reasonable from the assumptions about $\mathbf{E}^*$ above) we can obtain selective information about $\mathbf{K}_{Y\_unknown}$, independently of $\mathbf{K}_{Y\_known}$, by performing principal component analysis (see chapter 3.4) on $\hat{\mathbf{E}}$ (or $\hat{\mathbf{E}}'\hat{\mathbf{E}}$) after estimation of $\mathbf{K}_{Y\_known}$.

Once estimated, the loadings $\hat{\mathbf{P}}$ can be used to represent $\hat{\mathbf{K}}_{Y\_unknown}$ in subsequent unmixing together with $\hat{\mathbf{K}}_{Y\_known}$. Thereby the X-effect of the interferences are no longer mistakenly counted as analyte effects.

If there are two or more unidentified interferents, the PCA loadings $\hat{\mathbf{P}}$ will span their space, but due to rotation problems the individual loading vectors $\hat{\mathbf{p}}_a, a = 1, 2, \ldots$ cannot be expected to correspond to the individual unknown spectra $\mathbf{k}_{j,Y\_unknown}, j = 1, 2, \ldots$.

However, if there is only one major unmodelled interference in $\hat{\mathbf{E}}$, $\mathbf{k}_{1,Y\_unknown}$, then there will only be one major principal component. No rotation problem then exists, and $\hat{\mathbf{p}}_1$ is proportional to $\mathbf{k}_{1,Y\_unknown}$.

In the present illustration a principal component analysis of the matrix $\hat{\mathbf{E}}$ for the 12 calibration mixtures yielded one dominant principal component. Its loading vector $\hat{\mathbf{p}}_1'$ is given in Figure 3.27. This first component apparently differentiates between the two present 'interferents', ethanol and propanol. A few minor additional components could also have been extracted from $\hat{\mathbf{E}}$ but this is for simplicity ignored here. This 'unknown interference' spectrum $\hat{\mathbf{p}}_1$ can here be regarded as an estimated difference spectrum between the $\mathbf{k}_{prop}$ and $\mathbf{k}_{eth}$.

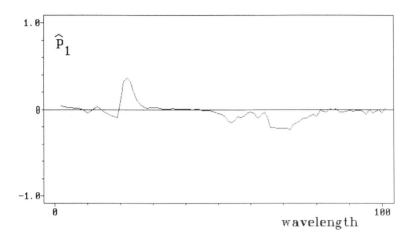

Figure 3.27 Loading vector for the first principal component in the residual $\widehat{\mathbf{E}}$ (Figure 3.26) after the over-simplified calibration model based only on methanol

This $\widehat{\mathbf{p}}_1$ spectrum was now used together with the aliased estimated in situ analyte spectrum $\widehat{\mathbf{k}}_{meth}$ in simultaneously determining the concentration of methanol (as well as the level of the unidentified interferent).

In the present case we can now perform unmixing of the 11 test mixtures, using $\widehat{\mathbf{k}}_0$ and the expanded $\widehat{\mathbf{K}} = (\widehat{\mathbf{k}}_{meth.}, \widehat{\mathbf{p}}_1)$. The resulting calibration coefficient vector $\widehat{\mathbf{b}}_{meth}$ is shown in Figure 3.28a. The figure shows that, contrary to Figure 3.25a, the indirect unmixing based on the extended mixture model now resulted in a $\widehat{\mathbf{b}}$-vector for methanol very similar to those obtained in all the other calibrations.

Predictions for the methanol concentrations for the 11 test mixtures based on the use of a model center $\widehat{\mathbf{k}}_0$ and two model spectra $\widehat{\mathbf{k}}_{meth}$ and $\widehat{\mathbf{p}}_1$ are given in Figure 3.28b and gave reasonably good results.

However, a curvature can be seen, resulting in understimation at the highest methanol levels. This curvature could actually have been modelled by including one more principal component from $\widehat{\mathbf{E}}$, resulting in three model spectra $\widehat{\mathbf{k}}_{meth}$, $\widehat{\mathbf{p}}_1$ and $\widehat{\mathbf{p}}_2$, in analogy to the PLS2 modelling (section 3.5.4.3), which required three, instead of two, PLS factors for satisfactory calibration.

Further statistical details of this extended mixture model and alternative calibration methods based on additional assumptions will be given in the next section (see also Martens and Næs, 1987).

———————————— Statistical extensions ————————————

3.6.3 THE EXTENDED MIXTURE MODEL

SUMMARY This section presents alternative calibration methods and results in the extended mixture model, with additional statistical assumptions applied. Different approaches are presented and their relative merits for

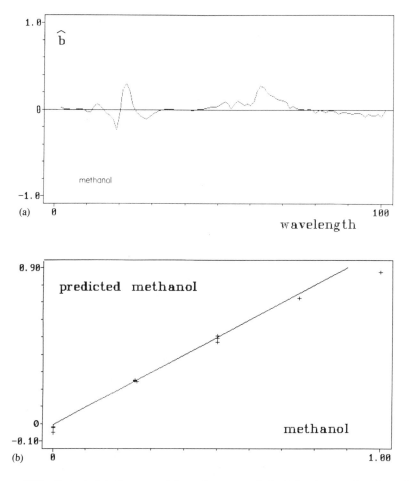

Figure 3.28 Extended indirect unmixing of spectra of alcohol mixtures, based on the model: NIR spectrum = f(offset + methanol + one unidentified PCA component). a) Calibration coefficient spectrum $\widehat{\mathbf{b}}_j$ for methanol for the extended indirect unmixing. b) Predicted vs 'true' methanol concentrations in test-set of 11 mixtures

different types of future prediction objects are discussed. Special emphasis is given to the relationship between the multivariate versions of the forward ('inverse') and reverse ('classical') approach.

As explained in section 3.6.1, the linear mixture model for a matrix of spectra **X** can be written as

$$\mathbf{X} = \mathbf{1}\mathbf{k}_0{}' + \mathbf{Y}\mathbf{K}' + \mathbf{E} \tag{3.96}$$

where $\mathbf{k}_0$ is possible background contribution, **Y** is the matrix of concentrations of absorbing constituents, **K** is the matrix of unit spectra and **E** represents random measurement noise.

In the same section, it was shown that good predictions can be obtained by least squares (LS) fitting as long as k_0 and all possible columns of K are known. If, however, not all of them are known, the LS predictor gives highly aliased and unreliable results. It was also shown that in indirect calibration with concentration data available only for a few of the absorbing constituents, the spectral estimates are aliased (or biased) and the derived predictor is misleading. At the end of section 3.6.2.2 it was, however, shown that if principal component analysis was performed on the residuals after estimation of K and the factor loadings or eigenvectors are used as additional columns in 'K', the LS predictor can still be used to give good prediction results.

This extended mixture model for X will here be written as

$$X = 1k_0' + YK' + TP' + E \tag{3.97}$$

where E still represents random measurement noise, YK' corresponds to the contributions from the analytes that we have calibration data for and TP' represents contributions from unidentified interferents.

In principle, there is no requirement that TP' in this model only represents interfering constituents. Therefore, any type of interference with approximately linear effect on the spectrum will fit into the same framework, i.e. can be represented by TP'. In section 3.6.3.3 below is given an illustration where model (3.97) is used and the light scattering interference is dominating.

Without further restrictions or assumptions on the P-parameters in this model, they are what is called *unidentified*. This means that many different choices of P will have the same effect on the spectra. For instance TP' has the same effect as $TZZ^{-1}P'$ for any invertible matrix Z. The model as presented here is then merely indicating that except for the random independent noise, the spectra fall in a linear subspace determined by K and P and the constituents in y are linearly related to directions in this space. When interpretation of P is important, factor rotation on $\hat{P}$ or $\hat{T}$ can be useful (see e.g. Harman, 1967).

Notice that if the number of constituents in Y and T together is much less than the number of wavelengths in X, and the residuals are small, the spectral matrix X is *collinear*. This is a clear indication that the extended mixture model can be of special importance in such cases (see section 3.6.3.3).

For a single object, the model (3.97) is equal to

$$x' = k_0' + y'K' + t'P' + e' \tag{3.98}$$

where x', y' and e' are rows of X, Y and E respectively. For simplicity, this model can also be written without transposes (see section 2.1.1.3) and is then equal to

$$x = k_0 + Ky + Pt + e \tag{3.99}$$

where now x is a column vector coresponding to a row in X etc. Throughout the rest of this chapter we use this simplified notation.

In the previous section, no further assumptions about the interferences **t** were made, and the predictor was obtained by putting estimates $\hat{\mathbf{K}}$ and eigenvectors $\hat{\mathbf{P}}$ into one single matrix. For the alternative methods to be covered in this chapter we will, however, require that

a) the expectation (or average over the population) of the unknown interferences t is independent of the value of **y**. Without loss of generality we may assume that this expectation is equal to 0.
b) the covariance matrix of the unknown interferences **t** in the population is independent of the value of **y**.

(It may be noted that if $(\mathbf{y}', \mathbf{t}')'$ comes from a (multi-) normally distributed population, a) and b) are fulfilled)

Thus, we may consider model (3.97) as a regression model with error terms **Pt+e** having a covariance matrix equal to

$$\Sigma = \mathbf{P} \operatorname{cov}(\mathbf{t}) \mathbf{P}' + \operatorname{diag}(\sigma_k^2). \tag{3.100}$$

If the dimension of **t** is the same as or larger than the dimension of **x**, the factor assumption for the error terms is the same as no restriction on Σ at all and is therefore a special case of our general treatment (see e.g. Brown, 1982; Brown and Sundberg, 1987; Sundberg, 1985). At the same time, it should also be emphasized that the factor structure on the residuals in the linear model, i.e. the assumption that residuals can be written **Pt+e**, has no practical effect on the derivation of predictors in direct calibration where all model parameters are assumed known. In indirect calibration, however, it can have a great effect on the estimation procedure applied and provides improved prediction results (see sections 3.6.3.2 and 3.6.3.3).

We refer to Osten and Kowalski (1985) for discussion of similar problems.

3.6.3.1 Direct calibration

As for direct calibration in section 3.6.1 we here assume that the model parameters in (3.97) and (3.100) are known.

Without further information it is natural to predict (or estimate) unknown concentration values **y** by generalized least squares fitting, i.e. by

$$\hat{\mathbf{y}}_{\text{GLS}} = (\mathbf{K}'\Sigma^{-1}\mathbf{K})^{-1}\mathbf{K}'\Sigma^{-1}(\mathbf{x} - \mathbf{k}_0) \tag{3.101}$$

since the covariance matrix of **Pt+e** is different from $\mathbf{I}\sigma^2$ (see section 2.1). As known from standard statistical texts, this predictor is unbiased and minimizes the expected squared difference

$$\mathbf{E}_{\hat{y}}(\hat{\mathbf{y}} - \mathbf{y})'(\hat{\mathbf{y}} - \mathbf{y}) \text{ for any value of } \mathbf{y}. \tag{3.102}$$

where $E_{\hat{y}}$ means that the expectation is only taken with respect to $\hat{y}$, not with respect to y. Note that in addition to the difference concerning prior information about model parameters, the principal difference between (3.101) and the approach to prediction based on least squares at the end of section 3.6.2 is the way the systematic unknown interferences are used. In section 3.6.2 they were put together with $\hat{K}$ in a new $\hat{K}$ matrix while in the present case their distribution is modelled in the covariance matrix Σ and used as weighting matrix in the generalized least squares predictor in (3.101). Therefore the two predictors may have quite different properties for different types of objects (see Martens and Næs, 1987). It can be shown that the predictor in (3.101) is superior for objects with typical values of interferences t, while the other one is superior for atypical values of t which are far from the average.

As explained, the GLS predictor is the unbiased predictor which minimizes (3.102) for any value of the concentrations y provided k_0, K and Σ are known. If, however, the most important property of a prediction method is that it has as good average precision as possible, these properties of the GLS predictor are not necessarily the most relevant. In this case one should instead use a predictor making e.g. the sum of squares

$$1/I_\mathrm{p} \sum_{i=1}^{I_\mathrm{p}} (\hat{y}_i - y_i)'(\hat{y}_i - y_i) \qquad (3.103)$$

where the sum is over the I_p prediction objects, as small as possible. For a large number of prediction objects this is equivalent to minimizing

$$E_{y,\hat{y}}(\hat{y} - y)'(\hat{y} - y) = \int E_{\hat{y}}(\hat{y} - y)'(\hat{y} - y)f(y)dy \qquad (3.104)$$

where $f(y)$ denotes the density function of y in the population and $E_{y,\hat{y}}$ means that expectation (averaging) is with respect to both the population of concentrations y and the predicted concentrations $\hat{y}$. (Notice the difference between this and the squared difference in (3.102) where the expectation is for fixed y-value.)

It can be shown that the predictor which minimizes (3.104) can be written as

$$\hat{y}_{\mathrm{BLP}} = v + DK'(KDK' + \Sigma)^{-1}(x - k_0 - Kv) \qquad (3.105)$$

where $v = E(y)$ is the known mean of the population and $D=\mathrm{Cov}(y)$ is its known covariance matrix. This method is usually known as the best linear predictor (BLP, Searle, 1974) of y and as we see it is dependent on available information about the mean and covariance matrix of the population of objects. This method is therefore preferable from an average quality point of view only if this type of information is available. If not, another method with less attractive properties must be used instead.

Alternative methods to BLP and the GLS predictor in the linear mixture model (3.64) can be found in e.g. Bibby and Toutenborg (1977). These methods are derived under different types of prior information and restrictions and possess different types of optimality or local optimality. As we consider the BLP and GLS predictors to be the most important in practice, the alternatives are not treated further here.

Thus, the basic message of this section is that alternative direct calibration methods can be derived in the extended mixture model and the choice of method depends on what type of optimality one wants and the amount of information about the population which is applied. To make a sensible selection in practice it is, however, necessary to know something about the statistical properties of the different choices and this is the theme in the following section.

Statistical properties of $\hat{y}_{GLS}$ and $\hat{y}_{BLP}$

As we saw above, the BLP incorporates information about the 'measurement instrument' represented by the extended linear mixture model (3.97), as well as information about the population of objects. On the other hand, the GLS predictor uses only information expressed by model (3.97). This different usage of information results in the important relation

$$E_{y,\hat{y}}(\hat{y}_{BLP} - y)'(\hat{y}_{BLP} - y) < E_{y,\hat{y}}(\hat{y}_{GLS} - y)'(\hat{y}_{GLS} - y) \qquad (3.106)$$

which shows that BLP has the best average prediction properties of the two.

But how are the statistical properties of the two predictors for different types of prediction objects or for different values of the concentrations y? From section 2.1.2.3 it is known that the GLS predictor is unbiased and has constant mean squared error equal to

$$\text{MSE}(\hat{y}_{GLS}) = \text{trace}(\mathbf{K}'\Sigma^{-1}\mathbf{K})^{-1} \qquad (3.107)$$

The predictor $\hat{y}_{BLP}$, however, is biased for all values of y except for $y=v$, i.e. for y equal to the mean of the population. The bias is equal to $-\mathbf{A}(y - v)$ (see Sundberg, 1985) where

$$\mathbf{A} = (\mathbf{D}^{-1} + (\mathbf{K}'\Sigma^{-1}\mathbf{K}))^{-1}\mathbf{D}^{-1} \qquad (3.108)$$

and this shows that the bias is greater the further y is away from the centre. In other words, the further an object is from the typical objects, the larger is the bias. The MSE curve of BLP as a function of y can also be derived and this is equal to

$$a + (y - v)'\mathbf{B}(y - v) \qquad (3.109)$$

for matrix $\mathbf{B}$ and scalar a (Næs, 1985a). Both a and $\mathbf{B}$ are difficult to interpret, but can be computed in each separate case if model parameters are known.

From comparison of (3.107) and (3.109) it follows that BLP is MSE-superior to GLS in an ellipsoid centered at $\mathbf{v}$ and outside the ellipsoid the opposite holds (Figure 3.29). First of all this shows that BLP gives the best results for typical objects near the centre of the population, while GLS is best suited for atypical objects. A consequence of this is that for instance when the purpose of the analysis is to detect objects with concentrations above a certain level or limit, one should use the GLS predictor. Alternatively one could use $\mathbf{v}$ and $\mathbf{D}$ values so that $\hat{\mathbf{y}}_{BLP}$ gives precise results near this limit.

Similar results can be found in Sundberg (1985) and Fujikoshi and Nishii (1984). Consequences of misspecification of model parameters are considered in Bibby and Toutenburg (1977).

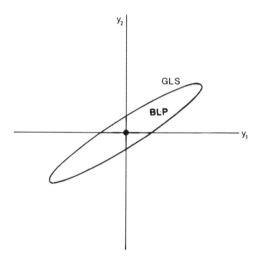

Figure 3.29 Illustration of the relative performance of the GLS predictor and the BLP. The BLP is superior with respect to MSE in an ellipsoid centered at $\mathbf{v}$ and the GLS predictor is superior outside this region

Functional relationship

The functional relationship between BLP and GLS can be written as

$$\hat{\mathbf{y}}_{BLP} - \mathbf{v} = (\mathbf{D}^{-1} + \mathbf{K}'\Sigma^{-1}\mathbf{K})^{-1}\mathbf{K}'\Sigma^{-1}\mathbf{K}(\hat{\mathbf{y}}_{GLS} - \mathbf{v}) \qquad (3.110)$$

or alternatively

$$\hat{y}_{GLS} - v = (I + (DK'\Sigma^{-1}K)^{-1})(\hat{y}_{BLP} - v) \tag{3.111}$$

As we see, the relation depends on D (the covariance matrix of the analytes) compared to

$$Cov(\hat{y}_{GLS}) = (K'\Sigma^{-1}K)^{-1}.$$

In particular, if the variances (i.e. diagonal elements) of Σ are very small, the predictors are indistinguishable. This is a very important result which shows that if GLS gives good results, this will also be the case for BLP. In other words, in cases where good calibration is possible, the difference between the predictors is small. In the univariate case, (3.110) is equal to (3.16).

3.6.3.2 Indirect calibration

Indirect *controlled* calibration (see section 3.2) will typically be a calibration where little or no information about the population of objects is available. From section 3.6.3.1 it follows that it is then most natural to use a GLS type approach. If, however, the model parameters are unknown (as they are in indirect calibration) a GLS predictor is impossible to use directly, but it can be estimated by simply replacing model parameters k_0 and K by estimates based on available calibration data. The resulting predictor is here called an estimated GLS or EGLS predictor.

In *natural* indirect calibration (see section 3.2), however, the calibration objects provide useful information about the population of objects. More specifically, we can use (x_i, y_i) $i = 1, \ldots, I$ to estimate v and D and these estimates can along with the estimates $\hat{K}$ and $\hat{\Sigma}$ be plugged into the BLP predictor to obtain an estimated best linear predictor (EBLP).

As we see from this, an EGLS predictor is sensible to use in controlled calibration cases while an EBLP is better suited for natural calibration. It should, however, be emphasized that the final decision about which method to use depends on their statistical properties in different situations and not on how they are derived.

Other approaches to indirect calibration in the multivariate mixture model can be found in Brown (1982) and Lwin and Maritz (1982).

Properties of EGLS and EBLP

First of all, it can be shown that the MSE of the EGLS predictor is sometimes infinite (see Lieffinck-Koeijers, 1988). Since a truncated predictor, i.e. a predictor of the type minimum $(\hat{y}_{EGLS}, a)$ where a is some fixed value, will always be used is practice, we think it is most fair (as long as squared error is used) to study such a truncated version of the EGLS predictor.

As for direct calibration, we would then like to know the performance of the (truncated) EGLS predictor and the EBLP for different values of the concentrations **y** and also their average prediction ability. For small sets of objects, however, such prediction results are difficult to derive for both these methods (for a special case, see Lieffinck-Koeijers, 1988). As, however, the number of calibration objects increases, the precision of parameter estimates increases and MSE's obtained by plugging such estimates into the MSE's for GLS and BLP can give reasonable approximations.

From this and the results for direct calibration we can conclude that at least for moderately large calibration sets, EBLP is best suited for interpolation, while EGLS is best suited for extrapolation. Besides, the average prediction ability of EBLP is better than that of the EGLS predictor. (Similar discussions can be found in Sundberg (1985) and in Fujikoshi and Nishii (1984)).

For small samples, the best advice we can give is to use resampling techniques as described in Chapter 4. Results from such analyses indicate that a similar relationship between EBLP and EGLS holds as for the large sample case (Næs, 1985c).

Model parameter estimates

The simplest and probably most common estimates of $\mathbf{k}_0$, $\mathbf{K}$ and Σ are

$$(\widehat{\mathbf{k}}_0, \widehat{\mathbf{K}})' = (\mathbf{Y}_e{}'\mathbf{Y}_e)^{-1}\mathbf{Y}_e{}'\mathbf{X} \tag{3.112}$$

$$\widehat{\Sigma} = (1/(I - J - 1))\mathbf{X}'(\mathbf{I} - \mathbf{Y}_e(\mathbf{Y}_e{}'\mathbf{Y}_e)^{-1}\mathbf{Y}_e{}')\mathbf{X} \tag{3.113}$$

where

$$\mathbf{Y}_e = (\mathbf{1}, \mathbf{Y}) \tag{3.114}$$

These estimates are unbiased for the underlying parameters and under normality assumptions $(\widehat{\mathbf{k}}_0, \widehat{\mathbf{K}})$ and $\widehat{\Sigma}*(I-J-1/I)$ are the maximum likelihood (ML) estimates of their corresponding parameters. For further properties we refer to Mardia et al. (1980).

The most common estimates of **v** and **D** are the object mean and covariance matrix of **y** as defined in section 2.1.2. These can be written as

$$\widehat{\mathbf{v}} = \frac{1}{I}\mathbf{Y}'\mathbf{1} \tag{3.115}$$

and

$$\widehat{\mathbf{D}} = \frac{1}{I - 1}(\mathbf{Y} - \mathbf{1}\bar{\mathbf{y}}')'(\mathbf{Y} - \mathbf{1}\bar{\mathbf{y}}') \tag{3.116}$$

With $I - 1$ replaced by I in $\widehat{\mathbf{D}}$, the two estimates are as above ML estimates under normality assumptions. We refer again to Mardia et al. (1980) for further properties.

With the mentioned ML estimates for model parameters it is easy to show that the EBLP is identical to the MLR predictor of $\mathbf{y}$ in the linear regression model where $\mathbf{y}$ is regressed on $\mathbf{X}$. In other words, with model parameters replaced by the ML estimates, EBLP is the natural generalization of the forward ('inverse') predictor presented for the univariate case in section 3.2. This result also represents an important link between mixture model methods and the usual forward ('inverse') regression methods obtained by regressing $\mathbf{y}$ on the $\mathbf{x}$'s. Likewise, EGLS with the same ML estimates is the natural generalization of the reverse ('classical') method.

It is important to realize that this link between forward regression methods and the EBLP implies that the EBLP properties described above also hold for other forward methods as e.g. the bilinear methods PCR and PLS.

Alternatives to the ML estimates could be robust regression (Huber, 1981) methods which are less sensitive to outliers and can also be more efficient in some cases.

So far, the linear factor structure of $\mathbf{Pt+e}$ in model (3.97) is not utilized to improve the prediction ability. A simple and straightforward way to do this is to do a factor analysis (in practice a PCA; Næs, 1985b) of $\hat{\Sigma}$ and replace $\hat{\Sigma}$ by an estimate of the form

$$\hat{\mathbf{P}}\widehat{\mathrm{Cov}(\mathbf{t})}\hat{\mathbf{P}}' + \mathrm{diag}(\hat{\sigma}^2_k) \tag{3.117}$$

which has the same structure as the model structure in (3.100). Using this estimator, improvements can be obtained when data are collinear. The main reason for this is that the factor estimate in (3.117) in many cases is a better estimate of the underlying covariance matrix Σ than $\hat{\Sigma}$ and inversion is then not as dangerous as for the full ML estimate. We refer to Næs (1985b) and Næs (1986) for demonstrations of the usefulness of the factor estimate in the EBLP and the EGLS predictor. In both cases the example is from near-infrared (NIR) spectroscopy and the number of factors or interferences is determined by cross-validation as will be discussed further in Chapter 4.

It is also important to realize that EBLP and EGLS computed with factor assumptions on the residuals are applicable even for $I < K$. This can be very important in spectroscopy where the number of objects may be small and the number of elements in the spectra quite large. For further discussion on this see Sundberg and Brown (1987).

3.6.3.3 Illustration of EBLP and EGLS

The data set in this example consists of $\mathbf{y}$ = concentration of fat and $\mathbf{X}$ = measurements of light reflectance at three selected wavelengths in the NIR region (between 1440 and 2400nm) for 38 individuals of fish (freeze-dried rainbow trout). The spectra were taken on a Technicon InfraAlyzer 400 and the fat percentage was measured by a traditional wet-chemical extraction and weighting method. The data is a subset of a larger data set published in Næs (1985b) and is here used as a simplified illustration of important relationships rather than as a demonstration of how to obtain optimal prediction of fat% in fish meat. The data set was divided into a calibration ($I = 20$) and test ($I = 18$) set with the calibration set having a somewhat larger span of fat percentage than the prediction set. Both sets are plotted in Figure 3.30. The estimation procedure used is the one described in the previous section. The $\hat{P}$ matrix was as in Næs (1985b) obtained by principal component analysis.

For this data set, the eigenvalues of the covariance matrix $\hat{\Sigma}$ in (3.113) were 0.0096, 0.00019 and 0.000013. As we see, the $\hat{\Sigma}$ is quite collinear with a condition index (see section 2.3) equal to 27.2. This means that there are clear systematic effects in the residuals after fitting of fat%. The predictors EBLP and EGLS should therefore be tested with both 1 and 2 factors in $\hat{P}$ and also compared with the use of the maximum which is equal to 3. The results from the predictions are shown in Figure 3.31 in terms of RMSEP and as we see, two factors in $\hat{P}$ gave optimal results for both methods.

The prediction equations for EBLP and EGLS with two factors in $\hat{P}$ are shown in Frame 3.7 together with their functional relationship. As we see EBLP represents a shrinking of the EGLS predictor towards the centre 40.70 by the factor 0.87.

Frame 3.7

$$\hat{y}_{EGLS} = 40.70 + 77.79(x_1 - 1.48) -$$
$$282.92(x_2 - 0.95) + 347.86(x_3 - 0.52)$$
$$\hat{y}_{EBLP} = 40.70 + 67.75(x_1 - 1.48) -$$
$$246.39(x_2 - 0.95) + 302.95(x_3 - 0.52)$$
$$(\hat{y}_{EBLP} - 40.70) = 0.87(\hat{y}_{EGLS} - 40.70)$$

The residuals from both EBLP and EGLS are plotted in Figure 3.32. We see that for values less than the average, the differences between EGLS and EBLP residuals are positive and for values larger than the average the opposite holds. This is in harmony with the functional relationship between them as shown in Frame 3.7. In other words, EBLP has a tendency of overestimation for small values of y (shrinking towards the middle, 'least squares effect') and underestimation for large values of y. This is also shown in Figure 3.33 where y is plotted vs $\hat{y}$ for EBLP. Except for the single point in the left corner the underestimation and over-estimation effect is clear (see Williams, 1987; Norris et al., 1976, for a similar effect).

For comparison we also performed a PLS regression analysis on the data and we obtained an optimal RMSEP (see Chapter 4) equal to 1.51 after 3 factors (=MLR), which was less accurate than the EBLP predictor with 2 factors (RMSEP=1.39). This indicates that for some type of data the explicit

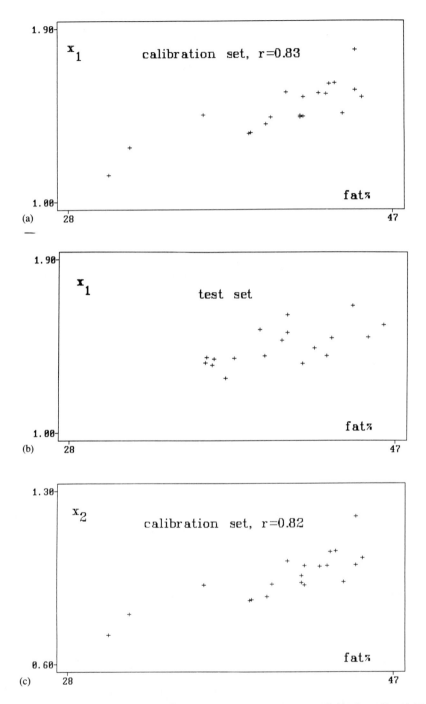

Figure 3.30 Illustration of the relation between fat percentage and the three *X*-variables for the calibration set and test set

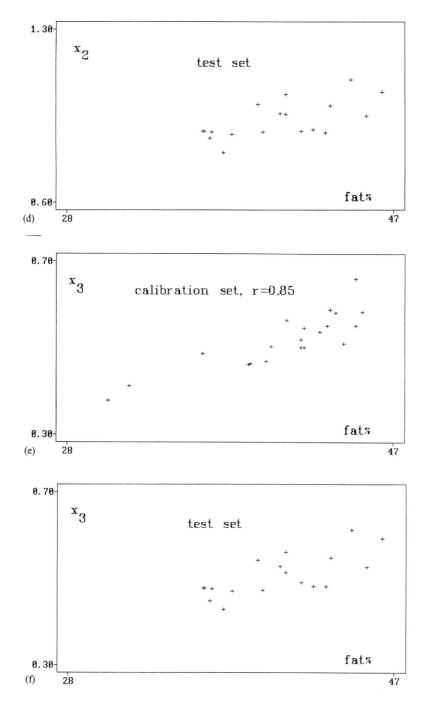

Figure 3.30 (*cont.*) Illustration of the relation between fat percentage and the three *X*-variables for the calibration set and test set

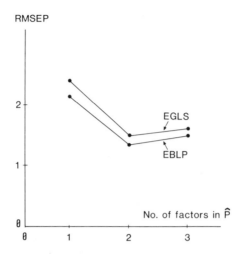

Figure 3.31 Prediction ability of EGLS and EBLP for different number of factors. The prediction ability is given in root mean square error of prediction (RMSEP) defined in Chapter 4

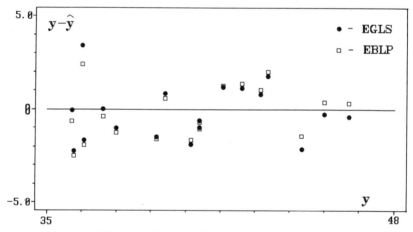

Figure 3.32 The residuals of EBLP and EGLS on the test objects

modelling of covariance structure as performed by EBLP and EGLS, can be advantageous over PLS which is one of the methods generally advocated in this book due to its simplicity.

It is also interesting to notice the difference between the prediction ability in Næs (1985b), when 9 wavelengths were used, and the results in Figure 3.31 where only 3 of the wavelengths are applied. The RMSEP obtained with 9 wavelengths was equal to 0.77 and this clearly tells us that the full set of data gave more information about fat percentage than did the subset. This means that other wavelengths contain further information about interferences

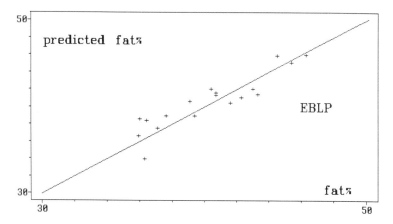

Figure 3.33 Predicted (EBLP) versus measured fat percentage in the 18 test objects

in the data making multivariate calibration and prediction more precise. For comparison, univariate calibration based on only one of the wavelengths among the 3 gave a RMSEP value equal to about 2.52. As we see this is more than 1 unit larger than for multivariate calibration based on 3 wavelengths.

——————————— End of statistical extensions ———————————

3.7   ALTERNATIVE CALIBRATION METHODS

SUMMARY   Alternative calibration methods are briefly presented and discussed. *Key words*: Fourier regression, Hruschcka regression, generalized standard addition method, curve resolution, rank annihilation, target transformation factor analysis, stepwise regression, ridge regression, projection pursuit regression. Limitations and strengths of the different methods are indicated along with references to more fundamental treatments.

3.7.1 ALTERNATIVE DATA COMPRESSION METHODS

We refer to section 3.3 for a description of the general structure of data compression.

3.7.1.1 Fourier regression

The Fourier transform regression is illustrated in Frame 3.1 and it uses a V matrix for $T=XV$ that is selected from a class of mathematically defined functions, rather than a V that is statistically estimated from experimental data. Thereby the number of calibration model parameters to be estimated is reduced.

The idea behind the method is that each spectrum x can be written as a sum of contributions from various mathematical functions. In Fourier regression these functions **V** are various trigonometric functions with different frequency, representing spectral phenomena ranging from slowly varying peaks and valleys to high-frequency noise ripple.

Each spectrum in **X** is thus transformed trigonometrically from the usual wavelength domain to a frequency domain **T**. By leaving the higher frequencies out from the calibration model $\mathbf{Y} = \mathbf{TQ'} + \mathbf{F}$, spectral redundancy and erratic features primarily due to noise can be eliminated.

The method is, however, mainly applicable in cases where some degree of smoothness is expected in the noise-free spectrum. An example such as that in section 3.4.5 (a fixed-channel spectrophotometer) is therefore not suitable to approach by Fourier regression.

For each spectrum **x**, the Fourier transform matrix **V** for A factors is defined according to

$$v_{k1} = 1/\sqrt{K}$$

$$v_{k,a+1} = \sqrt{(2/K)} \sin\left(\frac{2\pi ak}{K}\right),$$

$$a = 1, .., A/2 - 1; \; k = 1, \ldots, K \tag{3.118}$$

$$v_{k,A/2+a} = \sqrt{(2/K)} \cos\left(\frac{2\pi ak}{K}\right),$$

$$a = 1, \ldots, A/2 - 1; k = 1, \ldots, K.$$

There are maximally $A = K$ such terms (then $v_{kK} = 1/\sqrt{K} * (-1)^k$), half of them based on sines and half of them based on cosines. Using these weights, one can show that the columns of the resulting **V** matrix are orthonomial, i.e. $\mathbf{V'V=I}$.

Factors $\mathbf{t}_a$ (both sine and cosine factors) corresponding to high a-value measure the amplitude of the high frequency contributions to the spectrum and factors corresponding to small a-value measure the low-frequency contributions. The factors with large a-value are therefore supposed to represent random and irrelevant noise and they can be eliminated, i.e. $A/2$ is selected so that $A/2 \ll K/2$. The number of Fourier terms to incorporate can be determined by significance tests or cross-validation as described in Chapter 4.

Following the general procedure in section 3.3, the **T=XV** matrix is used in the regression equation

$$\mathbf{y} = \mathbf{1}q_0 + \mathbf{Tq} + \mathbf{f} \tag{3.119}$$

(uncentered univariate **y**), to estimate **q** and q_0. The predictor can then according to (3.43) be written as

$$\hat{y}_i = \hat{b}_0 + \mathbf{x}_i' \hat{\mathbf{b}} \tag{3.120}$$

where $\hat{\mathbf{b}} = \mathbf{V}\hat{\mathbf{q}}$ and $\hat{b}_0 = \bar{y} - \bar{\mathbf{x}}'\mathbf{b}$.

The easiest way to compute the Fourier factors $\mathbf{t}_i'$ from $\mathbf{x}_i'$ is by the so-called fast fourier transform (FFT). In that case all the K Fourier terms are obtained simultaneously, but this requires K to be a power of 2 (16,32,64,128, etc.); if not, the technique of 'zero filling' can be used. More detail is given in e.g. Rabiner and Gold (1975).

The Fourier technique and modifications are extensively used in near infrared spectroscopy (see e.g McClure, 1984; Davies and McClure, 1987; McClure et al., 1984) to filter out noise and compress the continuous spectral information. Some empirical comparisons have also been made with other calibration methods and the technique performs reasonably well. If the values of x_1 and x_K are very different, it is customary to subtract a ramp function 'baseline' from each spectrum, e.g. the straight line from $\bar{x}_1$ to $\bar{x}_K$ in the calibration set.

Fourier analysis can also be used as a separate spectral pre-treatment, for digital filtering and/or for domain transformation, like in FTIR, or e.g. for converting sound as a function of time into sound as a function of frequency. This is covered in more detail in Chapter 7.

3.7.1.2 Hruschka regression

If there are many wavelengths K, but only a limited number, A, of independent ways in which the spectra can vary, it may be possible to use the X-spectra of A selected calibration objects to construct the weight matrix V. These A objects must be selected in such a way that they include all the variation types present in the remaining $I - A$ calibration objects, as well as future objects.

In the simplest case these A objects represent the pure analytes with known spectra K, in which case the method corresponds to direct unmixing (section 3.6.1). If unidentified constituents or other interferents also have to be modelled, the number of objects A to be used in the model must be higher. We refer to the original paper by Hruschka and Norris (1982) and to Ottestad (1975), Næs (1987) and Kvalheim (1987) for techniques that can be used to select such objects.

The matrix of selected spectra is here denoted by W and the scores of Hruschka regression are found as the least squares solution for T in the equation

$$X = TW' + E \qquad (3.121)$$

This implies that the V matrix in the general framework is equal to

$$V = W(W'W)^{-1}. \qquad (3.122)$$

When the scores have been computed, the $\hat{q}$ and $\hat{q}_0$ can be found as for the other data compression methods yielding the predictor $\hat{y}$.

Since the factor scores are found by least squares fitting on W, it is evident that Hruschka regression requires very precise spectral readings W to obtain a sensible weight matrix V. It was originally developed for near infrared data which have exactly this property (Hruschka and Norris, 1982) and in that application it has shown to be able to give sensible results.

As expressed here, the Hruschka regression belongs to the family of forward calibration methods, where Y is modelled as function of elements from X. A reverse version also exists. For instance, Nyden et al. (1988), in their study of IR determination of human plasma protein mixtures, expressed every input spectrum x_i' as a linear combination of the X-spectra of a set of calibration objects with known analyte contents.

To overcome the problem of collinearity between different objects' spectra in the least squares fit of x_i' to X, Nyden et al did not resort to the selection of individual objects described above. Instead, they used a generalized inverse (section 2.1.1.13), which implied regressing each x_i' on principal component loadings from X, and propagating the results through the bilinear mixture model to obtain $\hat{y}_i$. Thus their approach is similar to PCR, but with the Y-loadings estimated in analogy to the mixture models (section 3.6).

3.7.2 ALTERNATIVE LINEAR MIXTURE MODEL METHODS

3.7.2.1 Generalized standard addition method (GSAM)

The GSAM was originally developed by Saxberg and Kowalski in 1980 and numerous articles have appeared since then by Kowalski and coworkers (see e.g. Jochum et al., 1981). The GSAM is primarily constructed for cases when physical or chemical 'matrix effects' are present, i.e. effects that cause different *in situ* constituent matrices K (see sections 3.6 and 1.2) to be needed for each separate object. In such cases the same prediction equation based on the ordinary least squares in the linear mixture model (see sections 3.6.1 and 3.6.2) may be impossible to use.

In sections 3.6.2 and 3.6.3 we presented one way of solving some of these 'interference' problems by modelling the error structure by factor models (extended mixture model), but the GSAM uses a very different approach. It consists of developing one separate calibration for each individual unknown object $i = 1, 2, \ldots$ by adding known combinations of the chemical constituents (standard additions) to the unknown object. In this way a matrix of spectra X_i and a matrix of concentrations Y_i are measured for each object i. These data matrices are then used to estimate the in situ K for this object. When $\hat{K}$ has been obtained, the predictor can be found by ordinary least squares fitting as in section 3.6.2.

The GSAM predictor can therefore as the reverse predictor in section 3.6.2 be written as

$$\hat{y}_i' = x_i'\hat{K}(\hat{K}'\hat{K})^{-1} \tag{3.123}$$

where the only difference is the acquisition of data to estimate $\hat{K}$.

The linear GSAM modelling assumes that the original in situ spectrum of the analyte(s) in the object, K, is of the same shape as the spectral effects induced by the artificially added analyte(s) and estimated as $\hat{K}$. This means that if the instrument response is non-linear or some unexpected baseline offset is present, the results obtained will be wrong.

The GSAM can, however, be extended outside the linear model framework and it has been used to compensate for drift and interferences (Kalivas and Kowalski,

1982). Design aspects have been considered thoroughly and formulae are developed which show how $\hat{y}$ changes with errors in the calibration.

3.7.2.2 Comparison with a reference set

The method discussed here is due to Haaland and Easterling (1982) and it is similar to GSAM in the sense that it makes a new calibration for each separate object. However, the in situ analyte response $\mathbf{K}$, although unknown, is now considered to be the same in all objects. The present method is not based on standard additions, but compares each new object with a reference object set. This reference object set consists of pure objects (i.e. objects with only one analyte) of the constituents.

We let b_j be the ratio between the concentration of the jth constituent in the new prediction object, y_j^p, and the concentration of the jth reference, y_j^r. The relation between

$$\mathbf{y}^p = (y_1{}^p, \ldots, y_J{}^p)' \tag{3.124}$$

and the reference data matrix

$$\mathbf{Y}^r = \begin{pmatrix} y_1{}^r & 0 \\ & \ddots & \\ 0 & & y_J{}^r \end{pmatrix} \tag{3.125}$$

is then equal to

$$\mathbf{y}^p = \mathbf{Y}^r \mathbf{b}. \tag{3.126}$$

When the linear mixture model applies we have

$$\mathbf{X}^r = \mathbf{Y}^r \mathbf{K}' + \mathbf{E}^r \tag{3.127}$$

$$(\mathbf{x}^p)' = (\mathbf{y}^p)' \mathbf{K}' + (\mathbf{e}^p)' \tag{3.128}$$

where $\mathbf{X}^r$ represents the spectra of the references and $\mathbf{x}^p$ is the spectrum of the new object. Substituting (3.126) into (3.128) and letting

$$\mathbf{Y}^r \mathbf{K}' = \mathbf{X}^r - \mathbf{E}^r \tag{3.129}$$

we obtain

$$(\mathbf{x}^p)' = \mathbf{b}' \mathbf{X}^r + (\mathbf{e}^*)' \tag{3.130}$$

where

$$(\mathbf{e}^*)' = -\mathbf{b}' \mathbf{E}^r + (\mathbf{e}^p)'. \tag{3.131}$$

Haaland and Easterling (1982) assume that the elements of $\mathbf{e}^*$ are uncorrelated and use weighted least squares (WLS) to find $\hat{\mathbf{b}}$. Then $\hat{\mathbf{b}}$ is used in (3.126) to obtain $\hat{\mathbf{y}}^p$. Its properties can easily be evaluated by LS methods (see section 2.1.2.3).

It can be shown that this approach, though seemingly quite different, is very similar to the estimated WLS estimator of $\mathbf{y}$, when the reference objects are used as calibration object set and calibration is performed as in section 3.6.2.

The method described is tested for infrared spectral data and it gave sensible results (Haaland and Easterling, 1982).

3.7.2.3 Self deconvolution

We can write the centered (or uncentered) mixture model

$$\mathbf{X} = \mathbf{YK'} + \mathbf{E} \tag{3.132}$$

where the rows in $\mathbf{X}$ represent unknown 'spectra', $\mathbf{Y}$ represents the constituent concentrations, $\mathbf{K}$ represents the constituent spectra and $\mathbf{E}$ *here* represents unsystematic residuals. Of course, if $\mathbf{K}$ is known for all constituents and interference phenomena, $\mathbf{Y}$ can be estimated by direct calibration. When $\mathbf{Y}$ is known for the analytes in a calibration set, the elements of $\mathbf{K}$ can be estimated and then used as if they were known. But what if neither $\mathbf{Y}$ nor $\mathbf{K}$ is known explicitly?

The present methods employ bilinear modelling (PCA) combined with extra information of various kinds to model such mixture data $\mathbf{X}$.

Number of factors

The first step in this type of approach is to determine the number of factors in $\mathbf{Y}$ or principal components in $\hat{\mathbf{T}}$. Due to noise $\mathbf{E}$, this is not always a trivial task. The best method to use in general is probably cross-validation as reported in Wold (1978) and Eastment and Krzanowski (1982), but significance tests as in Mardia et al. (1980) or in Malinowski and Howery (1980) can also be applied.

Ambiguities in the bilinear model

As discussed in section 3.4 the PCA decomposition model is

$$\mathbf{X} = \mathbf{TP'} + \mathbf{E} \tag{3.133}$$

The main problem with the PCA solution is, however, that $\hat{\mathbf{P}}$ and $\hat{\mathbf{T}}$ have a sort of inherent ambiguity that stops us from interpreting $\mathbf{Y}$ directly from $\hat{\mathbf{T}}$ and $\mathbf{K}$ directly from $\hat{\mathbf{P}}$. This is easily realized since

$$\hat{\mathbf{T}}\hat{\mathbf{P}}' = \hat{\mathbf{T}}\mathbf{CC}^{-1}\hat{\mathbf{P}}' \tag{3.134}$$

where $\mathbf{C}$ is any invertible matrix. This $\mathbf{C}$ matrix is typically a diagonal matrix (corresponding to scaling of the principal axes) or an orthogonal matrix ($\mathbf{C'C=I}$, corresponding to a rotation of the principal axes).

Can then a bilinear representation of $\mathbf{YK'+E}$ as (3.133) at all be useful for estimation of $\mathbf{Y}$ and $\mathbf{K}$ in unknown samples? The answer to this is definitely yes, provided additional information is available. In fact, there are a number of ways to do this, depending of how much and what type of additional information is available.

One-factor solutions

Let us first consider the simplest case, where only one type of variability ($A = 1$) is seen in the calibration data **X**. An example of this could be light absorbance spectra of solutions containing, as the only source of variation, a certain dye with unknown spectrum $\mathbf{k}_1$ at unknown, but varying concentrations $\mathbf{y}_1$, as in Figure 3.10a for litmus solutions at high pH. This results in only one major PCA factor $\hat{\mathbf{t}}_1 \hat{\mathbf{p}}_1{}'$. No rotational ambiguity thus exists, and $\hat{\mathbf{t}}_1$ would be proportional to the unknown, concentration $\mathbf{y}_j$, and $\hat{\mathbf{p}}_1$ would represent the unknown $\mathbf{k}_j$, except from an unknown scaling factor c_1.

Another example of a one-factor system would be amino acid chromatograms of mixtures of two proteins at constant total protein content (Spijøtvoll et al., 1982).

Figure 3.34a shows NIR spectra **X** from a set of mixtures of methanol (m) and propanol (p) which will be used for illustration. We shall assume that we know neither the pure constituent spectra $(\mathbf{k}_{\text{meth}}, \mathbf{k}_{\text{prop}})'$ nor the concentrations of the two $(\mathbf{y}_{\text{meth}}, \mathbf{y}_{\text{prop}})$. (The example is a subset of the calibration data from the example in section 3.6.2 (see also chapter 3.5).)

The constant spectral contribution, common to every object $i = 1.2.....I$, is subtracted as a common 'baseline' in the centering of the X-variables around $\bar{\mathbf{x}}$ prior to the PCA, so that only one type of variability remains in the centered X-data: The methanol-propanol mixtures fall more or less along a straight line in X-space (Figure 3.34b) and hence can be modelled by one single PCA factor.

The resulting PCA model is shown in Figure 3.34c in terms of $\bar{\mathbf{x}}$ and $\hat{\mathbf{p}}_1$. The resulting loading spectrum $\hat{\mathbf{p}}_1$ here represents a difference spectrum (see (3.86)) between methanol and propanol,

$$\hat{\mathbf{p}}_1 \approx (k_{\text{prop}} - \mathbf{k}_{\text{meth}})c_1{}^{-1} \qquad (3.135)$$

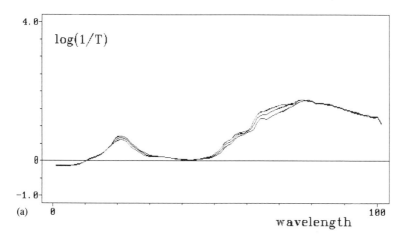

(a)

Figure 3.34 A one-factor self deconvolution problem: a) O.D. spectra of mixtures of methanol and propanol display only one major type of variability

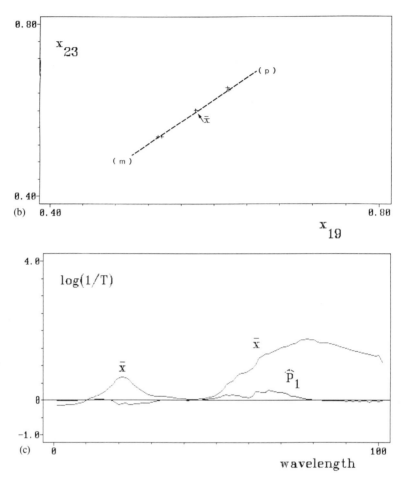

Figure 3.34 (*cont.*) A one-factor self deconvolution problem: b) The spectra plotted in two of the 100 wavelength channels show that the binary mixtures fall along a straight line in *X*-space (here illustrated for channels 19 and 23, corresponding to 1190 and 1210 nm). $\bar{\mathbf{x}}$ represents the center of the PCA model c) The resulting PCA model: $\bar{\mathbf{x}}$ and $\hat{\mathbf{p}}_1$

and the scores $\hat{\mathbf{t}}_1$ are correspondingly proportional to the (unknown but centred) methanol concentration, i.e.

$$\hat{\mathbf{t}}_1 \approx (\mathbf{y}_{\text{meth}}) * c_1. \tag{3.136}$$

Notice that if $\mathbf{y}_{\text{meth}}$ is known for one of the objects, c_1 can be determined and consequently $\mathbf{y}_{\text{meth}}$ can be estimated for all objects.

In general the presented analysis of *X*-data alone gave valuable information, but it does not give explicit information about each of the constituents without further information.

Intersecting different one-factor solutions

If there is more than one type of constituent variation in the data, it may sometimes be possible to find subsets of data that only show one single type of variation and therefore would yield causally interpretable information from analyzing $\mathbf{X}$ alone, without using $\mathbf{Y}$ or $\mathbf{K}$ information.

If we want to estimate the spectra $\mathbf{k}_j$ and concentrations $\mathbf{y}_j$ for each individual constituent, we can sometimes employ various types of closure as extra information. Under the closure condition that the constituent concentrations sum to a certain value, and assuming linear instrument response, mixtures of J constituents fall inside a $J - 1$ dimensional simplex with the spectra of the pure constituents as the extreme points (corners) of the simplex. A one-dimensional simplex is a line segment. A two-dimensional simplex is a triangle and a three-dimensional simplex is a tetrahedron etc.

Therefore the following method was called 'simplex intersect' (Martens, 1979; Martens and Bach-Knudsen, 1980):

In the methanol-ethanol-propanol mixtures example (see section 3.6.2), each set of binary ($J = 2$) mixtures fall more or less along one straight line in the K-dimensional X-space (see Figure 3.35 for a 2-dimensional illustration). These lines can be estimated by separate centered PC analyses on each of the (methanol-ethanol), (ethanol-propanol) and (methanol-propanol) mixture sets, each yielding 1 factor. By extrapolating these principal components, the point of closest fit between them in the X-space would give an estimate of each individual constituent spectrum $\mathbf{k}_{\text{meth}}$, $\mathbf{k}_{\text{eth}}$ and $\mathbf{k}_{\text{prop}}$ as points in X-space.

For instance, the methanol-propanol (m-p) model (consisting of the parameters $\bar{\mathbf{x}}_{\text{m-p}}$ and $\hat{\mathbf{p}}_{1,\text{m-p}}$) estimates $\mathbf{k}_{\text{prop}}$ to lie on the line

$$\bar{\mathbf{x}}_{\text{m-p}} + \hat{t}_{\text{m-p}}\hat{\mathbf{p}}_{1,\text{m-p}} \qquad (3.137)$$

and from the ethanol-propanol (e-p) model we likewise get that $\mathbf{k}_{\text{prop}}$ should be on the line

$$\bar{\mathbf{x}}_{\text{e-p}} + \hat{t}_{\text{e-p}}\hat{\mathbf{p}}_{1,\text{e-p}} \qquad (3.138)$$

The unknown constituent spectrum $\mathbf{k}_{\text{prop}}$ is thus estimated by finding the values of $\hat{t}_{\text{m-p}}$ and $\hat{t}_{\text{e-p}}$ that minimize the deviation between the two lines in the K dimensional space. The same intersection process can be done for the ethanol-propanol and methanol-ethanol models to estimate $\mathbf{k}_{\text{eth}}$, and for the methanol-ethanol and methanol-propanol models to estimate $\mathbf{k}_{\text{meth}}$.

An alternative way of estimating the constituent spectra by simplex intersect is to describe simultaneously all three binary mixture sets (see Figure 3.35) in a two-factor PCA model, and reconstruct the corners of the triangle by graphically extrapolating its sides in the $\hat{t}_2$ vs $\hat{t}_1$ plot.

The constituent spectra $\hat{\mathbf{k}}_j$ estimated for methanol and propanol by this so-called 'simplex intersect method' (Martens, 1979) are represented by (m) and (p) in the two-dimensional Figure 3.36, which also illustrates another method for estimating

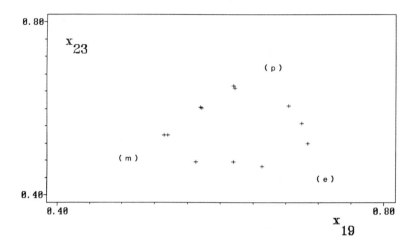

Figure 3.35 Intersecting different one-factor solutions yields estimates of individual constituents' spectra $\mathbf{k}_j$. Binary mixtures of mixtures of methanol-ethanol, ethanol-propanol and methanol-propanol are each approximated by a separate one-factor PCA model, and the location of the pure constituent spectra are obtained at the intersect or point of closest approach of these PCA models. This is here shown for two of the 100 X-variables, channels 19 and 23, 1190 and 1210 nm

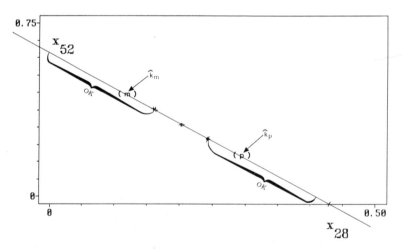

Figure 3.36 Nonnegativity restrictions of a PCA model for binary mixtures of methanol and propanol. Mixtures of methanol and propanol plotted for wavelength channels 28 and 52 (corresponding to 1240 and 1360 nm). Physically meaningful ranges within which the unknown constituent spectra $\mathbf{k}_{meth}$ and $\mathbf{k}_{prop}$ can be expected to lie are marked by symbols 'OK'. Points (m) and (p) represent the position of $\widehat{\mathbf{k}}_{meth}$ and $\widehat{\mathbf{k}}_{prop}$ obtained by the simplex intersect method (see text)

K (see below). These in situ estimates are similar to those obtained by regression on the known constituent concentrations (Figure 3.20), although in the present case Y-data were not used at all.

In the present case the simplex intersect was based on assumptions about closure in the (unknown) $\mathbf{Y}$ and a linear $X-Y$ relationship. An alternative to this approach is to base the simplex intersection on artificially induced closure in $\mathbf{X}$ itself, e.g. scaling each spectrum $\mathbf{x}_i$ to sum 100; which is discussed in more detail in Spijøtvoll et al. (1982).

Once estimated, the constituent spectra $\hat{\mathbf{K}}$ can be used for determining the concentrations $\mathbf{Y}$ in individual mixtures by linear unmixing as in section 3.6.2.

Generating one-factor residuals

In some cases it may not be possible to find subsets of objects with only one single type of spectral variation. Other alternatives then exist.

First of all, if the concentration $\mathbf{y}_j$ is known for all of the constituents except the analyte in question, one may project $\mathbf{X}$ (Martens, 1979) on this partially known Y-matrix, $\mathbf{Y}_{Y\_known}$, and perform PCA on the resulting residual matrix

$$\hat{\mathbf{E}} = \mathbf{X} - (\mathbf{Y}\hat{\mathbf{K}}')_{Y\_known} \tag{3.139}$$

If the data structures are linear and the observations precise enough to allow good estimation, this matrix $\hat{\mathbf{E}}$ would now only contain one type of variability, and the advantages of the one-factor PCA solutions may be utilized. The resulting loading $\hat{\mathbf{p}}_1$ will be proportional to the unknown, desired $\mathbf{k}_{y\_unknown}$. The scores $\hat{\mathbf{t}}_1$ are now proportional to that part of the unknown variability in the unknown analyte's concentration $\mathbf{y}_{y\_unknown}$ that is orthogonal to the known ones, $\mathbf{Y}_{y\_known}$ which was demonstrated at the end of section 3.6.2.

Limiting the rotation by non-negativity constraints

Another approach to limit the rotational ambiguity in PCA is to eliminate rotation ambiguity that represents physically meaningless solutions $\hat{\mathbf{T}}\mathbf{C}$ or $\mathbf{C}^{-1}\hat{\mathbf{P}}'$. One useful restriction on $\mathbf{C}$ is that the resulting concentration and spectral estimates $\hat{y}_{ij}$ and k_{kj} are zero or positive, not negative.

In the example with only methanol and propanol present the pure constituents are expected to lie along one factor, outside the range of mixtures (otherwise negative concentrations y_{ij} are implied in some of the mixtures). At the same time they have to lie inside a certain range, otherwise negative optical densities will occur in the $\mathbf{k}$'s in wavelength ranges where that is not expected. In Figure 3.36 the two wavelengths that had the maximum positive and negative ratio $\hat{p}_{k1}/\bar{x}_k$ (x_{28} and x_{52}, respectively) were chosen as axes for the illustration. These two X-variables are the ones that first reach negative values with increasing or decreasing scores along the factor and hence set the outer limit for the physically acceptable range for the $\mathbf{k}_{meth}$ and $\mathbf{k}_{prop}$ estimates.

The resulting ranges for $\mathbf{k}_{meth}$ and $\mathbf{k}_{prop}$ are given by the shaded areas in Figure

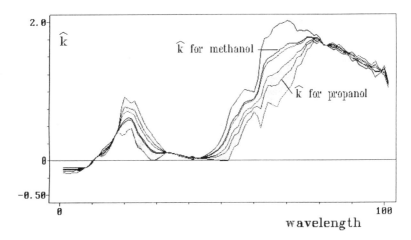

Figure 3.37 Estimation of $\hat{\mathbf{K}}$ Shaded area: The ranges of physically possible values of $\mathbf{k}_{meth}$ and $\mathbf{k}_{prop}$ estimated from the non-negativity restrictions in the methanol-propanol model (Figure 3.36). Solid curves: The $\hat{\mathbf{k}}_{meth}$ and $\hat{\mathbf{k}}_{prop}$ obtained by intersecting the methanol-ethanol, ethanol-propanol and the methanol-propanol simplex lines in the two-factor score plot (in analogy to Figure 3.35)

3.37. The constituent spectra estimated by the simplex intersect method (Figure 3.35) are also plotted, and are shown to fall nicely inside the allowable ranges.

(Since the present spectral data were taken by a fiber optic sensor with air as a reference, artificially negative absorbances happened to be obtained at the lowest wavelength range for all the mixtures. That is here ignored for simplicity. But it could be that the true baseline should have been 0.2 instead of 0.0 O.D. units throughout the spectrum. For normal calibration work that would be of no consequence, except that the allowable non-negativity range would just have been somewhat wider.)

How useful the non-negativity restrictions are in obtaining **K** and **Y** from X-data alone depends on the selectivity problem and the experimental design. If there are X-variables specific for the individual constituents, then the 'outer' limits (corresponding to non-negative constituent spectra **K**), will represent the constituent spectra themselves. In contrast, in the case of strongly non-selective spectral data the 'outer' limit will be quite unrealistic and of little help. The relevance of the 'inner limits' (corresponding to non-negative constituent concentrations **Y**) depends on how pure the mixtures in **X** are.

The non-negativity requirements can also be used for higher-dimensional PCA models. More details on self convolution can be found in e.g Lawton and Sylvestre (1971), Martens (1979), Spjøtvoll et al. (1982), Martens (1985) and Borgen and Kowalski (1985).

These different methods for studying constituent spectra **K** from mixture data **X** without any Y-data are summarized in Martens (1983).

3.7.2.4 Target transformation factor analysis (TFA)

TFA is closely related to the self deconvolution methods both in scope and analysis. The first step is exactly the same, namely principal component analysis and determination of the number of factors or components. The difference is the determination of concentrations and constituent spectra.

While the self deconvolution techniques are based on using mathematical constraints as extra information, the target transformations are based on the use of so-called target vectors, with which the principal component model is compared. This is similar to what is called factor rotation in the statistical literature, and can be done in various ways.

Again we write the PCA model for the X-data:

$$\mathbf{X} = \mathbf{TP}' + \mathbf{E} \tag{3.140}$$

External information can be of two kinds, concerning the objects $i = 1, 2, \ldots, I$ or concerning the X-variables $k = 1, 2, \ldots, K$. Such vectors of extra information about the objects or variables can be used in two different ways, either as rotation targets or as regressors. Here we concentrate on information about variables, but the same procedures can be used equally well for the other case. The method of TFA usually works on the *uncentered* $\mathbf{X}$- matrix.

The purpose is then to identify what constituents are spanned by the PCA model loadings $\widehat{\mathbf{P}}$, and then to use this for quantitative purposes. This is done by comparing the PCA loadings to spectra $\mathbf{K}$.

Fitting the targets to the PCA model

First we can use the $\mathbf{k}_j$ spectrum of various individual candidates as targets to be fitted to $\widehat{\mathbf{P}} = [\widehat{\mathbf{p}}_a, a = 1, 2, \ldots, A]$:

$$\mathbf{k}_j' = \mathbf{c}_j'\widehat{\mathbf{P}}' + \mathbf{e}_j' \tag{3.141}$$

with the LS solution

$$\widehat{\mathbf{c}}_j' = \mathbf{k}_j'\widehat{\mathbf{P}}(\widehat{\mathbf{P}}'\widehat{\mathbf{P}})^{-1} = \mathbf{k}_j'\widehat{\mathbf{P}} \tag{3.142}$$

and

$$\widehat{\mathbf{e}}_j' = \mathbf{k}_j' - \widehat{\mathbf{c}}_j'\widehat{\mathbf{P}}' \tag{3.143}$$

Candidates $\mathbf{k}_j$ that display small residuals $\widehat{\mathbf{e}}_j$ and are otherwise sensibly chosen can then be regarded as members of the mixture model $\mathbf{X} = \mathbf{YK}' + \mathbf{E}$.

For J constituents this can be written

$$\widehat{\mathbf{C}} = [\widehat{c}_{ja}, j = 1, 2, \ldots, J, a = 1, 2, \ldots, A] = \mathbf{K}'\widehat{\mathbf{P}}. \tag{3.144}$$

Thus, if $J = A$ and all J constituents have linearly independent spectra $\mathbf{K}$, $\widehat{\mathbf{C}}$ can be inverted and we have

$$\mathbf{X} \approx \hat{\mathbf{T}}\hat{\mathbf{C}}^{-1}\hat{\mathbf{C}}\hat{\mathbf{P}}' \approx \hat{\mathbf{T}}\hat{\mathbf{C}}^{-1}\hat{\mathbf{K}}' \qquad (3.145)$$

and

$$\hat{\mathbf{Y}} = \hat{\mathbf{T}}\hat{\mathbf{C}}^{-1} = \hat{\mathbf{T}}(\mathbf{K}'\hat{\mathbf{P}})^{-1} \qquad (3.146)$$

Fitting the PCA model to a set of targets

Alternatively, once $\mathbf{K}$, the spectra of all the J constituents have been identified by fitting the $\mathbf{k}_j$ to $\hat{\mathbf{P}}$, the $\mathbf{K}-\hat{\mathbf{P}}$ relationship can be reversed into:

$$\hat{\mathbf{P}}' = \mathbf{U}\mathbf{K}' + \mathbf{E} \qquad (3.147)$$

with the solution

$$\hat{\mathbf{U}} = \hat{\mathbf{P}}'\mathbf{K}(\mathbf{K}'\mathbf{K})^{-1} \qquad (3.148)$$

and consequently

$$\hat{\mathbf{P}}' \approx \hat{\mathbf{U}}\mathbf{K}' \qquad (3.149)$$

This $\hat{\mathbf{U}}$ can now be applied to the principal component scores $\hat{\mathbf{T}}$ to obtain the concentrations of the constituents. From the models $\mathbf{X} \approx \hat{\mathbf{T}}\hat{\mathbf{P}}'$ and $\mathbf{X} = \mathbf{Y}\mathbf{K}' + \mathbf{E}$ we get $\mathbf{X} \approx \hat{\mathbf{T}}\hat{\mathbf{U}}\mathbf{K}'$, so that

$$\hat{\mathbf{Y}} = \hat{\mathbf{T}}\hat{\mathbf{U}} \qquad (3.150)$$

Notice that even this approach requires the constituent spectra to be linearly independent so that $\mathbf{K}'\mathbf{K}$ is invertible.

General

This two-stage approach (first identifying $\mathbf{K}$ by regression on $\hat{\mathbf{P}}$, then 'calibration' by rotating $\hat{\mathbf{P}}$ into $\mathbf{K}$) has been applied with good results and a lot of theory and experience has been accumulated. Modified versions for finding the targets like iterative target transformation (Vandeginste et al., 1987) have been developed.

The method can be applied in many different applications; in spectral analysis where spectra are available for several objects as well as in chromatography where the time and wavelength are the two axes in the principal component analysis. We refer to Malinowski and Howery (1980) for an excellent reference on the subject.

It should be noted that quantitative predictions based on these target transformation approaches do require fully correct model specification. Like direct unmixing, no unmodelled interferents are allowed (unidentified constituents, physical interferences, nonlinearities etc.), otherwise erroneous predictions can result.

If these requirements are not fulfilled, it is advisable to use some 'open-ended' calibration method like PLSR (section 3.5) or extended mixture modelling (sections 3.6.2 and 3.6.3) instead. Then the ambitions are reduced from full causal modelling to a more pragmatic empirical interference compensation and mapping.

And it should be noted that when certain interferents' spectra $\mathbf{k}_j$ remain unknown, and the analytes' concentrations are not known in the calibration data, then no ordinary multivariate calibration method can resolve the selectivity problems. More data is then needed, unless the X-data come from a 'second-order' instrument.

3.7.3 RANK ANNIHILATION METHOD (RAM)

In contrast to most methods of multivariate calibration which apply a vector of spectral measurements for each object, the RAM works on data from second-order bilinear instruments (here bilinear means linear in two dimensions). These are instruments that for each object yield a two-dimensional bilinear data matrix $\mathbf{X}_i$ that is a sum of contributions from the constituents $j = 1, 2, \ldots, J$ of the form:

$$\mathbf{X}_i = \sum_{j=1}^{J} \mathbf{u}_j y_{ij} \mathbf{q}_j' + \mathbf{E}_i = \mathbf{U} \operatorname{diag}(y_{ij}) \mathbf{Q}' + \mathbf{E}_i \qquad (3.151)$$

where $\mathbf{u}_j$ are columns with information in one order, e.g. HPLC eluation profile at different retention times and $\mathbf{q}_j$ are rows with information in the second order, e.g. UV absorbance spectra at different wavelengths. The parameter y_{ij} is the concentration of the j'th analyte in object i.

The RAM is applicable to second-order instruments like HPLC-UV, or excitation-emission fluorescence. What is especially appealing about such rank annihilation modelling is that unidentified interferents do not have to be modelled, neither explicitly (as in mixture modelling) nor implicitly (like in bilinear modelling or extended mixture modelling). Provided that they have sufficiently unique patterns in $\mathbf{U}$ and $\mathbf{Q}$, they can be compensated for without modelling, due to the extremely rigid structure of the bilinear model in (3.151). This is done by comparing each $\mathbf{X}_i$ to the corresponding $\mathbf{X}_{\text{ref}}$ of a reference object modelled the same way as in (3.151).

The rank annihilation uses singular value decomposition (SVD, see section 2.1.1) and a sort of target transformation and ends up with estimates for the constituents that are present in both the calibration and prediction set.

The use of bilinear decomposition as in (3.151) for chromatography–spectrometry instruments is similar to developments in psychometrics (Harshman, 1970 and Carroll and Chang, 1970), as described by Martens (1979). The general rank annihilation called generalized RAM or GRAM was first published by Sanchez and Kowalski (1986), but for special applications it was developed already in 1978 by Ho and coworkers (see e.g. Ho et al., 1978).

3.7.4 TRADITIONAL LINEAR REGRESSION METHODS

This section is devoted to alternative methods that are developed to solve the collinearity problem (see e.g. section 2.3) in multiple linear regression (MLR) based on the forward calibration model

$$y = b_0 + \sum_{k=1}^{K} x_k b_k + f \qquad (3.152)$$

As pointed out e.g. by Dempster et al. (1977), the MLR predictor has seriously deficient performance when there is collinearity in $\mathbf{X}$. The way to solve the collinearity problem advocated for most cases in this book is based on regressing $\mathbf{y}$ on a reduced number of regressors from $\mathbf{X}$. For the data compression methods we select a small set of *linear combinations* of X-variables as regressors, $\mathbf{T} = \mathbf{XV}$ and use these in the regression equation $\mathbf{y} = \mathbf{Tq} + \mathbf{f}$. The next section concerns methods of this type that involve selection of a subset of *individual X*-variables instead. For comparison to the other forward calibration methods, each column in the weight matrix $\mathbf{V}$ now simply consists of 1.0 for one X-variable and the rest of the elements are zero (see Frame 3.1).

3.7.4.1 Wavelength selection or stepwise multiple linear regression (SMLR)

The SMLR approach may give better prediction ability than the full-spectrum methods—the data compression or mixture modelling—if there are many redundant X-variables and these have very different curvature in their relationships to $\mathbf{Y}$. The SMLR then allows us to eliminate those X-variables that are most nonlinear in their response, while their non-informative curvatures may contaminate the full-spectrum calibration models.

But if there is collinearity in $\mathbf{X}$ and these X-data in addition are somewhat noisy, then the full-spectrum methods are expected to give more precise predictions. And the resulting calibration models are normally easier to interpret from the full-spectrum methods, because the causally interpretable X-variables may not be the ones that happen to be drawn into the SMLR model that shows the best statistical performance.

There are a number of methods that exist to select *which* and *how many X*-variables to use in the final model: On the one side there are methods called forward selection methods. These start with one wavelength, incorporate further wavelengths one at a time and stop when a certain criterion is met. Another type is the backward elimination methods which start with the full spectrum and delete one by one of the wavelengths until the stopping criterion is reached. Some versions of forward methods also have the possibility to delete at later steps wavelengths already incorporated. Similarly, some versions of the backward elimination methods can incorporate wavelengths that have already been deleted. Finally, we have the best subset selection methods where all possible submodels are tested relative to an optimization criterion and the 'best' model is selected.

For each of the selection strategies a number of different selection criteria can be applied. Here we consider only three of them and refer to Weisberg (1985) and Hocking (1976) for more fundamental treatments.

However, it should be kept in mind that assumptions about independence and identical, well-behaved error distributions are made in the more formal statistical approaches (see section 2.1.2.3). If these assumptions are violated, the formal criteria may lead to over-fitting, i.e. bringing too many X-variables into the calibration model. It may then be sensible to use the more empirical validation methods like cross-validation instead (see Chapter 4).

Mallows C_p statistic

The C_p statistic is defined as

$$C_p = \frac{\text{RSS}_p}{s_f^{\,2}} + 2p - I \qquad (3.153)$$

where RSS_p is the residual sums of squares $\widehat{\mathbf{f}}\widehat{\mathbf{f}}$ for the actual model used (I objects with $p-1$ regressors plus the intercept, see section 2.1.2) and $s_f^{\,2}$ is the usual estimate of $\text{Var}(f) = \sigma^2$ estimated from $\widehat{\mathbf{f}}$ for the full regression model ($y = K$). The C_p can be considered as an estimate of the average prediction ability

$$1/\sigma^2 \sum_{i=1}^{I} \text{MSE}(\widehat{\mathbf{y}}_i) \qquad (3.154)$$

which is the rationale for using it. It is composed of a contribution measuring the fit of the p-dimensional submodel ($\text{RSS}_p/s_f^{\,2}$) and a penalty for many regressors ($2p$). Mallows (1973) suggested that good models have negative or small $C_p - p$ (see also Weisberg, 1985).

Coefficient of determination (squared multiple correlation coefficient)

The coefficient of determination, $R^2{}_p$ can be written as

$$R^2{}_p = 1 - \frac{\text{RSS}_p}{s^2{}_y * (I - 1)} \qquad (3.155)$$

This quantity varies between 0 and 1 and is interpreted as the square of the correlation between y and $\widehat{y}$. A value equal to 1 corresponds to perfect fit of the regression equation and a value equal to 0 reflects no linear relation at all. The value of R_p^2 increases with p and is therefore best suited for comparing models of the same size (i.e. with the same value of p). There are, however, adjusted versions that correct for this phenomenon (see e.g. Weisberg, 1985). If the error variance of the reference data $\mathbf{y}$ is known in advance, the R_p^2 may be corrected for that, as in the relative ability of prediction (RAP, see Chapter 4; Martens and Martens, 1986b; Hildrum et al., 1983).

Both C_p and $R^2{}_p$ are frequently used and are very useful in best subset selection, but can also be applied in forward selection and backward elimination.

The F-test and t-test

A subset model can be compared to a larger one by testing whether the

230

coefficients $\mathbf{b}_2$ of the added variables are 0. We let $\mathbf{b}_1$ be the regression coefficients of the subset model and let $\mathbf{b}' = (\mathbf{b}_1', \mathbf{b}_2')$ be the coefficient vector of the full model. If we compare a p-dimensional (including intercept) model with a larger one (K-dimensional, including intercept) the natural test statistic to use can be written as

$$F_p = \frac{(\text{RSS}_p - \text{RSS}_K)/(K - p)}{\text{RSS}_K/(I - K)} \tag{3.156}$$

and can be compared to a Fisher F-table with $K - p$ and $I - K$ degrees of freedom (Weisberg, 1985) to check significance.

If an F-test is used in forward selection the F-value for each variable which is not incorporated at an earlier step is computed and the variable with the most significant coefficient is incorporated. The whole process terminates when no added X-variable is significant. In backward elimination the variable with the smallest F-value is eliminated, provided it is not significant and this process terminates until all X-variables are significant.

It should be noted that for the F-test to be exactly valid, the usual regression assumptions described in section 2.1.2.4 plus normality of error terms f are required.

For only one element in $\mathbf{b}_2$, i.e. only one added variable, the reported F-test is equivalent to the ordinary t-test for testing significance of addition of one regressor variable. In fact, the absolute value of the t-variable is the square root of F.

Applications, comparisons and further discussion of wavelength selection methods in spectroscopy can be found in e.g. Isaksson and Næs (1987) and in Hruschka (1987).

3.7.4.2 Ridge regression (RR)

The RR method has great similarities with PCR, but can not be described by data compression. While PCR deletes the influence of the smaller eigenvectors, RR only decreases the influence of them. For centered $\mathbf{X}$-matrix the $\hat{\mathbf{b}}$-vector is equal to

$$\hat{\mathbf{b}}_{RR} = (\mathbf{X}'\mathbf{X} + k\mathbf{I})^{-1}\mathbf{X}'\mathbf{y} \tag{3.157}$$

or alternatively

$$\hat{\mathbf{b}}_{RR} = \hat{\mathbf{P}}(\text{diag}(\hat{\tau}_a + k))^{-1}\hat{\mathbf{P}}'\mathbf{X}'\mathbf{y} \tag{3.158}$$

where $\hat{\mathbf{P}}$ is the matrix of eigenvectors of $\mathbf{X}'\mathbf{X}$, the $\hat{\tau}_a$'s are the corresponding eigenvalues and k is the 'ridge parameter'. Comparing this equation with equation (3.50) the relation between RR and PCR becomes clear.

The ridge regression method is widely used and a lot of theory exists (see e.g. Hoerl and Kennard, 1970; Gunst and Mason, 1977, 1979). It has shown to be superior to PCR for sensible choice of k in some applications and inferior in others. Quite a lot of the literature on ridge regression concerns the difficult problem of estimating the ridge parameter k. For applications of the method in spectroscopy, see Fearn (1983) and Næs et al. (1986).

3.7.5 NONLINEAR CALIBRATION METHODS

In the preceding section we looked at models where y can be written as a linear function of $\mathbf{x}$, i.e.

$$y = b_0 + \sum_{k=1}^{K} x_k b_k + f \qquad (3.159)$$

Such models that are linear in the parameters are very nice to work with; they are computationally attractive, reasonably well understood theoretically, and they are quite flexible. For instance, the linear model can accomodate many types of curved $X-Y$ relationships, by letting the different X-variables compensate for each other's nonlinear relationship to $\mathbf{Y}$ (see Chapter 7).

But in some applications we may know that the $X-Y$ relationship definitely is nonlinear and difficult to describe by a linear model. In Chapter 7 preprocessing transformations of x (and/or y), which can make the linearity better are treated.

However, in some cases a nonlinear calibration regression may be preferable.

3.7.5.1 Nonlinear regression

The most general calibration model can be written

$$g(\mathbf{y}) = h(\mathbf{x}) + \text{residuals} \qquad (3.160)$$

with the solution

$$\hat{\mathbf{y}} = \hat{g}^{-1}(\widehat{h(\mathbf{x})}) \qquad (3.161)$$

The most common technique to use in estimating the parameters in nonlinear $g(\mathbf{y})$ and $h(\mathbf{x})$ are based on LS, as in the linear case. This means that the sum of squared residuals is minimized with respect to the parameters involved. If for instance the regression equation is

$$y = \alpha x^{\beta} + f \qquad (3.162)$$

one simply minimizes the sum of squares

$$\sum_{i=1}^{I} (y_i - \alpha x_i^{\beta})^2 \qquad (3.163)$$

with respect to both α and β.

Similarly, let us consider the case of a mixed additive and multiplicative effect. In for instance chromatography the total amount of sample applied to the column may vary uncontrollably from object to object. We can term this amount d_i and it will be a multiplicative factor. Likewise, in transmission spectroscopy a varying cuvette length d_i or in diffuse reflectance a varying effective optical path length d_i will be multiplicative factors for the absorbances (see e.g. Birth, 1978, 1982).

In all these examples let us assume that the X-data are best described by a mixture model

$$\mathbf{x}_i{}' = (y_{i1}\mathbf{k}_1{}' + y_{i2}\mathbf{k}_2{}' + \ldots + y_{iJ}\mathbf{k}_J{}')d_i + \mathbf{e}_i{}' \tag{3.164}$$

For small $\mathbf{e}$ this causal model can be inverted into the mixed additive and multiplicative model

$$y_{ij} = (x_{i1}b_{1j} + \ldots + x_{iK}b_{Kj}) * c_i + f_{ij} \tag{3.165}$$

However, the unknown variable $c_i = d_i{}^{-1}$ makes the ordinary linear model

$$y_{ij} = (x_{i1}b_{1j} + \ldots + x_{iK}b_{Kj}) + f_{ij} \tag{3.166}$$

unsuitable in this case and we may have to solve for both d_i and y_{ij} simultaneously using e.g. LS as in (3.163). This is numerically more complicated than in the linear case and it is generally impossible to find explicit formulae for the predictors.

The LS minimization can generally be performed by using e.g the Newton Raphson algorithm (see e.g. Henrici, 1964) from numerical analysis (or single optimization, Denning and Morgan, 1987).

Alternatively, an iterative minimization of $\hat{\mathbf{e}}_i{}'\hat{\mathbf{e}}_i$ can be applied, in analogy to the optimal scaling technique in psychometrics (see e.g. Young, 1981): Concentrations y_{ij} are then estimated from $\mathbf{x}_i$, using earlier estimates of d_i and multiplier d_i is estimated from $\mathbf{x}_i$ using earlier estimates of y_{ij}, etc.

A special type of nonlinear regression especially designed to correct for the multiplicative scatter effect in NIR analysis is developed by Norris (see e.g. Norris, 1983) and will be discussed in Chapter 7 which also discusses various data pretreatments.

3.7.5.2 Nonparametric regression

Nonparametric regression techniques are developed for cases where very little is known a priori about the regression equation

$$\mathbf{y} = h(\mathbf{x}) + \mathbf{f} \tag{3.167}$$

The idea is then to let the data to a larger extent reveal its functional form.

Since this makes less assumptions than other methods, flexibility is gained, but more observations are needed to obtain a reasonable precision and avoid over-fitting.

There exist several methods of this type, for instance projection pursuit regression (Friedman and Stuetzle, 1981), optimal transformations regression (Breiman and Friedman, 1985) and spline smoothing (Wold, 1974). Here we only give a brief discussion of projection pursuit (PP) regression.

The basis for this method is the wish to make few assumptions and simultaneously be able to look at the data in sensible 'data windows'. The method is then only based on the basic assumption of smooth relationship between y and $\mathbf{x}$. The regression equation is then approximated by a sum of empirically determined univariate smooth functions S_a (e.g., moving averages) of linear combinations of the predictor variables $\mathbf{x}$, $a = 1, 2, \ldots, A$:

$$h(\mathbf{x}) = \sum_{a=1}^{A} S_a(\boldsymbol{\alpha}_a{}'\mathbf{x}) \qquad (3.168)$$

where $\boldsymbol{\alpha}_a{}'\mathbf{x}$ denotes the usual scalar product (see section 2.1). The smooth functions S_a and the $\boldsymbol{\alpha}_a$ parameters are determined in an iterative manner. The $\boldsymbol{\alpha}_a$ vectors are usually assumed to have length 1 so that $\boldsymbol{\alpha}_a{}'\mathbf{x}$ can be interpreted as the projection of $\mathbf{x}$ along the $\boldsymbol{\alpha}_a$ direction.

One starts with $a = 1$ on centered variables. The first set of parameters, S_1 and $\boldsymbol{\alpha}_1$, are then determined as to minimize

$$\sum_{i=1}^{I} (y_i - S_1(\boldsymbol{\alpha}_1{}'\mathbf{x}_i))^2 \qquad (3.169)$$

with respect to $\boldsymbol{\alpha}_1$ and S_1. Afterwards, y_i is replaced by $y_i - \hat{S}_1(\hat{\boldsymbol{\alpha}}_1{}'x_i)$ and $S_1(\boldsymbol{\alpha}_1{}'\mathbf{x}_i)$ by $S_2(\boldsymbol{\alpha}_2{}'\mathbf{x}_i)$. Then S_2 and $\boldsymbol{\alpha}_2$ are determined the same way as S_1 and $\boldsymbol{\alpha}_1$. This results in an estimator $\hat{h}(\mathbf{x})$ of the regression function which is a sum of smooth functions of projections of the data. Since the method is based on projections, computer graphics are a good help in interpretation and can also be used in determining the degree of smoothing.

The projection pursuit regression mthod was applied to spectroscopic data in Isaksson and Næs (1988). The data applied were from a NIR analysis of a designed experiment consisting of water, fish protein and starch. The NIR spectra were taken on a Technicon InfraAlyzer 400 with 19 pre-specified wavelengths. A set of 37 samples were generated and submitted to multiplicative scatter correction (MSC, see Chapter 7) and then to principal component analysis.

The scores of the first two factors versus protein percent y are plotted in Figure 3.38. As we can see the relation is not too far from linear, which was to be expected since apart from minor constituents in the fish protein, the data (after MSC) have only two major independent factors. There is, however, an indication of nonlinearity in some parts of the region, especially

234

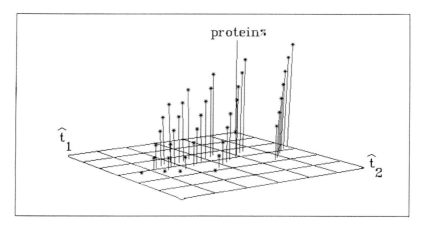

Figure 3.38 Plot of protein percentage vs. the two first principal components $\widehat{t}_1$ and $\widehat{t}_2$ for 37 calibration samples based on mixtures of fish protein, water and starch

for large values of protein percentage. Linear calibration tested on an equally generated sample set, gave an average prediction error (RMSEP, see section 4.3.1) equal to 2.40 percent protein and a prediction equation equal to

$$\widehat{y} = 34.94 - 91.19\widehat{t}_1 - 253.46\widehat{t}_2. \tag{3.170}$$

These three variables $(y, \widehat{t}_1, \widehat{t}_2)$ were then submitted to a PP regression and the results along the two first PP axes are given in Figure 3.39. As we see the first axis gave an almost perfect linear equation with coefficients equal to 0.34 and 0.94. These were (except form opposite sign) almost identical to the coefficients in (3.170) after scaling to length 1. The second axis, however, showed a clear nonlinear relation to y, i.e. S_2 was clearly nonlinear. The projection direction was equal to (0.04, 0.99) which shows that this axis is approximately parallel to the $\widehat{t}_2$ axis. The figure shows that S_2 has a shape quite similar to a second order polynomial and the two first PP axes then indicate that an equation of the type

$$\widehat{y} = \widehat{b}_0 + \widehat{t}_1\widehat{b}_1 + \widehat{t}_2\widehat{b}_2 + \widehat{t}_2^2\widehat{b}_3 \tag{3.171}$$

could be appropriate. Such a polynomial regression of y on $\widehat{t}_1, \widehat{t}_2$ and $\widehat{t}_2^2$ was then performed on the data and tested on the same set as the linear predictor. The prediction error (RMSEP) was in this case equal to 1.34% which was an improvement by 45% compared to the results from the linear predictor.

This result was even comparable with a full PCR predictor based on 9 factors which gave a prediction error (RMSEP) equal to 1.27%. Thus, the PP method showed us which type of equation could give better fit than the linear, so that improved predictions could be obtained. Due to the enhanced graphical possibilities in only two dimensions, it was argued in Isaksson and

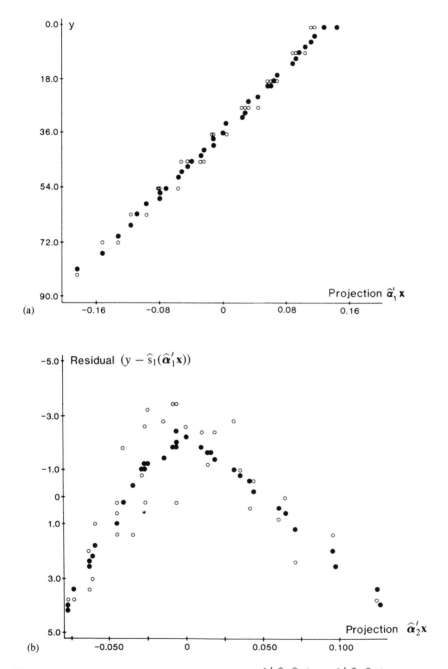

Figure 3.39 Plot of y versus the optimal directions $\widehat{\boldsymbol{\alpha}}'_1(\widehat{t}_1,\widehat{t}_2)'$ and $\widehat{\boldsymbol{\alpha}}'_2(\widehat{t}_1,\widehat{t}_2)'$ in projection pursuit regression. The smoothed values $\widehat{S}_1$ and $\widehat{S}_2$ are indicated by the solid circles, while the data are represented by the open circles

Næs (1988) that a predictor as in (3.171), though nonlinear, based on two factors $\hat{t}_1$ and $\hat{t}_2$ is preferable to a PCR with 9 factors.

Finally, it should be mentioned that combining several local linear models can solve strongly nonlinear $X-Y$ calibration problems. Breiman et al. (1984) have developed a hierarchical methodology for this (classification and regression trees (CART)). Thus, a hierarchy of linear full-spectrum calibration models (e.g. PLSR) can be used—the first levels are crude calibration models suitable for a wide range of samples and used primarily for classification purposes while the last levels represent finely adjusted local predictors with very limited range of applicability.

A modification of this approach is to perform a local calibration for each unknown object i: Based on its input spectrum $\mathbf{x}_i$ a subset of the most relevant calibration objects with known analyte concentrations are selected. This can be done by some pattern recognition/projection sequence in analogy to the Hruschka regression (section 3.7.1). A new calibration model is then estimated based on this local calibration set, and applied to predict y_i from input spectrum $\mathbf{x}_i$ for the unknown object.

Jensen et al. (1982) tested this, using a qualitative K nearest neighbours (KNN) interpolation technique in predicting botanical components from autofluorescence spectra of ground wheat, due to strongly nonlinear instrument response. E. Sanchez (pers. com. 1985) has instead suggested using a separate PLSR instead of the KNN interpolation on the local 'calibration set' for the unknown object, in order to attain improved local quantitative prediction ability.

———————————— End of statistical extensions ————————————

4 Assessment, Validation and Choice of Calibration Method

SUMMARY   The need for validation of calibration equations is demonstrated by an example from NIR analysis of protein in wheat. Different types of prediction error are discussed and alternative ways to estimate them are shown. Some of them are based on new prediction objects (external validation) and some are based on the calibration data themselves (internal validation). How to use these measures to determine the optimal complexity of a calibration model is demonstrated. Finally, how to check specific model assumptions is discussed.

4.1   GOOD AND BAD CALIBRATIONS

4.1.1 QUALITY FOR WHAT PURPOSE?

At least four groups of scientists have to consider quality in multivariate calibration:

The end user who applies the predictor formula for predicting **Y** from **X**, needs several types of statistical information. In particular, the end user is interested in the worst-case and average precision to be expected. Since multivariate calibration usually represents a local approximation, the end user also has to know the range of sensible applications for a given calibration equation. In addition, since the accuracy of individual prediction results may be somewhat different in the center and at the periphery of the normal range, the end user would also like to know the uncertainty of each individual future prediction.

The person who calibrates the instrument has to be interested in the statistical properties of different calibration methods, in order to select calibration data and a calibration method for a given purpose. Multivariate calibration is an interdisciplinary exercise, and it may be important to use concrete, domain-specific extra knowledge concerning the object types, the instruments and actual data involved. Therefore it may be important that the practitioner can do the calibration

238

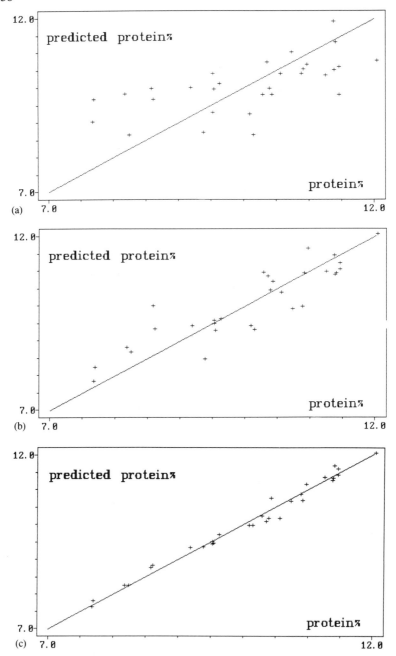

Figure 4.1 The need for assessment and validation. Underfitting and overfitting illustrated in an exaggerated form using PLSR for simulated whole-wheat NIR and protein data with $\sigma_{X,cal} = \sigma_{X,test} = 0.005$ O.D. units and $\sigma_{y,cal} = 0.5\%$ protein; $\sigma_{y,test} = 0$. Abscissa: 'True' protein content y_i. Ordinate: Protein content $\hat{y}_i$ estimated for calibration objects $i = 1,2,\ldots,30$ or test objects $i = 1,2,\ldots,109$: a) Calibration set, 2 factor PLS model. b) Calibration set, 4 factor PLS model. c) Calibration set, 15 factor PLS model

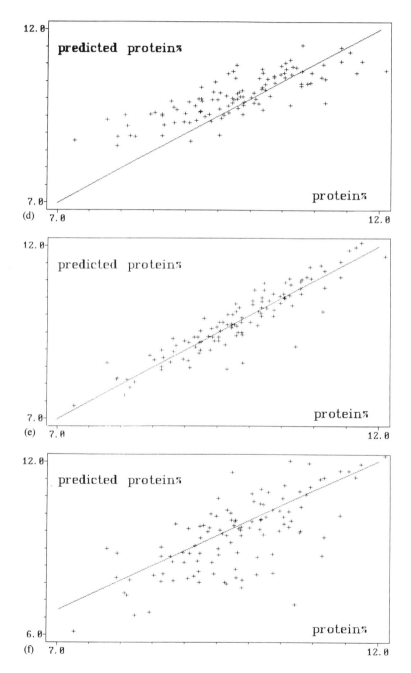

Figure 4.1 (*cont.*) d) Test set, 2 factor PLS model. e) Test set, 4 factor PLS model.
f) Test set, 15 factor PLS model (*for full caption see p.238*)

240

computations personally. On the other hand multivariate calibration is an advanced data analytic problem, and some statistical insight is required. As we shall see shortly, grave statistical mistakes can easily be made.

The instrument manufacturer may want to optimize the instrument's performance according to the criteria expected to be important for the customers and/or to satisfy the referees. The conventional univariate instrument specifications like detector repeatability and linearity at the individual wavelength channels etc. may not be very informative in this respect. The reason is that the multivariate calibration can be made to correct for systematic instrument variations (temperature-effects, wavelength shifts etc.). The manufacturer should therefore instead be able to describe the predictive ability of the instrument for given purposes.

The referee comparing different instruments needs well accepted methods for comparing their predictive abilities for well defined types (populations) of objects, as well as other important virtues.

4.1.2 VALIDATION IS IMPORTANT!

A simple example simulating determination of protein in whole wheat from NIR transmission spectra and showing the importance of validation is given in Figure 4.1. The data are described in Frame 4.1 and in the figure caption. The calibration method used is PLSR.

Frame 4.1

The artificial data were created from simplified 'perfect' data with exaggerated noise added:

1. *Background data*: By a Tecator Infratec transmission NIR instrument the optical density (O.D.=log(1/T)) was measured for 139 batches of whole wheat kernels at a subset of 100 wavelengths in the lower NIR wavelength range (see e.g. Figure 3.11). From these data 25 wavelength channels were selected to give the X-variables. Protein percentage (y) was measured by traditional Kjeldahl N analysis in every batch.

2. *Creation of error-free data*: The main variabilities in the 25 X-variables and in y were described in a centered PCR model. 'Perfect' (error-free) new X- and y-variables were then reconstructed from the five first PCR factors, using scores $\hat{\mathbf{T}} = (\hat{\mathbf{t}}_a)$ and loadings $\hat{\mathbf{P}} = (\hat{\mathbf{p}}_a)$ and $\hat{\mathbf{q}} = (\hat{\mathbf{q}}_a)$, $a = 1, 2, \ldots, 5$:

(*continued on next page*)

Frame 4.1 *continued*

$$\mathbf{X}_{\text{perfect}} = \mathbf{1}\bar{\mathbf{x}}' + \hat{\mathbf{T}}\hat{\mathbf{P}}'$$

$$\mathbf{y}_{\text{perfect}} = \mathbf{1}\bar{y} + \hat{\mathbf{T}}\hat{\mathbf{q}}$$

3. *Splitting into calibration and test set*: The 'perfect' data for the 139 objects were split into one set of 30 representative calibration objects and one set of 109 test objects.
4. *Simulated noise was added to the 'perfect' data*: Random normally distributed noise added at various levels to the data:

$$\mathbf{X}_{\text{cal}} = \mathbf{X}_{\text{perfect,cal}} + N(0, \sigma_{X,\text{cal}})$$

$$\mathbf{X}_{\text{test}} = \mathbf{X}_{\text{perfect,test}} + N(0, \sigma_{X,\text{test}})$$

$$\mathbf{y}_{\text{cal}} = \mathbf{y}_{\text{perfect,cal}} + N(0, \sigma_{y,\text{cal}})$$

$$\mathbf{y}_{\text{test}} = \mathbf{y}_{\text{perfect,test}} + N(0, \sigma_{y,\text{test}})$$

Underfitting: Using only 2 PLS factors in the calibration both the calibration (Figure 4.1a) and test results (Figure 4.1d) show unsatisfactory relation between predicted and actual protein concentrations. The reason is that interferences in the spectral data **X** remain unmodelled.

Optimal solution: Using 4 PLS factors (which the internal statistical validation in the calibration set (see section 4.3.2) showed to be optimal) gave reasonable predictions of *y* both in the calibration set (Figure 4.1b) and the test set (Figure 4.1e).

Overfitting: Using as much as 15 PLS factors yielded a very good fit in the calibration set (Figure 4.1c), but terrible predictions (Figure 4.1f).

4.1.3 DIFFERENT SOURCES OF PREDICTION ERROR

There are many different phenomena that can affect the quality of a calibration equation. In this section we discuss the most important of them and present an illustration of various error types. The data for this illustration are based on the same procedure as described in Frame 4.1. PLSR is used as calibration method and the optimal number of PLS factors is estimated by internal validation in the calibration set (cross validation, section 4.3.2.2).

Figure 4.2 shows the predicted vs. 'true' protein percentages for the same 109 test objects after various error sources have been added.

As required by the way the perfect (error-free) data were generated, Figure 4.2a shows that perfect prediction ability (other than round-off errors) was attained when

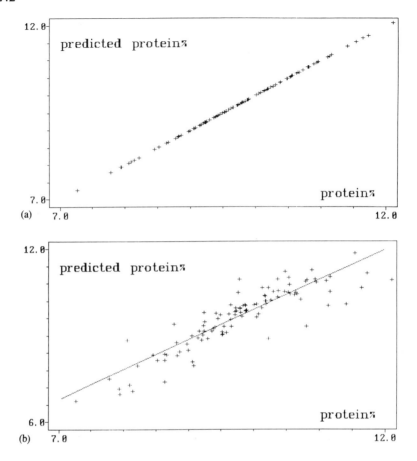

Figure 4.2 Effect of various error sources in calibration and prediction, illustrated for prediction of protein percentage in whole wheat grain by $\mathbf{X}$ = modified NIR transmission O.D. at 25 wavelengths. The $\mathbf{X}$ and $\mathbf{Y}$ data of the calibration and prediction objects have been changed in various ways. Abscissa: 'True' protein content, y_{test}, in 109 test objects. Ordinate: Predicted protein content $\hat{y}_{test}$ in the 109 test objects. The calibration formulae $\hat{y}_{test} = f(\mathbf{X}_{test})$ were estimated by PLSR based on data from 30 calibration objects. The solid diagonal indicates the ideal results ($\hat{y}_{test} = y_{test}$). a) Perfect data. Linear, noise-free calibration data and prediction data $(y, \mathbf{X})_{cal}, (y, \mathbf{X})_{test}$ were used. Root-mean-square error of prediction RMSEP = 0 (see section 4.3.1). b) Model error: non-linear instrument response. Same as in a), but with the noise-free calibration and prediction spectra $\mathbf{X}_{cal}$ and $\mathbf{X}_{test}$ expressed as transmission T instead of O.D.$=\log(1/T)$. RMSEP = 0.52%

the perfect calibration data $(\mathbf{X}_{cal}, y_{cal})$ were modelled by 5 PLS factors, the 5-factor model was applied to the perfect test data $\mathbf{X}_{test}$ and the results $\hat{y}_{test}$ compared to the perfect reference data y_{test}.

1) *Model errors*. The first problem illustrated concerns the fit of the calibration model. This is important because if the model used represents a bad fit to the

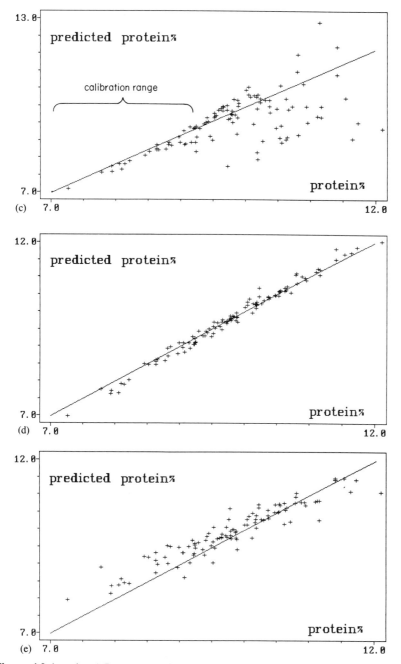

Figure 4.2 (*cont.*)   c) Representativity error. Same as b), but using only the 10 calibration objects that had protein concentrations in the range 7–9.5%. RMSEP = 0.88%. d) Noise in y_{cal} only. Same as a), but with random noise with $\sigma = 0.5\%$ added to **y** in calibration set. RMSEP = 0.09%. e) Noise in X_{cal} only. Same as a), but with random noise with $\sigma = 0.005$ O.D. units added to **X** in calibration set. RMSEP = 0.35% (*for full caption see p.242*)

244

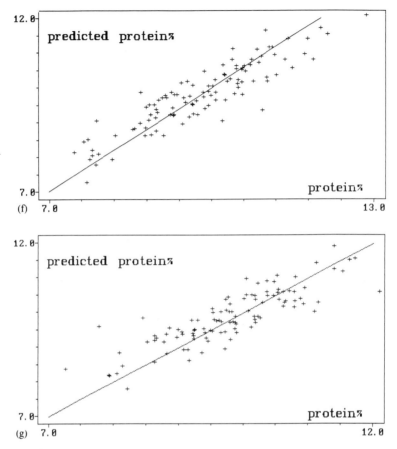

Figure 4.2 (*cont.*) f) Noise in y_{test}. Same as a), but with random noise with $\sigma = 0.5\%$ added to **y** in test set. Apparent RMSEP = 0.5%; of course the 'true' RMSEP is lower. g) Noise in X_{cal} and X_{test}. Same as a), but with random noise with $\sigma = 0.005$ O.D. units added to **X** in both calibration and test set. RMSEP = 0.49% (*for full caption see p.242*)

data, the calibration equation will never have the possibility of giving precise results.

While the error-free NIR data **X** expressed in O.D. units $\log(1/T)$ gave perfect predictions (Figure 4.2a), rather unsatisfactory predictions were obtained with the same error-free NIR data in 'transmittance' (T) (Figure 4.2b). Notice also that while the calibration data X_{cal} in O.D. units only spanned 5 factors due to the way the data were generated, 6 PLS factors were required in the non-linear case.

Since calibration modelling is usually a question of local approximation to some unknown, more or less nonlinear X–Y function, all calibration models must be expected to contain some degree of model errors.

2) *Lack of representativity.* The second error source arises when the calibration set does not cover the total range of variability in the population of future objects. In such cases one is in danger of bad prediction ability for objects outside the calibration range.

This is illustrated in Figure 4.2c where we only use objects in the narrow range 7 to 9.5% protein in the calibration set. The optimal 5-factor calibration model gave good predictions in that range. But as the figure shows, we cannot expect to get good predictions by extrapolation into the range 10–12% protein, unless the instrument response is very linear and the calibration data are very precise.

3) *Random noise in calibration and prediction data.*

Noise in y_{cal}. Figure 4.2d shows how noise in the calibration data y_{cal} can contaminate the estimation of the calibration parameters, resulting in lower prediction ability. But one may note that the precision of the predictions can still be much better than the precision of the data used for calibration: In Figure 4.2d the noise level added in y_{cal} was 0.5% on the average. Still, when compared to the perfect test data y_{test} the prediction error was found to be only 0.09% on average! This illustrates how multivariate calibration can cancel out much of the random noise in the calibration data when the number of calibration objects is high relative to the number of independent calibration parameters to be estimated.

Noise in X_{cal}. Figure 4.2e shows the corresponding effect of noise in the X-variables during calibration. In this case a random noise level of 0.005 O.D. units in X_{cal} made it impossible to estimate more than 4 PLS factors with sufficient precision. The average prediction error in y, y test, now became 0.35%. Notice the typical 'least-squares effect' (see e.g. section 3.6.3) due to serious noise in the calibration data: The high-protein objects are under-estimated and the low-protein objects are over-estimated.

Noise in y_{test}. Figure 4.2f illustrates the trivial, but important effect of error in the 'true' test values. The calibration data $(X,y)_{cal}$ were perfect, and so were X_{test}, so this figure should have been identical to Figure 4.2a if it had not been for the noise in y_{test}. The possibility of noise in the 'true' reference data should allways be taken into considerations when testing the predictive ability of a calibration model (see RAP, section 4.2.3).

Noise in X_{test} (*and* X_{cal}). Figure 4.2g shows predictions obtained when both X_{cal} and X_{test} contain the same level of noise ($\sigma = 0.005$ O.D. units), symbolizing the X-variables that cannot give perfect prediction of y. The noise in X_{test} considerably decreased the prediction ability (RMSEP=0.49%) compared to the prediction results from the same 4-factor PLSR solution on error-free test data X_{test} (RMSEP=0.35%, Fig. 4.2e).

So, while noise in the calibration data X_{cal} and y_{cal} can be countered by increasing the number of calibration objects, the noise in the individual future X-data, X test, constitutes' an *intrinsic* limitation in the analytical method.

The point 1) above is treated in more detail in section 4.5 and Chapter 7 and point 2) is discussed in Chapter 6.

Further discussion can be found in Fearn (1986) and in Williams (1987b).

4.2 PREDICTION ABILITY

4.2.1 PREDICTION ERROR WITH RESPECT TO WHAT?

We certainly do want good performance of our instruments—but good in what sense?

The prediction error of a future unknown object

When we predict by $\hat{y} = f(x)$ in a future unknown object, we assume that a corresponding 'true' y does exist conceptually, even though we have not measured it and may not ever be able to measure it with 100% precision. We want the difference between $\hat{y}$ and y to be small, and in order to accomodate both negative and positive differences equally well, we think of having $(y-\hat{y})^2$ small for that particular **y**. As we learnt in section 3.6.3, this difference can vary from object to object.

The average prediction error for the whole population of objects

We are not only interested in assessing the prediction for one single future object—there is a whole population of future objects to be predicted! What we want to minimize in many cases is some type of average prediction error for the whole population of objects for which the calibration applies. This can be described by the statistical concept of expected (e.g) squared difference $(y-\hat{y})$ symbolized by $E(y - \hat{y})^2$, which is frequently denoted mean squared error (MSE, see Section 2.1.2.2 and Section 3.6.3, equation (3.104)):

$$\text{MSE} = E(\mathbf{y} - \hat{\mathbf{y}})^2 \qquad (4.1)$$

Since MSE as defined here is a weighted average of squared differences with respect to the distribution, the most prevalent types of objects is given the most weight and the more seldom types of objects are given less weight.

Let us assume that we have made up our mind about what population of unknown future objects our calibration model is intended for, e.g. all hard red spring wheats grown in Canada, sampled and analyzed according to a certain procedure. The *true* expected prediction error is what we would like to assess. But this could only be known after we had determined y and $\hat{y} = f(x)$ on every one of these future objects. That of course makes calibration meaningless—the equivalent would be for a match factory to light every match for quality control. So we have to *estimate* MSE based on data. This can be done in two ways.

In some simple cases we can make estimates of MSE based on theoretical formulae from linear regression theory: By e.g. mathematical error-propagation methods we can calculate how a certain deviation in x_{ik} contributes with a certain deviation to the total error in $\hat{y}_i$ for that unknown object, i. But this has to be integrated with distributional information to give estimates of MSE.

However, for the more powerful and flexible calibration methods as presented in Chapter 3, the statistical theory has not yet been fully developed and such formulae are not available except in some very simple cases.

We therefore usually have to base the estimation of MSE on real comparison of data y and $\hat{y}$ from a limited number of objects from either the calibration set (internal validation) or from a separate test or prediction set (external validation). But this also involves statistical distribution considerations: To obtain a good estimate of the average prediction ability, the set of test objects on which it is based must be representative for the whole population of future unknown objects in question, otherwise the estimated MSE can be very misleading.

For other approaches than those presented here we refer to Lorber and Kowalski (1988), Akaike (1974) and UNSCRAMBLER user's manual (1987, 1991).

4.2.2 SQUARED ERRORS?

The prediction error can be defined in many different ways. The most common statistical expression is the expected square prediction error, MSE, defined above.

This squared 'error loss function' is mathematically tractable, both with respect to practical estimation and to theoretical statistical distribution considerations. It does, however, give relatively higher emphasis to large residuals than to small residuals, since $(y-\hat{y})$ is squared.

Other possibilities do exist, for instance based on the absolute error ($|y - \hat{y}|$), rather than squared error. Such measures may be more robust to outliers than the MSE. This book, however, recommends that outliers should be emphasized and handled explicitly, so these alternative measures of prediction errors are not treated here.

For non-statisticians, the squared prediction error may be difficult to interpret physically (the uncertainty is then given e.g. in the unit 'squared Mmol' or 'squared O.D. units'). The square root of the estimated MSE may then be preferable, because this is measured in the same unit as y itself (see section 2.1.2.2). The square root of MSE is denoted RMSE (root mean square error).

4.2.3 RELATIVE PREDICTION ERROR

The above mentioned measures of prediction ability are defined as being functions only of $\hat{f}$, the residual or difference between y and $\hat{y}$:

$$\hat{f} = y - \hat{y}. \tag{4.2}$$

This residual is given in the same unit as y itself and without comparing it to the total variability range of y it may be difficult to get an impression of the performance of the predictor.

In addition to 'absolute' measures like e.g. MSE or its root (RMSE), we therefore want to use relative measures of the type

$$\text{MSE}/\sigma^2_{\text{tot}} \tag{4.3}$$

where σ_{tot} is the total standard deviation of **y** in the full population. In practice this can be estimated e.g. from the calibration or test objects (see section 2.1.2.2):

$$s^2_{\text{tot}} = \sum_{i=1}^{I} (y_i - \bar{y})^2/(I - 1) \tag{4.4}$$

where s^2_{tot} is the sample estimate of σ^2_{tot} and $\bar{y}$ is the arithmetic mean of the calibration or test objects.

Functions of (4.3) are also relevant, e.g.

$$1 - \text{MSE}/\sigma^2_{\text{tot}} = (\sigma^2_{\text{tot}} - \text{MSE})/\sigma^2_{\text{tot}} \tag{4.5}$$

an estimate of which is relative explained variance:

$$(s^2_{\text{tot}} - \widehat{\text{MSE}})/s^2_{\text{tot}} \tag{4.6}$$

The expression (4.5) is close to zero for useless predictors and increases towards 1 for perfect predictors. But in practice, the 'true' test data **y** may contain error themselves, as illustrated in Figure 4.2f. The expression (4.5) becomes more meaningful if we correct it with the uncertainty standard deviation of this 'true' reference reading, σ_{test}:

$$(\sigma^2_{\text{tot}} - \text{MSE})/(\sigma^2_{\text{tot}} - \sigma^2_{\text{test}}) \tag{4.7}$$

If we have a good estimate of σ_{test}, s_{test}, we get the relative ability of prediction (RAP, see e.g Hildrum et al., 1983) from:

$$\text{RAP} = (s^2_{\text{tot}} - \widehat{\text{MSE}})/(s^2_{\text{tot}} - s^2_{\text{test}}) \tag{4.8}$$

Absolute and relative measures are illustrated in Figure 4.3. If we narrow the range of **y** for which we calibrate, this will usually improve the *absolute* prediction ability (MSE), since a better linearity is expected for a more narrow calibration range. But the *relative* prediction ability (RAP) appears to become worse, because it is compared to a more narrow total variability range (see also Helland, 1987).

4.2.4 THE EFFECT OF BIAS

Notice that in (4.1) we only consider the expected squared difference between y and $\hat{y}$, without differentiating between systematic and random prediction errors. Certain types of systematic prediction errors should be given special attention. One of these is the common 'bias', which is the average additive prediction error $E(\hat{y}) - y$, i.e. the difference between the expected value of $\hat{y}$ and y itself.

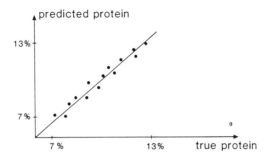

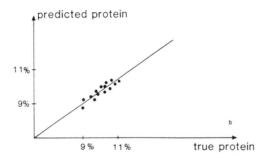

Figure 4.3 Absolute and relative prediction errors illustrated for calibration in two populations with the same error levels in the calibration data: Predicted vs. 'true' protein contents in wheat flour for test objects. Expected error standard deviation s_{test} = 0.2% in the 'true' protein data y. a) Wide range (7–13% protein; s_{tot} = 2%): RMSEP = 0.4%, RAP = 0.97. b) Narrow range (9–11% protein; s_{tot} = 0.7%): RMSEP = 0.3%, RAP = 0.89

With a sufficiently large and representative test set with known levels of y, the average bias can be estimated simply by

$$\widehat{Bias} = \sum_{i=1}^{I} (\hat{y}_i - y_i)/I \tag{4.9}$$

Theoretically, the MSE, residual variance σ_f^2 (see 2.1.2.2) and bias for a particular object are related according to:

$$MSE = \sigma_f^2 + Bias^2 \tag{4.10}$$

as also discussed in section 2.1.2.2.

Explicit correction for estimated bias may sometimes be useful: If the test set is sufficiently large, it may in some cases be sensible to correct the predictor offset b_0 for bias prior to employing it. This is particularly so if a long time has passed or changes in instrumentation (see e.g. Osborne and Fearn, 1986) or object type has occurred since the predictor was developed. The resulting prediction error is then expressed by the Standard Deviation of the residuals $\hat{f}$. (SEP).

In analogy, other types of updating, e.g. the common slope correction and more advanced corrections can sometimes be useful.

However, in our opinion updating should not be employed during validation: If the test set is actually used for assessing how various predictors will behave for future sets of objects (or how a certain instrument performs), the removal of systematic errors is not fair, because actual updating will be impossible for future objects since their true y-values will never be known.

In addition, the bias has to be estimated statistically, and if this estimate is uncertain because it is based on few or bad data, the updating can introduce extra uncertainty into the predictor. And the bias estimation, as reported in e.g. section 3.6.3, depends on the actual value of y and this makes updating dependent on the representatitvity of updating objects. At last, the bias estimate use Y- and X-information which possibly could have been used more efficiently in a full recalibration.

4.2.5 CONFIDENCE INTERVALS

Another way of assessing the quality of predictions rather than by MSE of a predictor is by using confidence intervals for the unknown quantities. For bilinear modelling such intervals have not been considered, but Brown (1982) treated the problem for the linear mixture model treated in section 3.6. He proposed an interval based on the EGLS predictor presented in section 3.6.3.2 and tested it on data. The proposal appeared to have some strange behaviour for certain values of $\mathbf{x}$. This was studied in some detail by Oman (1988) who also proposed another confidence interval for y which seems to have good properties.

This problem is not treated further in this book.

4.3 VALIDATION IN PRACTICE: ESTIMATION OF MSE

As discussed at the end of section 4.2.1 a MSE is in practice best estimated by direct comparison of $\mathbf{y}$ and $\widehat{\mathbf{y}}$ for a representative series of objects, either the calibration objects or a new set of test objects.

Let $\widehat{f}_{ij}$, the analyte lack-of-fit residual, be defined by

$$\widehat{f}_{ij} = y_{ij} - \widehat{y}_{ij} \tag{4.11}$$

and for a set of I objects let

$$\widehat{\mathbf{f}}_j = (y_{1j} - \widehat{y}_{1j}, \dots y_{Ij} - \widehat{y}_{Ij})' = \mathbf{y}_j - \widehat{\mathbf{y}}_j \tag{4.12}$$

where $\widehat{\mathbf{f}}_j$, $\mathbf{y}_j$ and $\widehat{\mathbf{y}}_j$ are I-dimensional column vectors.

4.3.1 EXTERNAL VALIDATION (PREDICTION TESTING)

The $\widehat{MSE}$ obtained by prediction testing (usually denoted mean squared error of prediction, MSEP) is found by simply computing the average squared difference between actual and predicted concentration value, i.e.

$$\text{MSEP} = I^{-1}(\widehat{\mathbf{f}_j'}\widehat{\mathbf{f}_j}) = I^{-1}\sum_{i=1}^{I}\widehat{f}_{ij}^{2} \tag{4.13}$$

for a set of objects not present in the calibration. Its square root is usually called root mean square error of prediction (RMSEP or only RMSP).

The prediction *variance*, on the other hand, is defined as

$$s_{\widehat{f}}^{2} = (I-1)^{-1}\sum_{i=1}^{I}(\widehat{y}_i - y_i - \overline{(\widehat{y} - y)})^2 \tag{4.14}$$

where

$$\overline{(\widehat{y} - y)} = \widehat{BIAS} \tag{4.15}$$

The square root of $s_{\widehat{f}}^{2}$ is sometimes called standard error of prediction (SEP). As can be proved, we still have the simple relation between MSEP, SEP and $\widehat{Bias}$ (in (4.9)) as when considering one object, i.e.

$$\text{MSEP} = \text{SEP}^2 + \widehat{Bias}^{2} \tag{4.16}$$

(To be exact, one has to divide by I instead of $(I-1)$ in (4.14) in order to obtain this. Replacing MSEP in the theoretical relative measure in (4.3) we obtain the relative prediction error

$$\text{MSEP}/s_{\text{tot}}^{2} \tag{4.17}$$

and the derived expressions

$$(s_{\text{tot}}^{2} - \text{MSEP})/s_{\text{tot}}^{2} \tag{4.18}$$

and

$$\text{RAP} = (s_{\text{tot}}^{2} - \text{MSEP})/(s_{\text{tot}}^{2} - s_{\text{test}}^{2}) \tag{4.19}$$

where s_{tot} is the usual measure of a variable's total estimated standard deviation. The quantity s_{test} is an estimate of the error level in the 'true' reference data y_{ij}.

One of the curves in Figure 4.4 illustrates the use of an external validation set of objects, using PLSR as calibration method. The data are taken from the simplified whole-wheat NIR O.D. determination of protein also used in Figure 4.1 (see Frame 4.1). A random, normally distributed noise level of $\sigma = 0.005$ O.D. units was added to the 'perfect' NIR data ($x_k, k = 1, 2, \ldots, 25$) for the 30 calibration and 109 test objects. A random normally distributed noise of $\sigma = 0.5\%$ was added to the 'perfect' protein data **y** for the 30 calibration objects, while the y-data for the 109 test objects were error-free.

The prediction error estimated from the 109 test objects starts at MSEP=0.92 (i.e. RMSEP=0.96% protein), which here represents the deviation of the test set from the calibration set's average, $\overline{y}$ (the 'zero-factor solution'). From this data analysis we also found that the corresponding start bias for the test set (the average absolute deviation from $\overline{y}$) was -0.25% protein.

252

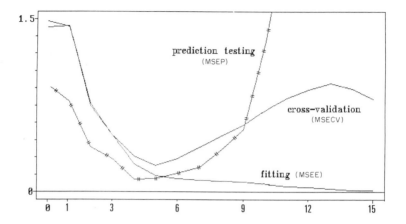

Figure 4.4 Validation of PLSR for predicting y=protein from **X** = O.D. at 25 NIR wavelength channels measured by transmission in whole wheat and simplified as described in the text. Various estimates of mean square error of prediction MSEP are given as functions of the number of PLS factors included in the calibration model. External prediction testing, using 109 new objects (MSEP). Calibration fitting, using the 30 calibration objects (MSEE). Internal cross validation, using the average of 10 separate calibrations and predictions, each time using a new subset of 3 calibration objects as local 'test set', and estimating the calibration model on the remaining 27 objects (MSECV)

As more and more of the structure in the data is modelled, the estimated prediction error decreased, reaching a minimum of MSEP = 0.107 (RMSEP = 0.33% protein) with 4 PLS factors. The corresponding bias with 4 factors was +0.046% protein. Hence the bias effect was negligible in this case.

Remember that the noise-free data were generated so that they give perfect fit after 5 PLS factors. Due to the noise then added to the data, the fifth factor here increased the prediction error, but not much (MSEP=0.116, RMSEP=0.34%). But when more factors than that were used, the prediction error increased sharply again; for instance, after 10 factors we obtained MSEP = 1.30 (RMSEP=1.14%, bias = 0.17% protein), which is higher than the original starting deviation!

Assuming that the test objects were representative for the future unknown objects, this illustrates the importance of finding the correct model complexity: modelling too few X-phenomena gives under-fitting, modelling too many phenomena gives over-fitting; both resulting in bad predictive ability (Figure 4.1).

Such external prediction testing is conceptually the easiest of the empirical validation methods to explain to users. But it is not the validation to be recommended in most routine calibration work, because it is usually rather wasteful and expensive: It requires a *large* and *representative* set of test objects in order to give relevant and reliable estimates of the future prediction ability. Multivariate calibration is often done because the traditional reference method for measuring **y** is too expensive, slow or otherwise undesirable so it is rather wasteful to keep aside a lot of the available data for testing purposes only.

On the other hand there is no sense in using only a few of the objects for testing: The temptation to assess future prediction ability from few, imprecise or unrepresentative test data should be resisted! It would be more economical to use all the available data both for calibration and for testing.

4.3.2 INTERNAL VALIDATION

Internal validation concerns validation from the calibration data themselves and an assessment based on internal validation is not the same as prediction testing. Ideally, the predictive ability can only be assessed by testing on new objects, but our experience is that in most sensible cases, cross-validation and leverage correction give sensible results with high information about the prediction ability. Let us, however, first look at an internal validation method that can be dangerous as also indicated in section 4.1.

4.3.2.1 Calibration fitting

When the predictor is tested on the calibration set itself we easily obtain the so-called mean square error of estimation (MSEE) defined as

$$\text{MSEE} = (\hat{\mathbf{f}}_j{}'\hat{\mathbf{f}}_j)/(I - \text{df}) \tag{4.20}$$

where df here denotes degrees of freedom used in the fitting of the regression equation. For MLR this is simply the number of parameters in the regression model ($K + 1$, see section 2.1.2.3) but for e.g. PLS the formula is more complicated. For simplicity we here use $A + 1$ in the PLS example, but this certainly underestimates df due to the use of $\mathbf{y}$ both in finding the factors $\hat{\mathbf{t}}$ and the loadings $\hat{\mathbf{q}}$. For some more details we refer to Martens and Jensen (1983).

As explained in section 2.1.2.3 the MSEE $= s_f^2$ is an estimator of the error variance of the error term f in the linear regression equation. Generally, it must therefore be considered more as a lower bound for prediction ability than a measure of the prediction ability itself. A real measure of prediction ability will have to bring into account the estimation error of the regression coefficients as well.

One of the curves in Figure 4.4 illustrates how MSEE develops as a function of the number of factors in PLSR assuming one degree of freedom for each factor. It starts at about 1.5 (corresponding to a total standard deviation in the calibration set of 1.2% protein). This is somewhat higher than the initial deviation in the test set, indicating that the 30 calibration objects on the average happen to span more variability than the 109 test objects. Contributing here is also the fact that in the calibration set the $\mathbf{y}$ data have had random noise of about $\sigma^2_{cal} = 0.25$ added, while in the test set $\mathbf{y}$ is error free.

Compared to the actual prediction error, the calibration MSEE was too low and continued to drop when the model was in fact over-fitted. This was also indicated in Figure 4.1f (compared to Figure 4.1c).

For most empirical calibration methods the mean of the fitted $\hat{f}$'s is zero so there is no difference between MSEE and the corresponding variance expression. The square root of MSEE is sometimes termed the standard error of estimation (SEE) or standard error of calibration (SEC) or simply root mean square error of estimation (RMSEE or RMSE).

Relative measures are here found similarly as in (4.19) above. It should be mentioned that in this case the $(s^2_{tot} - MSEE)/s^2_{tot}$ is closely related to the coefficient of determination R^2, which is frequently used for wavelength selection in stepwise regression analysis (see also section 3.7.4).

Such internal validation by calibration fitting is simple to understand and easy to compute, but it uses the same data both for model fitting and testing, and this may lead to serious under-estimation. It is therefore not usually recommended for validation purposes.

4.3.2.2 Cross validation

Cross validation (CV, see e.g. Stone, 1974; Snee, 1976) is a better internal validation method. Like the external validation approach it seeks to validate the calibration model on *independent* test data. But contrary to the external validation it does not waste data for testing only.

In full cross validation one repeats the calibration I times, each time treating one Ith part of the whole calibration set as prediction objects. In the end all the calibration objects have been treated as prediction objects and the estimated MSE (mean square error of cross validation, MSECV) can be computed in the same way as in (4.13).

Since full CV is based on repeated calibrations which may be somewhat time-consuming for the computer, an important alternative is to perform cross validation by only splitting the calibration set into $M(M < I)$ segments and then calibrating only M times, each time testing about a $(\frac{1}{M})$ part of the calibration set. (If a phenomenon in $\mathbf{X}$ is only present in a few objects, care must be taken to avoid that all these special objects fall into one cross validation segment.)

Full cross validation is also illustrated in Figure 4.4, where $M = 10$ cross validation segments were used. For the lowest-dimensional calibration models the cross validation and the internal calibration fit strongly resemble each other. But the cross validation estimate of MSE reaches a minimum and increases again. It shows that judging from internal predictive ability among the calibration objects, five PLS factors appeared to be optimal (MSECV = 0.23; RMSECV = 0.48% protein, corresponding well to the random noise level in $\mathbf{y}$ in the calibration data, $\sigma = 0.5\%$). The 4 and 6 factor solutions gave about the same predictive ability.

Note that the reason why the prediction error obtained in the test set appears lower than the cross validation prediction error is that the Y-data in the test set are error free.

This illustration showed that the external test set validation and the internal cross validation indicated about the same number of PLS factors.

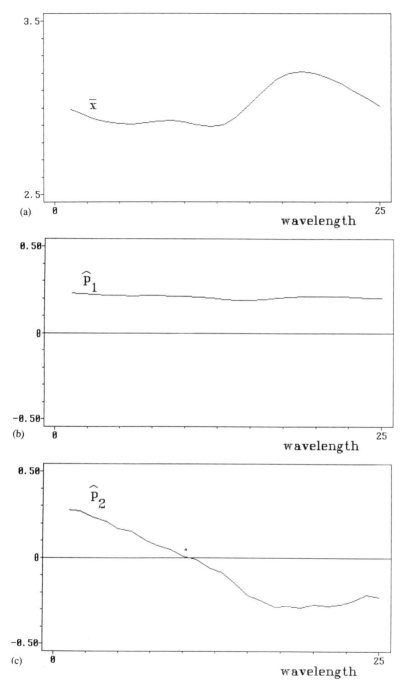

Figure 4.5 The PLSR model behind the validation results in Figure 4.4. 4.5a, b, c: The average NIR spectrum $\bar{\mathbf{x}}'$ and the first 2 loading vectors $\widehat{\mathbf{p}}'_a$, $a = 1,2$ are shown as functions of the wavelength channels $k = 1,2,\ldots,25$

256

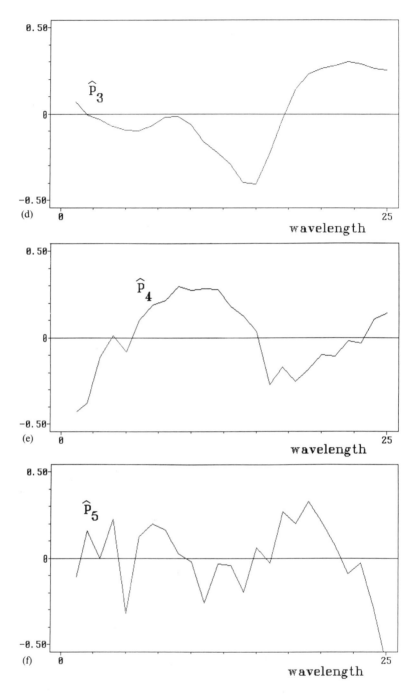

Figure 4.5 (*cont.*) The PLSR model behind the validation results in Figure 4.4. 4.5d, e, f:
The first 3 loading vectors $\widehat{\mathbf{p}}'_a$, $a = 3,4,5$ are shown as functions of the wavelength channels
$$k = 1,2,\dots,25$$

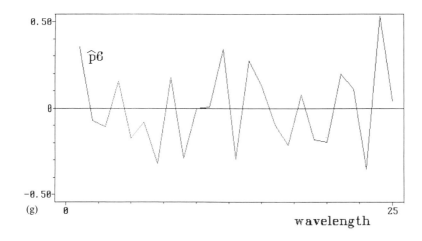

Figure 4.5 (*cont.*) The PLSR model behind the validation results in Figure 4.4. 4.5g: The loading vector $\widehat{\mathbf{p}}_a'$, $a = 6$ is shown as a function of the wavelength channels $k = 1,2,\ldots,25$

There are other ways of doing such cross validation for **y**. For bilinear modelling a modification is to cross validate only each individual factor (a), basing the estimation of this factor on the residuals from having estimated the former factors $(0, 1, \ldots, a - 1)$ using all the available calibration objects (Wold et al., 1983a,b). This can save computation time and give a somewhat better estimate of the optimal number of factors to use, at the expense of a direct, realistic estimate of the prediction error level.

Another type of validation in bilinear calibrations is the cross validation of the modelling of **X**. One way to do this (Wold, 1978) is to select a sequence of individual elements x_{ik} in the calibration data to be regarded as 'missing values' in the bilinear estimation algorithm. After having estimated the bilinear factor scores and loadings, these elements can be reconstructed as

$$\widehat{x}_{ik} = \overline{x}_k + \sum_{a=1}^{A} \widehat{t}_{ia}\widehat{p}_{ak} \tag{4.21}$$

and their deviation from the actual x_{ik} can be squared and summed. Like in ordinary cross validation for **y**, the model estimation and assessment are repeated several times, each time assessing a new subsegment of the data. It is important to select the segments of test elements (i, k) in special ways to avoid overfitting even in this type of cross validation assessment.

It may be noted that a different way of assessing the modelling of **X** is possible when the X-data are expected to yield smooth spectra. Figure 4.5 shows the PLS model resulting in Figure 4.4. It shows that the four first factors' loading spectra are smooth, while the fifth and the subsequent spectra seem to reflect random noise.

This corresponds well to validation results from the prediction testing and the cross validation.

Thus, visible inspection of calibration models for interpretability is a very important tool in model validation.

4.3.2.3 Mallows C_p

Another internal estimate of prediction MSE based on the calibration data is the C_p originally developed by Mallows (1973) and which is also dicussed in section 3.7.4. The C_p was developed for variable selection in stepwise regression and it can be written as

$$C_p = \frac{\text{RSS}_p}{s_f^2} + 2p - I \tag{4.22}$$

where RSS_p is the residual sums of squares after fitting of $p - 1$ variables plus intercept and s_f^2 is the usual estimate of σ_f^2 in the full regression model (see e.g. section 2.1.2.3). C_p is a natural estimate of $(I/\sigma_f^2)\text{MSE}$ if the calibration data set is representative for future objects. To obtain an estimate of MSE directly we then have to divide C_p by I (the number of objects) and multiply by s_f^2. Notice, however, that this operation has no effect on the relative ranking of models.

C_p is dependent on s_f^2 which is an estimate of σ_f^2 in the full model. Consequently, C_p is only applicable when $I > K$. The C_p method is developed for variable selection in stepwise regression, but it could also be applied for full-spectrum methods, with the modification that the number of regressors must be replaced by a function of the number of factors.

An important advantage with C_p over cross validation is the computation time. Several criteria developed in the same spirit exist and we refer to Weisberg (1985) and Akaike (1974) for results and references.

4.3.2.4 Leverage correction

In Cook and Weisberg (1982) is shown that the predicted residuals in full cross-validation, $\hat{f}_{i,\text{CV}}$, are equal to

$$\hat{f}_{i,\text{CV}} = \frac{\hat{f}_i}{1 - h_i} \tag{4.23}$$

where $\hat{f}_i$ is the ordinary fitted residual from the regression and h_i is the so-called leverage of object i. The leverage concept is treated in more detail in Chapter 5 and it is defined by

$$h_i = 1/I + \sum_{a=1}^{K} \frac{\hat{t}_{ia}^2}{\hat{t}_a' \hat{t}_a} \tag{4.24}$$

where $\hat{t}_a' \hat{t}_a = \hat{\tau}_a$ is the ath eigenvalue of $\mathbf{X}'\mathbf{X}$ and the $\hat{t}$'s are the principal components. The leverage can be interpreted as a Mahalanobis distance and measures how unique the independent variable vector x_i is relative to the rest.

Relation (4.23) implies that when ordinary regression is applied cross validation can be replaced by this leverage correction. The only thing we have to do is to find the residuals, divide them by $1 - h_i$, square them, sum up and divide by the number of objects I.

For other calibration methods, this simple relation does not fully hold. However, in connection with the bilinear methods of sections 3.4 and 3.5 the same idea has been used with some success (see e.g. Martens and Næs, 1987; Næs and Martens, 1988). In this case the identity in (4.23) is reduced to a useful approximation. One simply divides the Y-residuals $\hat{f}_i$ by the leverage

$$h_i^A = 1/I + \sum_{a=1}^{A} \frac{\hat{t}_{ia}^2}{\hat{t}_a' \hat{t}_a} \tag{4.25}$$

where $\hat{t}_a$ now refers to the orthogonal score vectors (either principal components or PLS scores) and A is the number of factors used. Notice that if $A = K$ the BLM leverage is equal to the ordinary regression leverage above. The leverage corrected residual mean square error is then simply defined as

$$\text{MSELC} = df^{-1} \sum_{i=1}^{I} \hat{f}_i^2 \bigg/ \left(1 - \sum_{a=1}^{A} \frac{\hat{t}_{ia}^2}{\hat{t}_a' \hat{t}_a}\right)^2 \tag{4.26}$$

where df, degrees of freedom, is, for instance, I. As we see we only need $\hat{f}$'s and the leverages from the full calibration and consequetly the MSELC is simple to compute.

Compared to MSECV, however, the MSELC will frequently be smaller than the MSECV (especially for large A). The reason for this is that while the CV computes a new $\hat{T}$-matrix for each deletion, the leverage correction is based on the $\hat{T}$-matrix for the full calibration set. In this way all the estimation variation involved in CV is not involved in leverage correction. This phenomenon can also result in overestimation of optimal A in PLSR. This is compensated for in e.g. the Unscrambler system (see e.g. Tyssø et al., 1987; UNSCRAMBLER user's manual) where a preliminary heuristically motivated degrees-of-freedom correction is applied $df = I - 1 - A$.

———————— Statistical extensions ————————

4.4 MORE ON DIFFERENT TYPES OF MSE

4.4.1 INTERPRETATION OF MSE'S

In section 3.6 we discussed different types of MSE and considered calibration methods with different properties relative to the different types of MSE's. In section 4.1 we repeated the arguments and in Figure 4.1 we gave an illustration.

Another distinction which should be dicussed in some more detail is the difference between external MSEP on the hand and the internal measures such as MSECV on the other. Since MSEP is computed for a separate test set, it is the prediction ability for these objects for the particular *fitted* calibration equation at hand. The internal measures presented are, however, merely measures of the statistical properties of the calibration method used (see e.g., Effron and Gong, 1983). In other words, the MSECV etc. are estimates of the average prediction ability of calibration equations derived by the actual calibration method and for the actual population. In this way, MSEP is more or less *one* particular realization of the set of mean squares averaged to obtain the internal validation MSE's. Therefore, MSECV and MSELC can be both larger and smaller than the realization MSEP depending on the MSEP's position relative to the average. For reasonable object size, however, the two strategies will seldom give results too far apart.

These arguments imply that the internal validation criteria can be difficult to interpret in controlled calibration situations (see e.g. sections 3.6.3 and 3.2) where the design is given in advance of the experiment. In most serious calibrations, however, where the number of objects is reasonably large and the calibration objects are selected as to cover the actual region as well as possible, the MSECV and MSELC will give good estimates of 'average prediction ability' where average now means average over a population similar to that of the calibration design.

4.4.2 THE BOOTSTRAP

To incorporate the dependence on the value of **y** or alternatively to assess the prediction ability for different values of **y** as mentioned in section 4.1 is generally impossible by cross validation since this technique computes averages over the available objects. An alternative to this full cross-validation exists, however, when the number of objects is large so that it is possible to validate the predictor at different intervals on the scale, but the Bootstrap presented in e.g. Effron (1982) may be a better alternative.

The MSE for a particular given y-value depends on the predictor used and on the conditional distribution of **x** given y. If we let $F_{\mathbf{x}|y}$ be this conditional distribution, the MSE for a specified y-value can be written as

$$E_{F_{\mathbf{x}|y}}(y - \hat{y})^2 \qquad (4.27)$$

The bootstrap estimate of this is simply defined by replacing the theoretical F by an estimate based on the calibration objects. In cases where **x** can be written as a function of y plus noise as in the linear mixture model, such an estimate is easily obtained. But its usefulness will depend on the representativity of the calibration set.

In practice the bootstrap estimate is computed by Monte-Carlo simulation according to the estimated distribution function. In Næs (1985c) a Bootstrap evaluation of EGLS and EBLP was performed and gave reasonable results.

——————————— End of statistical extensions ———————————

4.5   USING PREDICTION ERROR TO SELECT THE CALIBRATION MODEL

4.5.1 OPTIMAL MODELLING COMPLEXITY

The various statistical methods for indirect multivariate calibration have as discussed in sections 3.4, 3.5 and 3.6 one problem in common; namely how to choose the number of dimensions to be used in subsequent predictions.

Multivariate calibration aims at reducing the prediction error by modelling chemical and physical interference phenomena that would otherwise destroy concentration determinations. Each independent interfering phenomenon requires an independent factor dimension in the multi-dimensional calibration model. As illustrated in Figure 4.4, a minimal prediction error is attained at a certain number of calibration factors.

4.5.2 TYPES OF PREDICTION ERROR

The prediction error is composed of two main contributions, the remaining-interference error (underlying bias) and the estimation error (Figure 4.6a). The former is the systematic error due to unmodelled interference in the spectral data and the latter is caused by random measurement noise of various kinds.

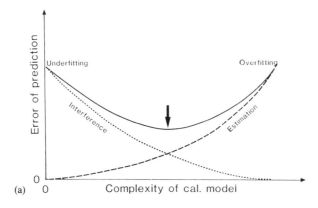

Figure 4.6   Validation: Conceptual illustration of prediction error as function of complexity of calibration model. a) Over- versus under-fitting. The arrow indicates optimum

The two contributions to the prediction error have opposite trends with increasing complexity of the calibration solution: Provided that the calibration objects are sufficiently representative for the new objects to be predicted, the remaining interference error should decrease with increasing number of interferences being modelled. But the statistical uncertainty error increases at the same time, due to the increased number of independent model parameters estimated from the available data.

This is important, because only a limited number of independent parameters can be estimated with high precision from a given set of calibration data. Conceptually, minimal prediction error is obtained when the remaining-interference error and the uncertainty error balance each other. Modelling too few phenomena is called underfitting and modelling too many phenomena is called overfitting (see also section 4.1.2, and Martens and Næs, 1984).

4.5.3 MODEL SELECTION

The problem is then how to select the number of dimensions in practice. The technique advocated here is to compute the prediction ability for the different number of factors by either prediction testing or internal validation for the calibration data. In this way one can draw a graph on the basis of the computed criteria as for instance in Figure 4.4. A suggestion is then to select the number of factors which gives the lowest point on the curve. Alternatively, it is usually advantageous and more robust not to be that strict. If for instance a similar value of RMSEP is obtained at a number of factors less than the 'optimal', A, this number of factors should usually be selected (see e.g. Osten, 1988).

Visual inspection of plots of the model parameters likewise represents an important tool in model selection. In multivariate calibration it is important both to attain good predictive ability and to understand the calibration models. In Figure 4.5 our prior information about what absorbance spectra to expect was limited; the visual assessment was limited to watching out for randomness in the loadings. In other cases (see e.g. the PLSR analysis of the alcohol mixtures in section 3.5) more information can be gained in this plotting phase as well.

4.5.4 HOW TO REDUCE PREDICTION ERROR

4.5.4.1 Using more data

Figure 4.6a shows that the prediction error is a 'sum' of the effects of under- and over-fitting. While the amount of unmodelled structure decreases, the amount of noise drawn into the model increases with increasing model complexity.

In Figure 4.6b is illustrated conceptually how prediction ability changes with the number of objects in the calibration. As we see the estimation error curve is better for large calibration sets. As this sample size increases we approach the best available of all predictors if the model assumptions are correct.

4.5.4.2 Using better data

Another way to improve prediction ability is of course to seek better data. This concerns first of all the noise level of the instrument used, but it is also related to the object preparation. It could for instance concern reduction of irrelevant variation due to object packing or other standardizations of measurement conditions (see again Figure 4.6b).

Sometimes irrelevant types of variation in the X-data can also be reduced by data preprocessing (Chapter 7).

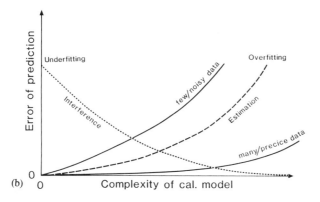

Figure 4.6 (*cont.*) b) the effect of increased amount of data and better data

4.5.4.3 Simplifying the population for calibration

Multivariate calibration is based on data approximation over a limited population of objects and the calibration model can not be expected to fit equally well to all types of objects. In some cases it may give improved fit and prediction results if the calibration population is split into several narrow sub-populations so that e.g. linear approximations become more satisfactory (see Chapter 7).

4.5.4.4 Using a better model

After a model selection strategy as above, it can very well happen that the results are too bad and the reason is the inappropriateness of the model. To get information about this and also information about potential improvements by using modifications or pretreatment of the data it is important to check assumptions of the method. For some important tools we refer to the next section.

4.6 MODEL CHECKING

Here we consider evaluation of *calibration* model assumptions. Model fit of a future prediction spectrum is closely related to detection of multivariate outliers and we refer to Chapter 5 for a treatment of this. Remedies such as transformations and nonlinear modelling are treated in Chapter 7 and section 3.7 respectively.

4.6.1 BILINEAR MODELS

The ordinary linear regression model is attained when the number of bilinear factors A is equal to K. The assumption of greatest importance to check in this case is therefore the linearity of y versus x. Although not explicitly stated in section 3.3, unweighted regression as used in the regression step of bilinear regression is best

264

suited for models with homogeneous error variances (i.e. equal error variance of f_i for each object) and this could also be investigated. In addition, it may be of interest to check the linearity assumption for $A < K$ which is an important property for obtaining good prediction results in the case of collinearity (see e.g sections 3.3, 2.3 and 4.5).

Linearity of **y** versus **x** is most easily investigated by plotting of residuals $\hat{f}$ or Studentized residuals r_i (residuals divided by their standard error, see Cook and Weisberg, 1982; and Chapter 5, equation (5.20)), versus $\hat{y}$, the measurement y or some of the individual variables x_k. Notable lack of fit will then usually appear as clear systematic tendencies (see Figure 4.7 and Weisberg, 1985). Plots of residuals vs. e.g. date of analysis or just the object number can also reveal useful patterns.

Homogeneity of the residual variances is similarly evaluated by simple plotting of residuals or Studentized residuals. Systematic tendencies as e.g increase or decrease of residuals as a function of $\hat{y}$ or y as in Figure 4.8 is a clear indication of heterogeneity.

A more formal statistical test of homogeneity of variances is found in e.g. Weisberg (1985) and more sophisticated tools to check linearity like added variable plots can be found in Cook and Weisberg (1982).

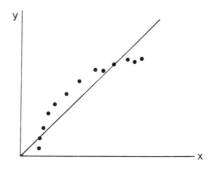

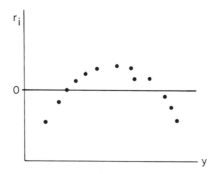

Figure 4.7 Plot (simulated) of unjust data x and y and of studentized residuals r_i versus the value of y

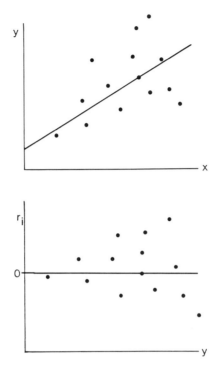

Figure 4.8   Plot (simulated) of unjust data x and y and of studentized residuals versus y

────────────── Statistical extensions ──────────────

4.6.2 THE LINEAR MIXTURE MODEL

The linear mixture model can be written as

$$\mathbf{X} = \mathbf{1k_0}' + \mathbf{YK'} + \mathbf{E} \tag{4.28}$$

In sections 3.6.1 and 3.6.2 it was assumed that the errors in $\mathbf{E}$ are uncorrelated and of comparable size while in section 3.6.3 the residuals were modelled by a linear factor structure $\mathbf{TP'}+\mathbf{E}$, where $\mathbf{TP'}$ represents systematic interferences and $\mathbf{E}$ represents uncorrelated errors with possibly different variance.

In any of these cases, the basic assumption is the linearity between $\mathbf{Y}$ and $\mathbf{X}$ and as in section 4.6.1 this can be checked by plotting of residuals $\mathbf{\hat{E}}$ (or studentized residuals) versus $\mathbf{Y}$, $\mathbf{X}$ or $\mathbf{\hat{X}}$. In the case of linear factor structure on the residuals, the linearity between $\mathbf{Y}$ and $\mathbf{X}$ should be done from residuals after fitting of both $\mathbf{YK'}$ and $\mathbf{TP'}$.

Formal tests of linear model fit can be used if the error variances of the covariance matrix Σ are known or we have replicates of $\mathbf{x}$ for some values of $\mathbf{y}$. For instance, the statistic

$$R_k = \sum_{i=1}^{I} (x_{ik} - \hat{k}_{k0} - \mathbf{y}_i' \, \hat{\mathbf{k}}_k)^2 / s_k^2 \qquad (4.29)$$

can be used to check linearity for each value of k or each wavelength. Here $\mathbf{k}_k$ is the vector corresponding to the kth row of $\mathbf{K}$, $\hat{k}_{k0}$ is the corresponding element of $\mathbf{k}_0$ and s_k^2 is either the known error variance σ_k^2 for wavelength k or the estimate based on the replicates. Note that neither of these 'estimates' assumes anything about linearity and if the fit of the model to data is bad, the estimated residuals represented by the numerator will be greater than that reflected by the variance s_k^2. Large values of R_k are then suspicious and it can be shown that as a useful approximation it can be tested for outliers against a chi-square distribution with $I - J - 1$ degrees of freedom. With linear factor structure on the residuals, the R_k and the test must be adequately modified to be most useful.

For the methods defined in section 3.6.3 the assumption of equal covariance matrix Σ for the systematic residuals **Pt+e** for different objects is critical. Checking of this can be done as in Oman and Wax (1984) by comparing estimates $\hat{\Sigma}$ from different subsets of the calibration data. Formal procedures for such comparisons are found in Mardia et al. (1980).

In general, linear calibration methods are best suited for normality assumptions on the variables, because in this case the optimal procedure is a linear method. To check for such normality, one can apply probability plotting techniques on the residuals or studentized residuals as shown in e.g. Weisberg (1985) or Atkinson (1987). The most common such technique is the so-called $q - q$ plot where the computed residual quantiles percentiles are plotted against their theoretical analogues in the normal distribution. A discrepancy between the plotted points and a straight line indicates deviation from normality.

——————————— End of statistical extensions ———————————

5 Outlier Detection

SUMMARY The reason for and importance of outlier detection and possible actions when they are discovered is discussed. Alternative detection criteria based on leverage and residuals are presented and tested with respect to a data based on litmus quantification. The emphasis is put on bilinear models, but the linear mixture model is also mentioned.

5.1   OUTLIER PROBLEMS

5.1.1 THE DANGER OF OUTLIERS

5.1.1.1 Anticipate errors!

Mistakes and unexpected phenomena are unavoidable in the real world. This is of course true in research into the unknown, but it is also true in routine laboratory analysis and in industrial process measurements. A number of phenomena can affect the quantitative analysis—operator mistakes, noise spikes, instrument drift, abnormal object types etc. Looking out for trouble is better than closing one's eyes to them. Detection of abnormal observations (or outliers) is therefore important.

In this respect the traditional univariate 'calibration curve' is dangerous. Multivariate calibration offers much more powerful tools for automatic outlier detection (Figure 1.1). There are several principally different types of errors that can be detected and handled—errors in X, errors in Y, errors in calibration data and errors in future prediction data. In addition, not all abnormalities represent useless errors—some may in fact give extremely valuable information.

Let us start the discussion on outliers by a mental analogy, based on our own 'spectrophotometer', the eye. Then we present a real spectroscopic example with outliers, and go through various computational methods to detect these.

5.1.1.2 Learning to recognize tea from coffee: An analogy

Assume that we have to learn how to prepare a good cup of tea. Of course we can go by the book on how many spoons of tea leaves should be used with how much boiling water for how long a time. But as a safety check it would be wise to watch the appearance of each cup of tea, because mistakes can occur in the kitchen.

How can we learn to recognize the colour of normal brown tea and to assess its strength ? We have to train our eyes, by comparing the visual appearance ($\mathbf{X}$) of various cups of tea with known tea concentrations ($\mathbf{y}$).

Let us consider various possible outliers among the tea cups: Assume first that one of the cups of tea in the 'calibration set' contained pure water, while all the other cups contained brown tea of various concentrations. This one cup of water would then be perceived as an outlier: Although it does belong to the general solvent/analyte model, it contains abnormally low level of the analyte, compared to the other cups. Conversely, if one cup of tea were extremely dark and undrinkably strong, then it would also be detected as an outlier, from its extremely dark brownness.

Such extreme calibration objects could strongly influence how we learn to recognize brown tea and evaluate tea strength. As long as the extreme objects are otherwise OK, the extreme span may be an advantage since it can facilitate our mental modelling (estimation) of the main colour variations. But the extremeness may stop us from noticing the finer nuances within the normal concentration span of tea, given the nonlinear detector and modelling system involved.

Irrelevant or erronous data in the training set could give us wrong impressions of how a good cup of tea may look. If for instance a cup of dark brown coffee were placed among the cups of tea, we had better disregard it as irrelevant when learning how to assess tea strength from the colour. A cup of weak brown tea erroneously described as containing strong tea could likewise create problems. The damage of such erroneous outliers in the training set is two-fold:

First of all we may get a wrong impression of how the colour of the tea in the cup varies with tea concentration, and we may therefore make erroneous predictions of tea concentrations when looking at future tea cups. Our mental predictor $\hat{\mathbf{Y}} = f(\mathbf{X})$ will become systematically wrong.

Secondly, the erroneous calibration data give us a rather ill defined mental model with large tolerances for how the appearance of a cup of tea may vary. This stops us from detecting strange cups of 'tea' in the future. In the extreme case we may later serve as brown tea anything from green tea or orange juice to coffee, because we have not learned to recognize a good cup of brown tea.

Some people, however, take milk in their tea, and others not. What then of our ability to recognize and assess the strength of the tea? If we had never seen a cup of tea with milk in our training period, we would later detect it as something abnormal, due to its turbidity.

On the other hand, if the training set contained various known tea strengths with and without milk added, then we could learn to disregard the milk addition as an interference: We would then develop a multi-factor mental calibration model, with a brownness-factor and a whiteness-factor. The increasing brownness would still indicate increasing tea concentrations, although seriously affected by the turbidity. But the increasing whiteness seen with increasing milk concentration would then compensate for the loss of brownness.

Alternatively, we may have developed two different mental calibration models for tea strength - one for tea with milk and one for tea without.

What then about outlier detection? In any case, extremely low or high whiteness would alert us to cups with too little or too much milk. The intensity of the brownness could still form the basis for detecting too weak or too strong cups of tea, but only in conjunction with the mental whiteness compensation. Off-colour mistakes from green tea or orange juice would still be recognizable and stopped before they reach our guests.

Just what type of calibration modelling our mind employs is irrelevant in this context. But the example illustrates that our body's sensory instruments allow quantitative multivariate pattern recognition. Without such outlier detection we could not survive very long.

5.1.1.3 Outliers in laboratory and in process

Outliers in general represent data elements that either are irrelevant, grossly erroneous or abnormal in some other way, compared to the majority of the data. In order to make optimal use of one's data it is important to identify such outliers.

Outliers always call for special attention. They often reflect 'bad' errors of some sort, and should be *eliminated* or *corrected*. But this is not always the case: Some outliers may in fact be very valuable and *informative* observations.

Both objects, variables and individual data elements can behave as outliers.

An outlying object, which is the most important of the outlier concepts, is an object whose data vector x_i and/or y_i strongly differ from those of the majority of objects calibrated for, either because of error in the data (noise, instrument drift, operator mistake, misprinting etc.) or because of a peculiar composition or physical state of the object analyzed. An 'outlying' or extreme variable x_k or y_j is here a variable whose data in a certain set of objects strongly differ from those of the majority of variables used in the calibration modelling, either because of high level of random noise, or because it alone reflects some unique property (a particular chemical or physical phenomenon, or a particular type of non-linearity). An outlying data element x_{ik} or y_{ij} may e.g. represent a noise spike or a misprinting.

In conventional analytical chemistry the detection of outliers is usually left to the subjective critical sense of the chemists, and this is a valuable stage that should not be ignored. But there are many different types of outliers, some of which are not easy to detect from one single measured variable. In music, one sound from a single string cannot reveal a disharmony! Some type of contextual modelling and pattern recognition is required, either in the mind of a person or in the computer.

This is especially important for routine analyses; modern multivariate computerized instrumentation produces too much data for just mental evaluation. To make things even worse, some people put unrealistic confidence in analytical results just because they are given in digital computer displays. To ensure correct usage of modern analytical instruments, outlier detection should therefore be included as an automatic and explicit part of the instruments' operation.

5.1.1.4 Outlier detection in calibration and prediction

During *calibration* it is important to detect and possibly remove or correct data from objects or variables that otherwise would decrease the prediction ability of the estimated calibration coefficients. During *prediction* of unknown objects it is important to have methods for detecting abnormalities; this increases the confidence of the predicted concentration results. Unexpected outliers detected can also give new scientific information.

With multivariate calibration modelling it is possible to perform outlier detections according to a variety of criteria. Let us repeat the essence of the calibration modelling: A mathematical model type is chosen for the $X-Y$ and $X-X$ relationships (and sometimes even for the $Y-Y$ relationships, if there are more than one analyte). The choice is based on assumptions about the data (expected interferences, noise level, distributions and linearities.) By some estimation method the parameters of the model are estimated statistically, using available calibration data. The resulting calibration model describes how the data from this type of objects measured with this type of X- and Y-instruments should be expected to behave, and what the normal, acceptable level of noise is.

Comparing data with this resulting estimated calibration model, individual objects, variables or data elements that behave abnormally can be detected. This checking can be performed both on the calibration data and data from prediction objects later on.

But the rather simple outlier warnings to be given in this chapter should be used with critical sense. Especially during calibration they should be supplemented by graphical inspection of the resulting model parameters and residual statistics. This can reveal multiple outliers that passed unnoticed due to the so-called *masking* effect. Alternatively, more advanced statistical tools can be employed in order to reveal *multiple* outliers, as outlined by Cook and Weisberg (1982).

For comprehensive reviews of the whole field of outlier detection in linear models, we refer to Cook and Weisberg (1982), Belsley et al. (1980) and Beckman and Cook (1983).

5.1.2 WHAT SHOULD BE DONE WITH OUTLIERS

5.1.2.1 Good and bad outliers

Outliers are important and unavoidable in the scientific process. We must learn to live with them and try to learn from them. Therefore outliers should be dealt with critically: Once detected, we should correct or ignore outliers if they seem to damage our modelling.

Whenever possible one should try to understand the reason for every outlier detected. Sometimes this is easy. For instance, the objects with the highest or lowest analyte level can often become tagged as outliers; if the instrument response is reasonably linear, such outliers may in fact be very informative and should be retained in the calibration set. The objects with the highest or lowest levels of the various interferents will likewise have a good chance of being detected as outliers—these are 'good' outliers.

Bad outliers are usually those which we cannot understand and which strongly affect the calibration model. The most obvious action when such 'bad', apparently erroneous outliers are detected is to look for simple mistakes like misprintings, and correct them. But often the interpretation of suspicious outliers may take more time—for instance sending the object to repeated chemical and instrumental analysis.

Meanwhile, a sensible strategy is then to delete the bad outliers from the data set and repeat the calibration modelling, to be sure that they do not influence the calibration or destroy predictions. With outliers due to operator mistakes or failure of the equipment this is a sensible strategy. But this is primarily recommended when we have lots of good data and the number of outliers is low. Excessive 'pruning' of a data set for outliers should be avoided except in very initial exploratory data analyses.

Strategies for more or less automatic down-weighing or elimination of outliers are treated in the literature on robust statistics (see e.g. Huber, 1981). But this approach is not considered in the present book, since only the user has sufficient contextual background knowledge to distinguish between good and bad outliers (see also section 2.3).

5.1.2.2 Outliers, classes and split models

Outlier detection, classification and 'split models' are closely related topics. Multivariate calibration concerns how to find simple, local mathematical approximations of the underlying unknown X–Y and X–X relations in data. Making an overall linear calibration for inhomogenous populations of objects can be useful for a general overview, but may give unsatisfactory local approximations and tag many objects as 'outliers'.

Assume that the objects come from more than one natural population, for instance meat- and fish products to be analyzed for fat content ($\mathbf{y}$) from NIR reflectance spectroscopy ($\mathbf{X}$). The X–Y and X–X relationships will normally be quite different for these two types of objects.

If we intend to calibrate for meat products only, then a fish object or two among the calibration meat objects would probably be tagged as outliers relative to the meat calibration model. For optimal modelling and prediction in meat objects, this calibration modelling should be repeated with the fish outliers eliminated. Likewise, a meat object or two among many fish objects for calibration would be detected as outliers relative to the fish calibration model, and should be removed prior to recalibration.

The outlier detection capability of these two calibration models could then be used for classifying new objects based on their X-data: A normal future meat object would then be classified as belonging to the meat model while being an outlier relative to the fish model. Hence its fat content should be predicted from the meat calibration model. Likewise, a future unknown fish object would be classified as being a fish product and not a meat product.

If there were about equal numbers of fish and meat products in the calibration set, it would be possible to calibrate for both types simultaneously. But a rather complicated calibration model would probably result, with sub-optimal prediction ability and high outlier tolerances.

In such cases it may lead to better results to split the calibration set into more homogenous sub-classes (e.g. meat products and fish products separated) and then develop a specialized calibration for each of these local sub-classes. This splitting can usually be based on prior information about the objects, combined with visual inspection of preliminary calibration results (factor score plots etc.). But suggestions for model splitting can also be obtained by some automatic clustering procedure, e.g. FOSE (see e.g. Gunderson et al., 1988 and CART (Breiman et al., 1984)).

5.1.3 A DATA SET WITH OUTLIERS

The following example will be used for illustration of the various outlier detection methods.

Assume that we want to determine the concentration of the analyte litmus in aqueous solutions at pH 10 from O.D. spectra in the 400–800 nm wavelength range. We have 17 calibration objects and a small set of 6 test objects.

First we study the calibation set: Figure 5.1a shows the O.D at the absorbance maximum (576 nm) of the analyte vs. 'true' analyte concentration given in percent (relative to the concentration of a reference level of 1.25 mg/ml). The spectral data were obtained by reading the O.D. in beakers containing the solutions, using a Guided Wave Model 200 spectrometer.

The data are similar to those reported in section 3.5 and Chapter 8, but have been modified to illustrate various types of outlier problems. The figure shows that most of the objects contain analyte levels between zero and 50 relative percent. Most of the calibration solutions fall nicely along a straight line with object 3 lying as an outlier at the extension of this line. Calibration objects 1, 2 and 7 fall clearly off the normal calibration line.

Figure 5.1b shows the full O.D. spectra of the 17 calibration solutions of litmus where the foremost spectra represent the lowest analyte concentrations ($y = 0$). Figure 5.1c maps the 17 objects for two of these 50 wavelength channels. With our eyes some of the outliers can be seen directly in these plots. But how can they be detected *automatically*? And are there other outliers? This will be illustrated in the following treatment of outlier detection in bilinear methods in the next section. Table 5.1 gives some details of the input data.

The prediction set consists of six normal solutions of blue litmus (aqueous solutions, pH 10) in the normal concentration range ($0 < y_i < 50$ relative percent). These objects should not give rise to outlier warnings.

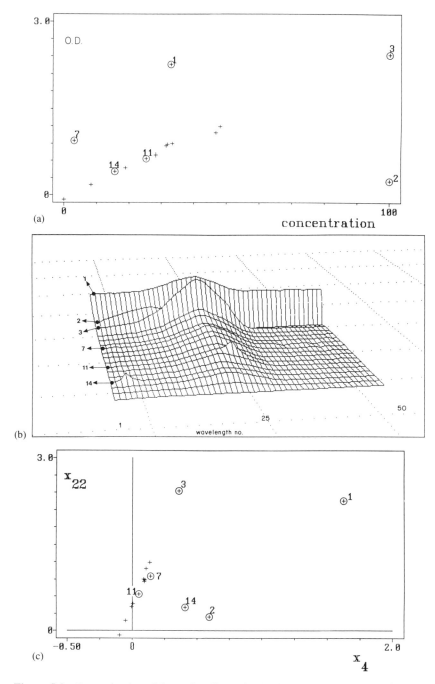

Figure 5.1 Determination of the analyte litmus in alkaline aqueous solutions. a) O.D. at the analyte absorbance maximum (576 nm, channel 22) vs. 'true' analyte concentration for 17 solutions. b) O.D. spectra of the 17 solutions at 50 channels between 400 and 800 nm. The objects are numbered from the back, so that the one with the highest O.D. corresponds to no.1. in a). c) O.D. at absorbance maximum (channel 22, 576 nm) vs absorbance at a lower wavelength (channel 4, 532 nm)

Table 5.1 Data set with outliers: Aqueous solutions of a blue coloured analyte, litmus at pH 10

Calibration set:
  Seventeen objects, of which 11 are normal, containing various analyte levels and no interference.
  Outliers in calibration set:

Object	Outlier type
1	Physical interference: Light scatter, turbidity from white ZnO powder added.
2	Chemical interference: Changed molecular state, analyte colour changed from blue to red by acid added (pH 4), for abnormally high analyte level (100 percent).
3	Extreme composition: Abnormally high analyte level (100 percent), otherwise normal.
7	Error in reference data **y**: 'True' analyte concentration is not true. Misprint y = '3' instead of '33' percent.
11	Error in instrument data **x**: Measurement noise at channel x_{25}, i.e. on top of the analyte absorbance peak.
14	Error in instrument data **x**: Measurement noise at channel x_4, i.e. off the analyte absorbance peak.

Test set:
  Objects 18–23 (no outliers).

5.2 OUTLIER DETECTION IN BILINEAR MODELS

5.2.1 OUTLIERS IN CALIBRATION

The bilinear model for one analyte is as described in section 3.3 equal to

$$\mathbf{X} = \mathbf{1}\bar{\mathbf{x}}' + \mathbf{TP}' + \mathbf{E}$$
$$\mathbf{y} = \mathbf{1}\bar{y} + \mathbf{Tq} + \mathbf{f}$$

$$(5.1)$$

The basic tools for outlier detection in this model are based on leverage and residuals and below we shall discuss each of them separately and also consider what types of abnormalities they are able to reveal.

For illustration the 17 calibration objects presented in the previous section were first submitted to bilinear PLS calibration for total analyte concentration. Figure 5.2a shows how the PLS model describes the analyte concentration with increasing number of factors: The cross-validation curve (10 segments) shows strong signs of outliers in the calibration set, with highly irregular deviations from the fitted curve. This indicates that we may have to correct or remove some outliers before we develop the final calibration model.

Without special types of prior information about the rank of the data, one would normally choose to look for outliers relative to the fitted model that has the best apparent predictive ability in e.g. cross validation, which in this case is the 3-factor solution (see Chapter 4). (But why three factors, when we only expect one?)

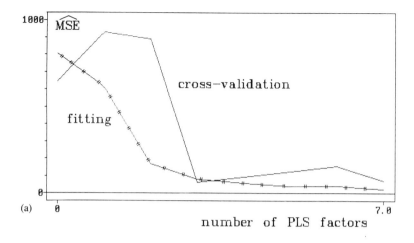

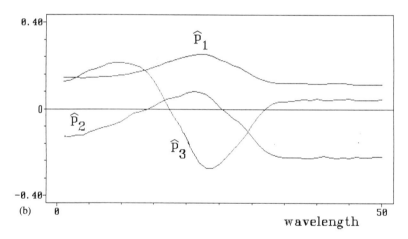

Figure 5.2 Preliminary calibration for litmus in alkaline aqueous solutions. a) Apparent predicting ability vs. model complexity: Mean square error $\widehat{MSE}$ for litmus concentration vs. number of PLSR factors $a = 1,2,\ldots,7$. The two curves represent the fitted calibration error (dotted line) and the estimated validation error (estimated by full cross validation, using 10 cross validation segments). b) Loading spectra of the PLS calibration model, scaled according to size, $\widehat{\mathbf{p}}_a{}' * \sqrt{\widehat{\mathbf{t}_a{}'\mathbf{t}_a}}, a = 1,2,3$

This will hopefully reveal the most serious outliers. Then we can correct or eliminate these outliers and repeat the calibration modelling. Figure 5.2b shows the corresponding 3 loading vectors $\widehat{\mathbf{P}} = (\widehat{\mathbf{p}}_a, a = 1, 2, 3)$, plotted according to their relative importances for modelling $\mathbf{X}$ (i.e. multiplied by $\sqrt{\mathbf{t}_a{}'\widehat{\mathbf{t}_a}}$. The figure shows all three loading spectra to be smooth. They are also approximately equal in size and hence about equally important in the modelling of $\mathbf{X}$. So at least these three factors do not seem to represent random noise.

5.2.1.1 Data with extreme leverage in calibration

Leverage of an observation is a concept developed in ordinary regression theory (see e.g. Cook and Weisberg, 1982, for an excellent discussion) and concerns the position of the observation's 'independent variables' (or X-variables) relative to the others. Here we will treat a modification of the original concept applied to the bilinear model above.

Our definition is equal to

$$h_i = 1/I + \widehat{\mathbf{t}}_i' \, (\widehat{\mathbf{T}}'\widehat{\mathbf{T}})^{-1}\widehat{\mathbf{t}}_i \qquad (5.2)$$

where $\widehat{\mathbf{t}}_i' = (\widehat{t}_{ia}, a = 1, 2, \ldots, A)$ is the vector of factors for object i, and $\widehat{\mathbf{T}} = (\widehat{\mathbf{t}}_i, i = 1, 2, \ldots, I)$ is the factor score matrix for the whole calibration set. (In Chapter 4, this h_i was denoted h_i^A to emphasize the dependence on A, but here this special notation is avoided for simplicity.)

The term $1/I$ represents the contribution from the intercept and could be deleted from the definition, but is here included to fit the definition of h_i made in standard statistical texts.

Formally speaking, h_i is defined as a Mahalanobis distance (Cook and Weisberg, 1982; Mardia et al., 1980), with $\widehat{\mathbf{T}}'\widehat{\mathbf{T}}$ acting as the weighting covariance matrix. With $A = K$ this leverage h_i corresponds to the usual leverage definition, which gives the full Mahalanobis distance of object i to the calibration centre. For regression methods optimized for predictive ability, however, the leverage h_i corresponds to a truncated Mahalanobis distance, where the small and uncertain types of variability in the X-data have been ignored (see e.g. section 3.4.6). The basic reason for this is that bilinear methods use only A factors in regression and it is important to have criteria giving specific information about this A-dimensional systematic variability space. (Detection criteria for further dimensions are covered in the next section.)

From Cook and Weisberg (1982) we know that $1/I < h_i < 1$. A low leverage for an object (near $1/I$) shows that the object is close to the centre of the calibration set in $\mathbf{X}$ with respect to the A-dimensional factor space, and consequently this calibration object has usually very little importance for the calibration solution. But a leverage near 1 shows that the object is far from the mean and this may have had a very high importance on the resulting A-dimensional calibration model: The interpretation of such a high leverage point could be an observation with extreme analyte concentration or extreme value of an interferent also modelled by the A-dimensional bilinear model.

A calibration object with high leverage may have strong effect on the resulting X- and Y-loadings $\widehat{\mathbf{P}}$ and $\widehat{\mathbf{q}}$, since they are found by regressing $\mathbf{X}$ and $\mathbf{Y}$ on $\widehat{\mathbf{T}}$, and such least-squares estimations are outlier sensitive. In addition, it may even have a major effect on the weights $\widehat{\mathbf{V}}$ (via loadings $\widehat{\mathbf{P}}$ and/or $\widehat{\mathbf{W}}$) and hence on how $\widehat{\mathbf{T}}$ is derived from $\mathbf{X}$. So, if a high-leverage observation in the calibration set contains errors or some other undesired abnormality in its y- or X-data, such an object may be dangerous and decrease the predictive ability of the resulting calibration model. On the other hand, if its high leverage is caused by the fact that the object is particularly informative, it had better not be eliminated!

The average value of h_i in the calibration set is $(1 + A)/I$, and a warning may be given by e.g. $h_i >$ constant$(1 + A)/I$ where the constant is for instance 2 or 3 (see e.g. Velleman and Welsch, 1981).

For methods like PCR and PLSR that give orthogonal scores in the calibration set $(\widehat{\mathbf{T}}'\widehat{\mathbf{T}} = \text{diagonal})$ formula (5.2) simplifies to

$$h_i = 1/I + \sum_{a=1}^{A} (\widehat{t}_{ia}^{\;2}/\widehat{\mathbf{t}}_a'\widehat{\mathbf{t}}_a) \tag{5.3}$$

Here we see that leverage is simply a weighted sum of X-scores, where the weights for each factor are the inverse of the sum of squares of $\widehat{t}_{ia}$ in the calibration set. In this way we easily see that each factor is weighted so that they give contributions of comparable size to the leverage: An extreme object with respect to one of the small factors is equally important as an extreme one with respect to one of the larger factors in $\mathbf{X}$.

While outlier detection based on h_i has the advantage of summarizing the scores of many factors $a = 1.2.....A$, it does not tell us from which of the factor(s) the outlier condition came. If individual factors have been found to represent distinct phenomena (as a result of a controlled experimental design, or after axis rotation), it may be useful to add outlier tests on individual factors. This can be done by the partial leverage h_{ia} for each factor, which for orthogonal factors is simply defined as

$$h_{ia} = \widehat{t}_{ia}^{\;2}/\widehat{\mathbf{t}}_a'\widehat{\mathbf{t}}_a \tag{5.4}$$

The average partial leverage is by definition $1/I$ in the calibration set, so a pragmatic outlier detection limit could here for instance be $h_{ia} > 3/A$.

Let us consider the present example of alkaline litmus, where we expected to find one single type of spectral variation (varying analyte concentrations) and a linear instrument response. If there were no abnormalities in the data, one would expect the objects to lie as points along a straight line with increasing analyte concentration, in the K-dimensional $\mathbf{X}$-space. Points that fall along a straight line in the K-dimensional $\mathbf{X}$-space should also do so in the A-dimensional $\widehat{\mathbf{T}}$-space, if this type of variability has been modelled by the A factors.

In the present example we have chosen to look closer at the model with $A = 3$ factors. Figure 5.2c shows the scores of the 17 calibration objects with respect to the three factors included in the preliminary calibration model. It shows that most of the objects in fact fall along a straight line inside this model space.

But some objects, particularly nos 1, 2 and 3, clearly differ from the rest in this respect. In the initial PLSR modelling the 3-factor PLS solution yielded very strong leverage warnings for object 1 (the only turbid solution, i.e. abnormal systematic interference, $h_i > 0.99$) and object 2 (the only red-coloured solution, i.e. abnormal systematic interference, $h_i > 0.99$) and a somewhat weaker warning for object 3 (containing twice as much analyte as any of the others, i.e. abnormal analyte concentration, $h_i = 0.67$).

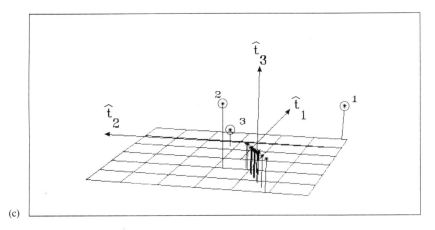

(c)

Figure 5.2 (*cont.*) Preliminary calibration for litmus in alkaline aqueous solutions. c) Scores of the first of the PLS calibration factors, $\hat{\mathbf{t}}_a$, a = 1,2,3

Other examples of the use of h_i can be found in e.g. Martens and Næs (1987). This estimate of the 'importance' of an object to the calibration can be modified in various ways to enhance those factors that have particularly strong modelling power for **Y**, e.g. by basing the leverage computation on the **Y**-weighted scores $\hat{t}_{ia}\hat{q}_{ja}$ instead of $\hat{t}_{ia}$ alone.

Analogously we could consider the leverage of the different X-variables. In PCR this can be expressed in terms of the orthogonal loadings $\hat{\mathbf{P}}$:

$$h_k = \hat{\mathbf{p}}_k{}'\hat{\mathbf{p}}_k \tag{5.5}$$

and in PLSR from the orthogonal loading weights $\widehat{\mathbf{W}}$:

$$h_k = \hat{\mathbf{w}}_k{}'\hat{\mathbf{w}}_k \tag{5.6}$$

A value of h_k near zero for an X-variable shows that the variable has not affected the calibration model very much. A leverage near 1 indicates a very high importance for the model; this variable may have grabbed a regression factor alone. The average value of h_k is A/K, and a warning may be issued for e.g. $h_k > 2A/K$ or $3A/K$.

For the present three-factor solution this limit would be 2*3/25=0.24, while the maximum leverage was $h_k = 0.14$ (for wavelength channel $k = 23$). Thus none of the X-variables were found to be particularly extreme in the 3-factor PLSR model. This is in keeping with the fact that neither of the three phenomena drawn into the model (varying levels of blue litmus, red/blue colour change and turbidity) affect only one X-channel.

5.2.1.2 Unmodelled residuals in spectral data X

In multivariate calibration the number of relevant variables available from our analytical instruments (e.g. a scanning spectrophotometer) is often much higher than the number of factors A required by the BLM calibration models. This yields over-determined equations, and X-residuals may be obtained. The X residuals will reflect whatever lack-of-fit there is between the calibration data X and the resulting model defined by the $\hat{V}$ matrix; random noise in X, unmodelled interferences, unmodelled nonlinearities etc.

The calibration residuals in X are simply defined as

$$\hat{E} = X - 1\bar{x}' - \hat{T}\hat{P}' \tag{5.7}$$

For each object this X-residual can be written

$$\hat{e}_i' = x_i' - \bar{x}' - \hat{t}_i'\hat{P}' \tag{5.8}$$

and for each element in the data table

$$\hat{e}_{ik} = x_{ik} - \bar{x}_k - \sum_{a=1}^{A} \hat{t}_{ia}\hat{p}_{ka} \tag{5.9}$$

The X-residuals can be used in many different ways for a variety of purposes. First of all they can be used to identify objects or observations or single data-points that do not fit the rest of the data. They can also be studied in more detail to tell if the problem stems from erratic measurements or systematic errors due to some specific interference.

Generally, an observation with large residuals or otherwise strange residual pattern $\hat{e}_i$ indicates an observation with abnormal interferent not present in the other objects and not modelled by the bilinear model. A variable with abnormal residual pattern $\hat{e}_k$ can be interpreted as one with higher noise level than the rest, either for one or more of the objects. A single large residual value $\hat{e}_{ik}$ indicates an abnormal value of x for exactly that particular object and variable.

The simplest, and often the best, way to treat the residuals is to study them graphically. This can be done in many different ways, and many powerful techniques exist (see e.g. Weisberg, 1985), but the most straightforward way is to plot each residual against observation number or wavelength number.

Figure 5.2d shows the residual spectra $\hat{e}_{ik}$ after 3 PLSR factors for 10 of the calibration objects in the present illustration (including all the objects with special problems (Table 5.1), plotted against wavelength channel $k = 1, 2, \ldots, 50$. It shows that most of the residual spectra fall within ± 0.01 O.D. units at most of the wavelengths. But, object 11 shows a large positive residual peak at channel 25, and object 14 shows a large positive residual peak at channel 4. This is not surprising, since a narrow 'error peak' had been added to their X-data at just these wavelength ranges.

280

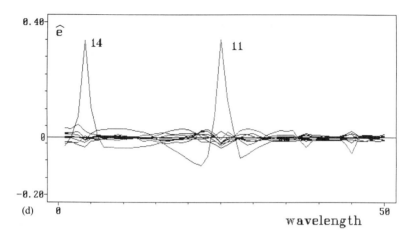

Figure 5.2 (*cont.*) Preliminary calibration for litmus in alkaline aqueous solutions. d) Spectral residuals $\hat{e}_i{}'$ after 3 PLS factors for the 17 calibration objects. The two outliers (objects 11 and 14) with large residuals are marked explicitly

Perhaps more surprising are the negative shoulders on both sides of channel 25 in object 11; these wavelength ranges do not contain particular errors. The reason is probably that the error spike in object 11 falls inside a wavelength range where PLS factors 1 and 3 have strongly non-zero loadings $\hat{p}_{ka}$. The positive error spike around channel 25 therefore affects the resulting scores for these factors (too high for factor 1 and too low for factor 3), which again results in the unexpected negative residuals outside the error spike but inside the loading peaks. The error peak in object 14 falls in a wavelength range where the factors do not have such important loadings, and hence causes less pronounced shoulder effects. This illustrates that the residuals obtained cannot be interpreted uncritically as if they were original measurements; they have been modified through the bilinear modelling.

For further identification of such **X**-residual spectra, they can be correlated to the corresponding residual spectra of various potential candidates obtained by fitting the candidate spectra to the same bilinear model.

The plotting procedures can be very informative. But in routine operation and elsewhere when the number of objects is high there is a need for more compact and automatic errror warnings. Below we shall study ways of summarizing the information from e.g. Figure 5.2d.

Total residual standard deviation in **X**

In the calibration set containing objects $i = 1, 2, \ldots, I$ the total residual standard deviation $s(\hat{e})$ (attempted estimate of the root mean square variability of the *x*-residuals) after A factors is defined by

$$s(\hat{e})^2 = \sum_{i=1}^{I} \sum_{k=1}^{K} \hat{e}_{ik}{}^2 / df \qquad (5.10)$$

where df represents the remaining degrees of freedom left for the residuals $\hat{e}_{ik}$.

The assessment of df depends on the number of calibration objects and X-variables, on how the calibration modelling was performed (the number of alternative models tried), and on what type of calibration method was employed (the number of truly independent model parameters estimated from $\mathbf{X}$ and from $\mathbf{Y}$).

Thus the exact df can be rather intractable. Fortunately, exact assessment of df is not particularly important for outlier detection.

Martens and Jensen (1983) approximated it roughly by

$$df = (IK - K - A(\max(I, K))) \tag{5.11}$$

for PLS regression. Mandel (1971) gives alternative results for PCR.

(Since $s(\hat{e})$ is an attempted estimate of variances of E-terms, the $\hat{ }$ could have been omitted from the definition. However, because of the generally vague definitions of, and assumptions about, $\mathbf{E}$ in bilinear modelling, we have chosen to keep the $\hat{ }$ in this chapter about BLM in order to distinguish from other more stringent estimates of error terms in, for instance, Beer's law models (see (3.113)). For the same reason, this will also be done for $s(\hat{e}_i)$, $s(\hat{e}_k)$, $s(\hat{f}_j)$ defined later in this chapter.)

Residual standard deviation of individual calibration objects in X

An individual calibration object's residual standard deviation $s(\hat{e}_i)$ (attempted estimate of the square root of the average variance of the e-variables for object i) is defined by

$$s(\hat{e}_i)^2 = \sum_{k=1}^{K} \hat{e}_{ik}^2 / (df/I) \tag{5.12}$$

where df is the degrees of freedom used above and I is the number of calibration objects.

If $s(\hat{e}_i)^2$ is much larger than $s(\hat{e})^2$, which is the average residual variance in the calibration set, this is an indication of an abnormal object. How much larger than the average should then this be before the object is considered suspicious? There is no simple exact answer to this, but it can be useful in practice to use an F-test approximation for the quantity

$$(s(\hat{e}_i)/s(\hat{e}))^2 \tag{5.13}$$

and conclude that the object is suspicious if the quantity is larger than a certain percentile in the F distribution with df/I degrees of freedom for the numerator and df degrees of freedom for the denominator.

More sensitive tests can be obtained with the quantity

$$(s(\hat{e}_i)/s(\hat{e}_{-i}))^2 \tag{5.14}$$

where $s(\hat{e}_{-i})^2$ is the average residual variance for all the calibration objects except object i.

A rough but practical approach is to consider the object as abnormal if $s(\hat{e}_i) > \text{constant} * s(\hat{e})$ where the constant is e.g. 2 or 3.

In the present example, objects 11 and 14 were thus tagged as outliers, based on $s(\hat{e}_i)^2$—in keeping with their unmodelled error-peaks (See Figure 5.1b and 5.2d).

Residual standard deviation of individual X-variables

In analogy, the residual standard deviations for the different X-variables are defined in the calibration set by

$$s(\hat{\mathbf{e}}_k)^2 = \sum_{i=1}^{I} \hat{e}_{ik}^2 / (\mathrm{df}/K) \qquad (5.15)$$

where df is still the same as before. A large $s(\hat{\mathbf{e}}_k)^2$ compared to $s(\hat{\mathbf{e}})^2$ after A factors indicates an X-variable with high residual level for its a priori weighting. An F-test approximation may again be useful. But usually the rougher test $s(\hat{\mathbf{e}}_k) >$ constant$*s(\hat{\mathbf{e}})$ where the constant is e.g. 2 or 3 suffices.

The reason for abnormally high $s(\hat{\mathbf{e}}_k)^2$ could be that x_k spans a systematic variability type that has not been included in the A-dimensional calibration model— either because it was masked by larger noise in some X-variables or because it is irrelevant for modelling **Y**. Alternatively, x_k may have too high random noise level compared to its a priori weighting.

In the present data set the X-variable channels 4 and 25 were thus detected as outliers, due to the large residuals in objects 11 and 14.

Ideally all the X-variables in a bilinear calibration should usually be scaled to about the same noise level (see Chapter 7). If different X-variables display very different residual variances $s(\hat{\mathbf{e}}_k)^2$, this means that the noise level in some X-variables will to some degree contaminate the estimation of bilinear factors before other variables have been emptied for information. The X-variables could then be re-scaled in proportion to the inverse of these residual levels (after a possible leverage-correction; see below). With these new a priori weights the bilinear regression modelling should then be repeated. This can sometimes improve the predictive ability and simplify the interpretation.

Residuals of individual data elements in **X**

Abnormalities in individual data elements $\hat{e}_{ik}$ can also be detected in tests. It may be useful first to correct $\hat{e}_{ik}$ for the number of degrees of freedom used in estimating e_{ik}, by using e.g.

$$\hat{d}_{ik} = \hat{e}_{ik} / \sqrt{(\mathrm{df}/IK)} \qquad (5.16)$$

Again an F-test approximation can be useful, e.g. testing $\hat{d}_{ik}$ against $s(\hat{\mathbf{e}}_k)$. Another approach is to issue warnings for

$$|\hat{d}_{ik}| > \text{constant} * s(\hat{\mathbf{e}}_k)$$

where the constant is 3, 4 or 5 (the test should be rather severe, because there are very many elements to be tested, and the probability of spurious warnings therefore increases).

In the present example wavelength 25 of object 11 and wavelength element 4 of object 15 were automatically tagged as very strong outliers, in keeping with the error spikes added to these data elements x_{ik}.

5.2.1.3 Unmodelled residuals in analyte data Y

In the calibration set (as well as in possible validation set) we know the reference data y_{ij} and hence we can obtain corresponding residuals $\hat{f}_{ij}$. In analogy to the 'spectral' residuals, $\hat{e}_{ik}$, these y-residuals can reveal certain types of outliers, i.e. objects with errors in the 'true' target data y_{ij} or objects for which the X–Y calibration relationship is different than for the other calibration objects.

We define

$$\hat{f}_{ij} = y_{ij} - \bar{y}_j - \sum_{a=1}^{A} \hat{t}_{ia}\hat{q}_{ja}$$

or (5.17)

$$\hat{f}_{ij} = y_{ij} - \bar{y}_j - \hat{\mathbf{t}}_i{'}\hat{\mathbf{q}}_j$$

In the present example used for illustration we only have one analyte (total litmus; $J = 1$). Therefore it is relatively simple to assess the **Y**-residuals graphically. Figure 5.2e shows the obtained residuals $\hat{f}_{ij}$ for the 17 calibration objects using 3 PLS factors, plotted against the 'true' concentration, y_{ij}. It shows a single major **Y**-residual of -30 O.D. units for object no. 7; the other residuals are between 0 and 5 O.D. units.

Remember that objects 1,2,(3),11 and 14 were previously tagged correctly as outliers from diagnostics developed in **X**-space. In addition, object no. 7 now seems to be an outlier. In fact, this object actually contains a -30 O.D. units error in its reference data y_7; see Table 1).

How can outlier detection based on the **Y**-residuals be automated? A summary for each individual Y-variable j over objects $i = 1, 2, \ldots, I$, $s(\hat{f}_j)$ ($=$ RMSEE$_j$ $= \sqrt{\text{MSEE}_j}$, Chapter 4, $= s_f$, section 2.1.2.3. See Section 5.2.1.2 for a discussion of the $\hat{}$.) can be computed from

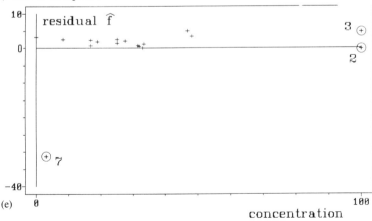

Figure 5.2 (*cont.*) Preliminary calibration for litmus in alkaline aqueous solutions. e) Analyte fit for the 3 PLS factors calibration modelling. Analyte residual $\hat{f}_i$ after 3 factors vs. 'true' analyte level y_i for the 17 calibration objects. The outliers 2, 3 and 7 are marked explicitly

$$\text{MSEE}_j = s(\widehat{f_j})^2 = \sum_{i=1}^{I} \widehat{f_{ij}}^2/\text{df2} \qquad (5.18)$$

where df2 is the number of degrees of freedom remaining after fitting the calibration regression of $\mathbf{y}$ on $\widehat{\mathbf{T}}$. This number varies with the number of calibation factors used, A, but also with the type of regression method used and with the structure in the X-data and their relationship with the Y-data (see Chapter 4). As a simple estimate one can use df2 $= I - 1 - A$ which is exact for $A = K$. (In that particular case it would be more natural to use the symbol s_f as in Section 2.1.2.3, see also discussion after (5.11).)

For $A < K$ this probably underestimates df2 somewhat. In PLSR, where the Y-data are used twice (in obtaining both $\widehat{\mathbf{V}}$ and $\widehat{\mathbf{Q}}$), an expression as e.g. df2 $= I - 1 - 2A$ may then possibly be more realistic; when K/I is high and the X-variables contain much noise, an even more conservative estimate of df2 may be required. But again, an exact degrees-of-freedom estimate is not critical for satisfactory outlier detection.

A large $\widehat{f_{ij}}$ residual compared to the RMSEE$_j$ indicates an outlier. In the test we may want to correct $\widehat{f_{ij}}$ for degrees of freedom, e.g.

$$\widehat{d_{ij}} = \widehat{f_{ik}}/\sqrt{(\text{df2}/I)} \qquad (5.19)$$

Again an F-test approximation can be useful, e.g. testing $\widehat{d_{ij}}^2/\text{MSEE}_j$ vs a percentile in the F-distribution, using df2$/I$ and df2 degrees of freedom.

More sensitive tests can be obtained by replacing MSEE$_j$ with a similar quantity computed for all the calibration objects except object i. More sophisticated techniques can be found in Cook and Weisberg (1982) for ordinary regression.

A simpler approach is to issue warnings for

$$|\widehat{d_{ij}}| > \text{constant} * \text{RMSEE}_j$$

where the constant is e.g. 2 or 3.

With this approach, object no. 7 was automatically detected as an outlier. In this case, the reason for outlier warning in $\widehat{f_{ij}}$ was a misprinting error in the 'true' Y-data. In other cases, the outlier warning in $\widehat{f_{ij}}$ could be due to a different X–Y relationship for a certain object, compared to the other objects.

Quite often, objects detected as outliers in the X-space due to errors in the X-data will also be flagged as outliers with respect to $\widehat{f_{ij}}$. The reason is that $\widehat{t_i} = \mathbf{x}_i' \widehat{\mathbf{V}}$ and hence $\widehat{y_{ij}} = \widehat{t_i}' \widehat{q_j}$ are affected by these X-errors, and this creates large residual $\widehat{f_{ij}}$.

5.2.1.4 Leverage corrections of the residuals

An object that has particularly high leverage h_i in the calibration set will necessarily have relatively low residual $\widehat{e_i}'$ and $\widehat{f_i}$, irrespective its true error level (see e.g. Weisberg, 1985). This was evident for the ZnO-containing litmus solution for $\widehat{e_i}$ in Figure 5.2d and for $\widehat{f_i}$ in Figure 5.2e. Object 1 did not give high residuals. Outliers with high $\widehat{e}$ or f will often not be detected by testing $\widehat{e_i}$ and $\widehat{f_{ij}}$ alone. One way to correct for this is to divide the residual by some factor that approaches zero as h_i increases and hence amplifies the residual of an object. This can be done in different ways.

The most common technique is to divide the **Y**-residual $\widehat{f}_{ij}$ by an estimate of the standard deviation of the ordinary regression residual in order to obtain the so-called Studentized residuals (see Chapter 4). The standard deviation can according to Weisberg be written as $\sigma_f\sqrt{(1 - h_i)}$ and consequantly the Studentized residuals for MLR are defined as

$$r_{ij} = \widehat{f}_{ij}/s_f\sqrt{1 - h_i} \qquad (5.20)$$

where s_f is the usual RMSEE defined in Chapter 4 or above. The Studentized residual can for simplicity be compared to a $N(0, 1)$ distribution (Normal distribution with mean 0 and variance 1) scale to reveal outliers. This approximation is best for normally distributed errors f.

To put even more emphasis on observations with large leverage, one can square the $\sqrt{1 - h_i}$ quantity and obtain the so-called leverage corrected residuals (see UNSCRAMBLER user's guide, 1987):

$$r_{\text{lev}} = \widehat{f}_{ij}/s_f(1 - h_i) \qquad (5.21)$$

This more severe leverage correction serves to simulate actual prediction residuals in **Y** (see section 4.3.2.4). These leverage corrected residuals can also be used as the basis for quick-and-dirty model validation (see section 4.3.2.4).

The r_{ij} and r_{lev} can also be used in bilinear regressions with the leverage definition and residual definitions given in the previous chapters.

In the present data set the leverage correction did not yield any new outlier detections in **Y**.

Similarly the leverage correction can also be performed on **X**-residuals, both for individual elements $\widehat{e}_{ik}$, and for objects $\widehat{\mathbf{e}}_i$. These leverage-corrected **X**-residuals can then be used instead of the original residuals to reveal outliers.

In the present example, object 2 (the red-coloured analyte object) yielded outlier warnings for such leverage-corrected **X**-residuals, both for the total spectrum $\widehat{\mathbf{e}}_i$ and for $\widehat{e}_{ik}$ in the particular range $19 < k < 30$.

Even the X-variables' residuals $\widehat{\mathbf{e}}_k$ can be leverage-corrected, with respect to h_k instead of, or in additional to, h_i, for instance in connection with re-computing of the a priori weights. But this will not be treated in this book.

5.2.1.5 Influence measures and influence plots

Another concept which is important in regression analysis is an observation's influence on the solution. Such measures of influence can be useful for assessing the importance of an outlier for the estimation of the regression parameters. Outliers with large influence are the most dangerous and should be given special attention.

In ordinary regression analysis the Cook's D_i measure (see e.g. Cook, 1977) for influence is frequently used, but other possibilities also exist (see Belsley et al., 1980). Cook's D_i is defined by

$$D_i = \frac{(\hat{\mathbf{b}}_{(-i)} - \hat{\mathbf{b}})'\mathbf{X}'\mathbf{X}(\hat{\mathbf{b}}_{(-i)} - \hat{\mathbf{b}})}{s_f^2(K + 1)} \qquad (5.22)$$

where $\hat{\mathbf{b}}$ and $\hat{\mathbf{b}}_{(-i)}$ represent the obtained MLR regression coefficients with all I objects and without object no. i, respectively (here $\mathbf{b}_0$ is incorporated in $\mathbf{b}$, see section 2.1.2.3). The parameter estimate s_f represents the residual lack-of-fit standard deviation of $\mathbf{y}$.

This can equivalently be written as

$$D_i = \frac{1}{K + 1} \frac{h_i}{1 - h_i} r_i^2 \qquad (5.23)$$

As we see D_i is a simple function of the studentized residual r_i and the leverage h_i and is simple and quick to compute. The D_i can be calibrated against an F-scale as shown in e.g. Weisberg (1985).

This MLR influence measure (5.21) can also be applied for bilinear models to provide useful information ($\hat{\mathbf{T}}$ must be used instead of $\mathbf{X}$ and $\mathbf{q}$ must replace $\mathbf{b}$). But in this case the relationship between (5.22) and (5.23) is not straightforward. In Næs (1989) is given a more detailed discussion of influence measures in bilinear models and also conditions for the identity to hold. In the same paper h_i as defined above along with the usefulness of $\hat{\mathbf{f}}$ is discussed (see also Critchley, 1985).

Generally speaking, large leverage alone or large studentized residual alone is not necessarily enough for the observation to be influential. At least a moderate contribution from each of these quantities is required for the influence to be large.

For instance, in the previous example, object no. 3 had rather low influence on the 3-factor PLSR model, in spite of its relatively high leverage ($h_i = 0.67$). The reason is that its residual $\hat{f}_{ij}$ was rather low, since represented a straight extension of the X–Y relationship of the other normal calibration objects at lower analyte concentration (see Figure 5.1a) and so also for the X–X relationship (see Figures 5.1c and 5.2c).

Conversely, object 7 also had low influence on this 3-factor PLSR model, in spite of its high residual $\hat{f}_{ij}$ (Figure 5.2e), the reason being its low leverage ($h_i < 0.1$).

Thus, both the residual and the leverage affect the influence of an observation in regression: The object must 'disagree' with the other objects, and must have been described by the calibration model.

It is useful to vizualize the influence by plotting the $\mathbf{y}$-residual $\hat{f}_{ij}$ (or its squares, or suitably studentized) against the leverage h_i for the objects. This can formally be done also for bilinear models and can be useful in searching for influential cases. Comparing such plots for $a = 0, 1, 2, 3, \ldots$ gives an efficient overview of how the calibration model develops with increasing factors.

As stated above, bilinear methods like PCR and PLSR also yield $\mathbf{X}$-residuals and statistical variance summaries of these. Plots of the residual $\mathbf{X}$-variance vs. leverage can therefore give overviews of how the X-data are modelled by the BLM. The theoretical merits of such plots in bilinear models are not clarified, but they have proven useful in practice.

Figure 5.2f shows such an 'influence plot' for the 17 calibration objects in the present example. The figure shows that objects 1 and 2 have leverages h_i very close to the theoretical maximum, 1.0, and low residual variance $s(\widehat{e}_i)^2$. They have had high influence on the modelling. Further inspection of the results reveal that they represent bad outliers.

Object 3 has moderate leverage and still relatively low residual variance, since it truly belongs to the same class as the majority of the other calibration objects.

Objects 11 and 14 show relatively high residual variances, due to their error

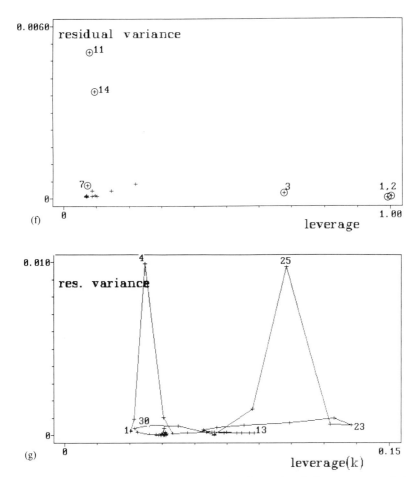

Figure 5.2 (*cont.*) Preliminary calibration for litmus in alkaline aqueous solutions. f) Influence plot for the objects' X-data in the 3-factor PLSR calibration model. Residual spectral variance $s(\widehat{e}_i)^2$ vs leverage h_i for the 17 calibration objects. All the detected outliers are marked explicitly. g) Influence plot for the X-variables in the 3-factor PLSR calibration model. Residual spectral variance $s(\widehat{e}_k)^2$ vs leverage h_k for the 50 X-variables. Adjacent wavelength channels are connected by straight lines. The outlier wavelength channels (4 and 25) are marked explicitly, together with some other channels to indicate how the spectrum behaves in the plot

peaks (see Figures 5.1b and 5.2d). But since they have not been modelled by the first three PLSR factors, their leverage is low, and their effect cannot be seen in the factor loadings (Figure 5.2b). So they are not very dangerous outliers, but we consider them sufficiently bad that we want to eliminate them before we recalibrate (see the next section).

Figure 5.2g shows the corresponding influence plot for the X-variables; here the residual variance $s(\hat{e}_k)^2$ is plotted against h_k for the same 3-factor PLSR solution. Adjacent wavelength channels $k = 1, 2, \ldots \ldots, K$ have been connected by lines, and a few of the channel numbers k are identified explicitly for interpretation.

The figure shows that none of the X-variables come anywhere near a dangerous leverage; the highest value is about $h_k = 0.14$ (for wavelength channel x_{23}, which is near the analyte absorbance maximum). The most conspicuous patterns in the figure are the high residual variances for narrow wavelength ranges around channels 4 and 25. Of course, these reflect the noise spikes added at these channels for objects 15 and 11, respectively. So here we could have given lower a priori weights to X-variables 4 and 24 (or regions 3, 4, 5 and 24, 25, 26), and repeated the calibration with objects 14 and 11 in the calibration set. But that will not be pursued here.

5.2.1.6 Recalibration with outliers deleted

Having automatically detected six outliers, and manually identified five of them (objects 1, 2, 7, 11 and 14) as had outliers, we now re-calibrate without these. The remaining 12 calibration objects now truly should reflect one single phenomenon, the varying concentration of blue litmus.

Figure 5.3a shows the prediction error obtained by cross validation; a very good internal prediction ability was obtained using only 1 PLS factor solution, but with a very small second factor visible.

During this second calibration we got one outlier warning: Object 3 was again shown to have relatively high leverage, $h_i = 0.71$ for $A = 1$ factor, and .82 for $A = 2$ factors (using 3 factors it had a leverage of 0.95).

Of course this warning was again issued because object 3 contains more than twice the analyte concentration of any of the other objects. Apart from that its data are correct.

Figure 5.3b shows the average spectrum $\bar{x}$ plus the two factors accepted: their loading vectors are scaled to reflect their importance in modelling the X-spectra ($\hat{p}_1\sqrt{\hat{t}'_1\hat{t}_1}$ and $\hat{p}_2\sqrt{\hat{t}'_2\hat{t}_2}$). It shows that the dominant spectral variation phenomenon has a loading vector $\hat{p}_1$ that is similar to the average spectrum $\bar{x}'$, which is not unexpected for an interference-free single-analyte system like this.

The second loading vector, although smaller and apparently more noisy, also displays an apparently non-random pattern of peaks and valleys. This second factor also improved the apparent predictive ability and we include it in the final PLSR calibration model as a correction for some unidentified interference type, even though we do not fully know what it represents. It displays a minimum at the analyte peak maximum and maxima at either flank, so it could be a loading vector correcting for nonlinear instrument responses, e.g. when the peak maximum

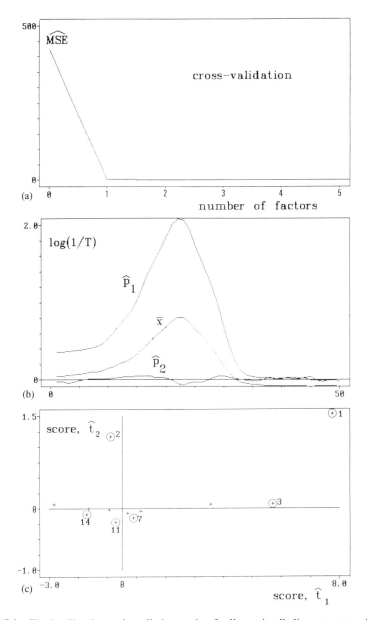

Figure 5.3 Final calibration and prediction testing for litmus in alkaline aqueous solutions. Objects 1, 2, 7, 11 and 14 were moved from the calibration set to the test set. a) Apparent predicting ability vs. model complexity: Mean square error ($\widehat{\mathrm{MSE}}$) for litmus concentration (estimated by cross validation) vs. number of PLS regression factors $a = 1,2,\ldots,7$ for the 12 remaining calibration objects. b) Calibration model: Average spectrum $\bar{\mathbf{x}}$ and two first loading spectra $\widehat{\mathbf{p}}_a$, $a = 1,2$ for the final PLS calibration model. The loadings are scaled according to size, $\widehat{\mathbf{p}}_a * \sqrt{\widehat{\mathbf{t}_a'\mathbf{t}_a}}$. c) Prediction scores of the two PLS calibration factors, $\mathbf{t}_2$ vs $\mathbf{t}_1$ for the 6 preliminary calibration outliers (numbered) and the 'unknown' test objects ('+')

approaches saturation at concentrations where the off-peak wavelengths still give linear response.

Now we are ready to predict the litmus concentration $\mathbf{y}$ from spectra $\mathbf{X}$ in new objects, using this resulting 2-factor calibration model.

5.2.2 OUTLIERS IN PREDICTION

Outlier detection during prediction for the bilinear calibration methods is primarily based on $\mathbf{X}$-residuals and the prediction leverage. Except for objects for which y_{ij} has been measured for control, the $\mathbf{Y}$-residuals do not exist and consequently measures based on $\widehat{\mathbf{f}}$ can not be used.

5.2.2.1 Objects with extreme leverage in prediction

Let us now apply this new 2-factor PLS calibration model to new X-data. We now regard the 5 erroneous outliers and the one good outlier (object 3) from the original calibration set as future unknown objects. In addition, we have data from 6 normal unknown objects, in total 12 'new' objects for which we want to predict $\mathbf{y}$ from $\mathbf{X}$.

Figure 5.3c shows the new objects' scores in the two-factor prediction model, $\widehat{\mathbf{t}}_2$ vs $\widehat{\mathbf{t}}_1$. It shows the normal, low-analyte objects to the left and the high-concentration object (no.3) to the right, reasonably close to factor 1. The normal objects (marked by +) lie close to the factor 1 axis, ranging from the lowest analyte concentrations at the left towards the high-concentration object (no.3) at the right.

What about outliers here? Notice first of all that object 7, which was tagged as an outlier during calibration because of its $\mathbf{Y}$-residuals, seems to be quite well positioned amongst the normal calibration objects, as expected since its error was in the 'true' reference data $\mathbf{Y}$, not in its $\mathbf{X}$-data. But objects 1, 2 and 3 are visually detected as outliers in this score plot. How can this be expressed mathematically?

The leverage h_i summarizes extremeness in all the factors applied in the modelling. In analogy to the leverage of the calibration objects it is for prediction objects defined as

$$h_i = 1/I + \widehat{\mathbf{t}}_i{}'(\widehat{\mathbf{T}}'\widehat{\mathbf{T}})^{-1}\widehat{\mathbf{t}}_i \qquad (5.24)$$

Here $\widehat{\mathbf{t}}_i = (\widehat{t}_{ia}, a = 1, 2, \ldots, A)'$ is the vector of regression factors for *prediction* object i, while $\widehat{\mathbf{T}}$ is again the regression factor score matrix for the *calibration* set. Like the leverage of calibration objects (see above) this leverage of prediction objects is defined as a truncated Mahalanobis distance, with $\widehat{\mathbf{T}}'\widehat{\mathbf{T}}$ acting as the weighting covariance matrix.

For methods like PCR and PLSR that gives orthogonal scores in the calibration set $(\widehat{\mathbf{T}}'\widehat{\mathbf{T}} = \text{diagonal}(\widehat{\mathbf{t}}_a{}'\widehat{\mathbf{t}}_a))$ this formula simplifies as above to

$$h_i = 1/I + \sum_{a=1}^{A} \widehat{t}_{ia}{}^2 \Big/ \widehat{\mathbf{t}}_a{}'\widehat{\mathbf{t}}_a \qquad (5.25)$$

For outlier detection, this prediction leverage can be tested against the average leverage of the I objects in the calibration set, e.g. $h_i >$ constant$*(A + 1)/I$ where the constant is e.g. 3.

Remember that for calibration objects the leverage could never be higher than 1.0. But for the prediction objects this limitation does not apply: New input spectra x_i with strange structures can more or less randomly generate large factor scores.

In our litmus example, objects 1 and 2 (light scattering and red colour) obtained very high leverages ($h_i = 62$ and 34, respectively). Prediction object 11 (error spike near the analyte response maximum, x_{25}) obtained slightly abnormal leverage (1.4).

Applications of h_i for prediction objects are given in Næs and Martens (1987) and Næs and Isaksson (1988).

5.2.2.2 Unmodelled residuals in spectral data X

The residual spectrum $\hat{e}_i{}'$ for a particular prediction object is obtained by

$$\hat{e}_i{}' = x_i{}' - \bar{x}' - \hat{t}_i{}'\hat{P}' \tag{5.26}$$

Figure 5.3d shows the obtained residuals of the 12 objects in the litmus prediction example, using the 2-factor PLSR model. As expected (Table 5.1), the abnormal objects 1 and 2 now reveal large unmodelled spectral features, and objects 11 and 14 reveal minor positive residual peaks at the positions where their errors had been introduced around $k = 25$ and $k = 4$, respectively. The rest of the objects predicted show spectral residuals close to zero O.D. units after fitting to this 2-factor calibration model.

The obtained X-residuals for the new objects can be tested automatically for abnormality in analogy to section 5.2.1.2. The average residual variance in X, $s(\hat{e}_i)^2$, is now defined by

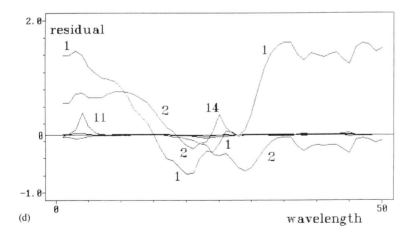

(d)

Figure 5.3 (*cont.*) Final calibration and prediction testing for litmus in alkaline aqueous solutions. Objects 1, 2, 7, 11 and 14 were moved from the calibration set to the test set. d) Spectral residuals $\hat{e}_i{}'$ after 2 PLS factors for the six outliers (numbered) and some of the normal 'unknown' test objects

$$s(\widehat{\mathbf{e}}_i)^2 = \sum_{k=1}^{K} \widehat{e}_{ik}^2 / (K - A) \qquad (5.27)$$

Each individual element $\widehat{e}_{ik}^2$ or the average residual variance, $s(\widehat{\mathbf{e}}_i)^2$, is now compared to the corresponding average noise levels $s(\widehat{\mathbf{e}}_k)^2$ and $s(\widehat{\mathbf{e}})^2$ from the calibration.

In the litmus example, objects 1, 2, 11 and 14 were automatically tagged as outliers, based on their $s(\widehat{\mathbf{e}}_i)^2$. In addition, data elements $x_{11,25}$ and $x_{14,4}$ were detected as particularly abnormal single elements in the new X-data.

Figure 5.3e summarizes the two outlier tests based on prediction data by showing the spectral residual level $s(\widehat{\mathbf{e}}_i)^2$ against the leverage h_i. The plot was scaled for increased detail, and does not correctly show the two extreme objects $i = 1$ and $i = 2$ (light scattering and red colour); their coordinates are as much as (62,1.2) and (34,0.2) respectively.

The figure also shows the interesting difference between objects 11 and 14: Both had the more or less similar error spikes, but located in different parts of the spectrum (Figure 5.1b). For object 11 the error spike was located around $k = 25$, which is inside the wavelength range spanned by the PLS factors (see Figure 5.3b). For object 14 the error spike was located around $k = 4$, which is outside this main variability range. Hence, the error spike had less influence on the score estimation in object 14 than in object 11. This in turn yielded higher leverage h_i; for the former, particularly due to the contributions from factor 2.

Since we know the true values y_{ij} for the present prediction objects, we can finally study the resulting prediction abilities in more detail. Figure 5.3f shows the predicted vs. the 'true' litmus concentrations in the 12 objects submitted to prediction testing. Grossly erroneous analyte predictions were obtained for outliers 1 and 2. On the other hand, the six normal prediction objects (marked '+') obtained very precise analyte prediction.

Object 3 gets almost perfect analyte prediction. This may be a deceptive artefact, because this object is a calibration object, and even had relatively high leverage during the final calibration! But extrapolating the good linear trend seen for the six normal prediction objects, good prediction ability even at high analyte concentrations seems reasonable in this case.

As expected, object 7 shows high apparent prediction error, but this is an artefact; the predicted analyte concentration is in fact precise, while the 'true' reference value is 30 percent too low. Objects 11 and 14 also give good analyte predictions, in spite of having been tagged as outliers. The reason is that their error spikes are rather small and concern only a couple of the X-variables used in the calibration model.

5.2.2.3 Enhancement of structures in residual spectra

It is possible to make outlier detection based on the X-residuals even more sensitive. The residual spectrum $\widehat{\mathbf{e}}_i{}'$ is expected to contain most of the random noise in $\mathbf{x}_i{}'$

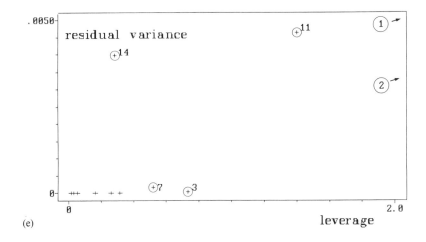

(e)

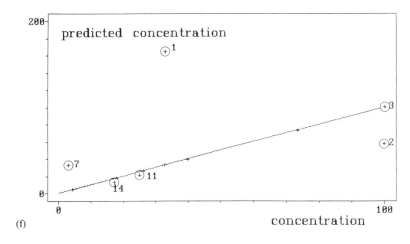

(f)

Figure 5.3 (*cont.*) Final calibration and prediction testing for litmus in alkaline aqueous solutions. Objects 1, 2, 7, 11 and 14 were moved from the calibration set to the test set. e) Influence plot for prediction, using the 2-factor calibration modelling. Residual spectral variance $s(\widehat{e}_i)^2$ vs leverage h_i for the six outliers (numbered) and the 'unknown' test objects ('+'). f) Predictive ability of the final 2-factor factor PLS calibration model. Predicted analyte concentration $\widehat{y}_i$ vs 'true' analyte level y_i for the six outliers (numbered) and the 'unknown' test objects ('+')

plus the contributions from X-factors intentionally left out from the calibration model because they are considered too irrelevant or too uncertain for the modelling of Y.

In Næs and Isaksson (1988) is discussed how even such observations can contribute to the prediction error. The residuals $\widehat{e}$ are, however, usually strongly dominated by the first factor after truncation $(\widehat{t}_{i,A+1})$ and abnormalities in e.g $\widehat{t}_{i,A+3}$ will not be detected using only $\widehat{e}$ and leverage. In such cases leverage contributions for the factors $a > A$ can be useful. It may, however, be dangerous to go too far out because the larger the value of a, the more is the factor contaminated by noise.

5.2.2.4 Other outlier detection methods

A number of alternative outlier tests can be envisioned. A simplified test type that may be easy to understand for non-statisticians assumes that the new observations must never exceed certain fixed boundaries. For instance, one may test $\hat{y}_{ij}$ vs upper and lower fixed bounds, like $0 \geq \hat{y}_{ij} \geq 100\%$.

Useful fixed bounderies can be defined from the variation range in the calibration set. For instance, the individual scores $\hat{t}_{ij}$ can be tested against fixed bounds representing the minima and maxima obtained in the calibration set. Similar tests can be done on h_i, $\hat{e}_{ik}$, $s(\hat{e}_k)$, etc.

If extra information is available about the normal variability of the X-data, this can be used to enhance the outlier detection further. For instance, if it is known that x_{ik} comes from a scanning spectrophotometer and adjacent wavelengths $k = 1, 2, \ldots$ should give similar readings, this can be used explicitly to test for broad peak-and-valley like features and for spikes, for instance after Fourier analysis (Chapter 7) of $\hat{e}_i$. In this book we have shown how this type of information instead can be used visually (see e.g. Figure 5.3d).

Conversely, if certain objects $i = 1, 2, \ldots$ are expected to resemble each other, e.g. because they represent a time series from a process, specialized outlier warnings can be constructed both for $\hat{t}_i$ and h_i, and for $\hat{e}_i$ and $s(\hat{e}_h)$ etc., again e.g. by Fourier analysis over $i = 1, 2, \ldots, I$.

———————————— Statistical extensions ————————————

5.3 OUTLIER DETECTION IN ADDITIVE MIXTURE MODELS

The additive mixture model can for each object be written as

$$\mathbf{x}' = \mathbf{k}_0' + \mathbf{y}'\mathbf{K}' + \mathbf{e}' \qquad (5.28)$$

(or $\mathbf{x} = \mathbf{k}_0 + \mathbf{K}\mathbf{y} + \mathbf{e}$, see section 3.6.3)

or for the whole data set

$$\mathbf{X} = \mathbf{1}\mathbf{k}_0' + \mathbf{Y}\mathbf{K}' + \mathbf{E} \qquad (5.29)$$

In section 3.6 different assumptions on the error terms were made. Here we will concentrate on the same assumptions as used in section 3.6.3, but similar diagnostics as those considered can also be developed in other cases. More specifically, we assume that $E(\mathbf{e}) = 0$ and $\mathrm{cov}(\mathbf{e}) = \Sigma$ is independent of $\mathbf{y}$.

5.3.1 DETECTION OF OUTLIERS IN CALIBRATION

In calibration, the estimation of **K** is done by using ordinary (or weighted) LS regression on each X-variable $k = 1, 2, \ldots, K$ separately (see e.g. section 3.6.3). This means that for each wavelength we are in an ordinary linear regression situation. The model is then equivalent to the model considered in section 5.2.1 when $A = K$, i.e. the full regression model, and thus the same outlier diagnostics and influence measures as desribed can be applied in the same way for this linear model.

We refer to Oman (1984) for another and very interesting approach in the univariate case.

5.3.2 OUTLIERS IN PREDICTION (GOODNESS OF FIT)

For bilinear methods, leverage and **X**-residuals were the basic quantities for the evaluation of outliers in prediction. Here we will concentrate on similar measures using a different approach.

In this connection **K** is thought of as a fixed matrix reflecting the nature of the measurement 'instrument'. To be useful in practice, however, estimates of the spectra **K** and matrix $\Sigma = \text{cov}(\mathbf{e})$ based on the calibration objects are needed.

First of all, in so-called linear *fixed* models, i.e. models where no distributional assumptions are made for **y** (see section 3.6.3), the only types of outliers which can occur are those with bad fit to the space spanned by **K**, i.e. those observations with large value of the projection of **x** orthogonal to **K**. We call such observations 'multivariate outliers of type A'. Conceptually, these correspond to outliers with large residuals $\hat{\mathbf{e}}_i$ for the BLM models, i.e. observations that do not fit the A-dimensional space modelled by the BLM factors.

As described in section 3.6.3 we are, however, not only interested in fixed models when predicting **y**. If the unknown object is considered to be selected from a population, it was shown that improved overall prediction ability (reduced MSE) is expected when this information is incorporated in the predictor. The model is then often termed a linear *random* model and in this model additional outliers may exist. Suppose for instance that **e** is small so **x** fits well to the **K**-space but let **y** be a value far out in the population of **y**'s. This will result in an observation **x** inside the **K**-space, but still abnormally positioned (inside this space). We call such observations 'multivariate outliers of type B'. Conceptually, these correspond to outliers with high leverage h_i for the BLM models, i.e. observations far out in the modelled BLM-space. For an illustration, see Figure 5.4.

According to Rao (1959) and Næs and Martens (1987b) a sensible statistic to use to discover outliers of type A (assuming model parameters known) is the GLS residual or the weighted sum

$$\text{RGLS} = \hat{\mathbf{e}}_{\text{GLS}}' \Sigma^{-1} \hat{\mathbf{e}}_{\text{GLS}} \tag{5.30}$$

where $\hat{\mathbf{e}}_{\text{GLS}}$ is the vector of residuals after GLS prediction of **y**. Similarly, it

296

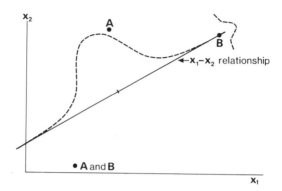

Figure 5.4 Illustration of multivariate outliers of type A and type B for two X-variables with a simple linear relationship ($A = 1$ bilinear factor = straight line). Type A has large residual deviation from the bilinear model, type B no residual deviation but extreme position along the model factor. The two distribution curves illustrate the natural variability of observations along and orthogonally to the model line, for detection of outliers of type B and A, respectively

was shown by Næs and Martens (1987b) that a sensible statistic to use to reveal outliers of type B is

$$\text{RDIF} = \hat{\mathbf{e}}'_{\text{DIF}}\mathbf{S}^{-1}\hat{\mathbf{e}}_{\text{DIF}} \tag{5.31}$$

where

$$\hat{\mathbf{e}}_{\text{DIF}} = \mathbf{X}\hat{\mathbf{b}}_{\text{BLP}} - \mathbf{X}\hat{\mathbf{b}}_{\text{GLS}} \tag{5.32}$$

and

$$\mathbf{S} = \text{cov}(\hat{\mathbf{e}}_{\text{BLP}}) \tag{5.33}$$

Under normality assumptions RGLS and RDIF are independent and chi-square distributed with $I - K$ and K degrees of freedom, respectively. It follows that RGLS and RDIF larger than certain percentiles in these distributions are suspicious and indicate outlying observations. The expression for $\mathbf{S}$ can be found in Næs and Martens (1987b).

In Næs and Martens (1987b) it is also shown that

$$\text{RBLP} = \hat{\mathbf{e}}'_{\text{BLP}}\mathbf{S}^{-1}\hat{\mathbf{e}}_{\text{BLP}} = \text{RGLS} + \text{RDIF} \tag{5.34}$$

which is a weighted sum of residuals from BLP prediction weighted by their covariance matrix. This RBLP is the natural residual criterion in a random model and by the relation in (5.34) it is able to detect both types of outliers (both A and B).

In practice the parameters in RBLP etc. must be replaced by estimates based on the calibration objects. The robustness to estimation error is discussed in Næs and Martens (1987b).

For further discussion on outliers in prediction we refer to Brown and Sundberg (1985) and Næs (1986b).

———————————— End of statistical extensions ————————————

6 Data Selection and Experimental Design

SUMMARY Basic discussion about the importance and the role of experimental design and data selection is given. General principles and illustrations based on NIR data are presented. Object selection based on spectral values is briefly summarized.

6.1 INTRODUCTION

6.1.1 DESIGN IS CENTRAL IN CALIBRATION

Calibration for simple mixtures is usually easy and cheap, but calibration of complex measurements, such as diffuse reflectance of powders, slurries or intact biological material, requires a lot of representative empirical data and careful multivariate calibration. This can be time-consuming and also quite expensive. So multivariate calibration is only worth the work for reasonably large series of analyses.

We first have to decide upon analytical procedures for measuring Y and X, and to consider their expected performances. Thus the question is: 'Given the noise in the reference method Y, and given the selectivity problems and noise in the X-data, is it at all possible to attain the required prediction ability?' If the answer is definitely 'no', then there is no sense in wasting time trying to make a calibration.

If the answer to the preliminary assessment is 'yes' or 'may be', then the problem is: 'How should we calibrate?'. This requires an assessment of what are the phenomena that have to be modelled,- analyte and interferences, nonlinearities, interactions etc. This can be based on prior knowledge about the object material and the instrumentation, or possibly on data analytic inspection of a large amount of available X-data.

On this basis follows the actual *experimental planning*—how to get good calibration data. We must decide which variables to measure, which experimental factors to control, how to combine them, and how many replicates to use. We must also decide whether to include a number of more or less *randomly* chosen observations in order to pick up uncontrollable factors and unexpected surprises. In some extreme cases in for instance NIR spectroscopy, so little information is available for design that a randomly selected sample may even be the best we can use. This is termed random or *natural* calibraton as opposed to *controlled* calibration which is usually to be preferred. In Figure 6.1 are shown cases of frequency diagrams for a controlled and a random calibration experiment.

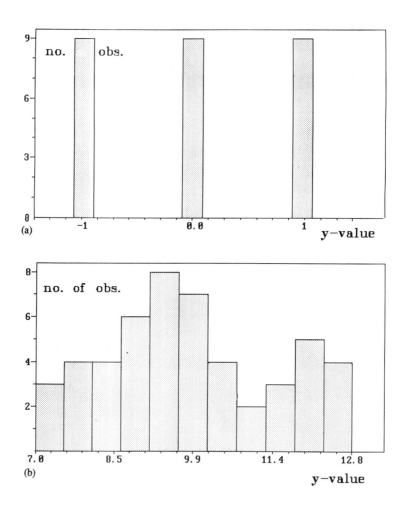

Figure 6.1 a) Frequency histogram for one of the constituents y_j in the controlled calibration experiment shown in Figure 6.2b. b) Frequency diagram for constituent $y =$ protein percentage in the random calibration experiment used in the wheat example in section 3.5

In multivariate calibration, experimental design is required in order to ensure a satisfactory description of the analyte-instrument relationship and of the instrument interferences in light of the unavoidable measurement noise and model errors. Good calibration plans can save us time and money: They increase the efficiency of our calibration work, by ensuring that the important information is obtained, while useless or unnecessary measurements are avoided.

Ideally, good planning does require good understanding of our instruments and good understanding of the type of objects to be calibrated for. But even when our understanding is very incomplete, elements of conscious planning can help us get useful calibrations. Thus we can learn while we work, through a cycle of planning—measurements—calibration—practical prediction—outlier detection and interpretation—more planning—more measurements—improved re-calibration etc.

6.1.2 IMPORTANT ASPECTS OF DESIGN

The design of a calibration experiment has two basic aspects:

1) To span the important types of systematic variability, including interferences, so that they can be assessed and hopefully modelled
2) To minimize the effect of random noise of various sorts.

Both should be given attention.

Fortunately, unexpected interferences can often be detected later on through their abnormal leverage contribution and/or their abnormal spectral residual contribution during prediction, as discussed in Chapter 5. But to avoid recalibration or updating, it would have been better to include the various types of interferences from the beginning.

It should be noted that the word *interferences* in this context means both chemical/physical phenomena affecting the X-data, and effects like nonlinear instrument responses and inter-constituent interactions (see Chapter 1).

Nonlinearities, and in particular distinctions between different types of them, are treated in detail in Chapter 7 and we shall here only briefly mention that even *nonlinear* structures between **Y** and *individual* X-variables can often be modelled well by linear or bilinear models. In such cases the nonlinearities behave like any other type of interferences in **X**, being modelled e.g. by additional factors in PLS or PCR (see e.g. Martens, 1983; Næs and Isaksson, 1988). In any case, to give satisfactory results, even these extra nonlinearity dimensions must be spanned properly by objects in the calibration design.

In general, the wider the range of objects calibrated for in one calibration model, the higher is the chance of running into difficult nonlinearity and heterogeneity problems. So from a mathematical model point of view, one should develop calibration models for *narrow* ranges of objects.

On the other hand, our local approximation models have parameters that need to be *estimated* statistically, and that calls for *wide* ranges of calibration objects to be modelled together. Sampling problems and measurement noise in **X** and **Y** can create damaging errors in these parameter estimates—depending on the signal/noise ratio in the calibration data. We can reduce the estimation error by bringing down the noise—improving our sampling and measuring techniques and/or by including more calibration objects. But this brings the cost of calibration up.

So in designing multivariate calibration experiments we must strike a compromise between optimizing the model structure (narrow range) and optimizing the parameter estimation (wide range)—and usually under constraints of cost, time and availability of objects.

6.1.3 DEFINING THE TARGET POPULATION OF A CALIBRATION

Since a multivariate calibration only gives local approximations, it will give different predictive abilities for different types of objects, even within the normal ranges of analyte and interferents included in the calibration set (see Chapter 4).

For instance in the case of NIR protein determinations in a certain type of wheat, we may want to calibrate for a typical protein range, say 10–14%. Even within this range our predictions would probably be most precise for the most typical protein levels calibrated for, say near 12%, and considerably more uncertain at 10 and 14%, and even more so for objects with 16% protein or with other qualities outside the normal range calibrated for (see sections 4.1 and 3.6.3).

The general goodness of a predictor $f(\mathbf{X})$ is a *statistically weighted average over the different types of future object qualities*. So it is dangerous to develop a calibration model and to specify its predictive ability if we do not know which distribution of future objects to target on.

Once the target population has been more or less identified, it is important to get calibration objects that together are representative for these future objects, both for model estimation and for validation assessments.

6.1.4 PRELIMINARY ASSESSMENT OF PROBLEM COMPLEXITY

A list of some of the phenomena expected to affect the chosen X-measurements can usually be made.

In e.g. high-speed analysis of wheat flour by diffuse NIR reflectance, where we want to determine the water, protein and hardness, the following phenomena are expected to affect:

1) Chemical constituents that we want to calibrate for, e.g. water and protein contents of wheat flour
2) Chemical constituents and physical phenomena that cause higher-order qualities that we want to calibrate for, e.g. starch content and genetic starch type determining the 'hardness' of wheat
3) Chemical and physical interferents inherent in the objects, e.g. celluloses content and particle size
4) Physical interferences arising in the measurement process, e.g. object packing density and temperature.

The list of such potential X-phenomena can easily become long. A combinatorial explosion is experienced if all possible combinations of all these phenomena at several levels should be measured.

So the next step is to simplify the list.

First of all we choose which phenomena we can safely *ignore* because they are constant or insignificant. For instance, in analysis of solvents, the bulk solvent may sometimes be regarded as a constant background when analyzing minor solutes, even when the solvent does add a signal, like in NIR spectroscopy. And of course constituents with no effects at all on the X-signal can safely be ignored. But there may be indirect effects lurking in the form of inter-constituent interactions: For instance, NaCl has no NIR signal itself, but modifies the NIR water signal so much that it can be determined in e.g. meat products by NIR spectroscopy! It would be unwise to ignore such a constituent in the design.

Are there other ways to simplify the design?

Phenomena that always vary linearly with each other can be treated as one *group* of phenomena, as exemplified by two constituents in binary mixtures: When one goes up, the other one goes down. This is true when two or more interferents are strongly correlated (positively or negatively)—they can be treated as one single interference effect. So we usually do not have to consider every single chemical compound present in our objects; for a reasonably narrow population of objects, most of the constituents will vary in groups, for instance in biological material due to common regulatory genetics. Knowing what groups to expect requires domain-specific knowledge.

The major remaining non-constant and significant X-effects (or groups of X-effects) should ideally by described by the calibration model. They must therefore vary independently of one another in the calibration data. So the next phase is to plan how to obtain a calibration set with sufficient information.

In some cases when prior information about interfering phenomena is incomplete, it may be advantageous to incorporate some *randomly* selected objects as well (see, for instance, chapter 3.6.3), to enhance the chance of having all important sources of variation present in the calibration.

6.2 CHOOSING THE VARIABLES TO BE MEASURED

Before we consider experimental design for the calibration objects in more detail, we have to decide which variables to measure and to what precision.

6.2.1 THE REFERENCE VARIABLES Y

When the variable(s) y_j to calibrate for have been defined, we have to decide exactly how we shall measure it. To save a lot of frustration and wasted time trying to do the impossible, it is important to assess the noise level to be expected in these reference measurements with the given sampling, object preparation and measuring technique.

This noise level of the reference method, $\sigma_{j,\text{ref}}$, can be estimated from earlier calibration experiments of the same type, by theoretical considerations or by taking replicates for various objects and averaging the variance estimates.

The errors summarized by $\sigma_{j,\text{ref}}$ will contaminate the resulting parameters in the calibration model. Still, it is indeed possible to attain predictions $\hat{y}_j$ with higher precision than that of the reference data in the calibration as shown in Figure 4.2. The reason is that the different objects $i = 1, 2, \ldots, I$ in the calibration set act as each other's replicates.

But this appealing property should not be over-rated. What precision to require from the reference method, and how many calibration objects to require, can best be assessed by analyzing a preliminary set of $(\mathbf{X}, \mathbf{Y})$ data.

However, a crude upper bound can be assessed directly. The uncertainty in $\bar{y}$ due to random uncertainty $\sigma_{j,\text{ref}}$ in the I independent calibration objects is $\sigma(\bar{y}) = \sigma_{j,\text{ref}}/\sqrt{I}$. This uncertainty in $\bar{y}$ will necessarily contaminate every prediction $\hat{y}_{ij}$. So the noise level of the reference method $\sigma_{j,\text{ref}}$ must at least be so small that

$$\sigma_{j,\text{ref}} \ll \text{RMSEP}_j \sqrt{I}$$

where RMSEP_j is the required precision of the predictions. Assume for instance that we know that we later shall afford to analyze only $I = 25$ calibration objects, and that a maximum prediction uncertainty of $\text{RMSEP}_j = 0.2\%$ will be tolerated. Then we get the requirement $\sigma_{j,\text{ref}} \ll 1\%$.

Of course, in practice other errors will usually dominate RMSEP_j—estimation uncertainty in other model parameters, unmodelled interferences, propagation of noise in $\mathbf{x}_i$ etc. So this bound is usually trivial, but may at least stop the statistical novice from trying the impossible.

6.2.2 THE INSTRUMENT VARIABLES X

Then we have to decide which X-variables to measure for determining $\mathbf{Y}$. The X-variables chosen must be expected to have a highly reliable (although not necessarily causal, selective or linear) relation to the analyte(s) $\mathbf{y}_j$ (see section 2.2).

In addition, the choice of X-instrumentation, sampling procedure, object preparation and measurement procedure is a question of price, availability and reliability—plus of course the ability to distinguish the analyte from interferences and noise.

Traditionally, such interferences had to be removed from the objects prior to the X-measurements, which can be laboursome, expensive and introduce artifacts, but which requires only a single-channel X-instrument. With multivariate calibration their effects can instead be removed from the data, after the X-measurements, but this requires several properly chosen X-variables.

This multivariate data analytic selectivity enhancement (see e.g. Martens et al., 1987) requires the analyte's and the interferents' X-responses to be sufficiently different, and the analyte's signal to be sufficiently strong, compared to the measurement noise.

If this is impossible to attain, then a combination of the two strategies may work—applying some simple analyte concentrating and interference removal steps before measuring the multivariate X-data, and then removing the effects of the rest of the intereferences by multivariate calibration modelling.

Which X-variables to use?

In the simplest case all the X-variables come from one multi-channel instrument. But if necessary, $\mathbf{X}$ can consist of variables from several different instruments. The practical standardization and maintainance problems may then increase, but a great improvement in selectivity can be gained. Simple combinations like adding a temperature sensor or an extra turbidity detector to one's scanning spectrophotometer should be contemplated.

In general it is better to measure too many than too few of the available X-variables, and to use them at least in the preliminary phases of a calibration project. This increases the probability of picking up and correcting for important selectivity problems, especially unexpected interferents. Contrary to traditional regression methods like MLR, the full-spectrum calibration methods like PLS and PCR, redundancy among the X-variables is generally a very small problem (see sections 3.3, 3.4 and 3.5). But irrelevant X-variables should certainly be eliminated.

Example: We have measured absorbance at 5 wavelengths (x_1, x_2, x_3, x_4, x_5), all with the same precision, in solutions of an analyte A and two interference phenomena B and C. With '+' and '0' symbolizing positive absorbance and zero absorbance, respectively, their pure spectra can be described:

Wavelength channel		1	2	3	4	5
Analyte A:	$\mathbf{k}_A' =$	0,	+,	0,	0,	0
Interferent B:	$\mathbf{k}_B' =$	0,	+,	+,	+,	+
Interferent C:	$\mathbf{k}_C' =$	0,	0,	0,	0,	+

Should we measure and use all 5 wavelength channels as X-variables in this case, or just one of them, or what?

First of all, channel x_1 is just a baseline variable and will not improve the predictions if the baseline has been properly subtracted. With a reasonably large amount of data it could, however, be measured and included in the model to give improved graphical understanding and outlier detection. But with limited amounts of calibration data its noise could be drawn into the calibration model and result in lowered prediction ability.

Channel x_2 is necessary since that is where the analyte responds according to $\mathbf{k}_A$. But $\mathbf{k}_B$ shows that interferent B upsets the simple determination of A at this channel and therefore must be corrected for by a multivariate calibration model (a two-factor model). Channels x_3, x_4 and x_5 respond to interferent B and could be used in this selectivity enhancement—one, two or all three of them.

However, at channel x_5 the response of interferent B is in turn upset by a secondary interferent, C, which does not in itself interfere with the analyte channel x_2. So with x_5 included, an extended calibration model would be needed (a three-factor model). If enough calibration objects were available with precise enough data, this resulting calibration model would probably work well (a 3-factor model). But it is unnecessarily complicated, requiring unnecessarily many calibration objects and unnecessarily precise input data.

By ignoring x_5 and x_1, the remaining x_2, x_3 and x_4 will still give the desired outlier detection and selectivity enhancement, with a simple two-factor calibration model. If we knew this in advance, we would not have had to measure and store x_5 and x_1. If we were uncertain in advance, we should have recorded them and it is hoped that we would have found them from data analysis so that we could choose to eliminate them.

Is the noise in **X** *low enough?*

How precise should the X-measurements be in order to give the desired prediction precision RMSEP? The total prediction error is a consequence of both calibration errors in the parameter estimate $\hat{\mathbf{b}}$, and prediction error propagation.

The random noise in the X-data, characterized by $s(\mathbf{e})$, will systematically affect the parameter estimation of the calibration. How strong this model contamination by X-noise is, depends on the number of independent calibration objects and how many independent model parameters are estimated from these data.

The general level of prediction error caused by the estimation error in $\hat{\mathbf{b}}$ is somewhat difficult to assess (see e.g. sections 3.4 and 3.5), since it will vary from object to object, depending on how spectrum $\mathbf{x}_i{}'$ differs from the center $\bar{\mathbf{x}}'$. Generally, however, the estimation error in $\hat{\mathbf{b}}_j$ and hence the uncertainty of $\hat{y}_{ij}$ will decrease with the number of independent calibration objects, I. So, with noisy X-data it is necessary to include a large number of calibration objects—how large depends on the structure in the data. This is best assessed after calibration. But in general, if the X-data are too noisy, the calibration model parameters can still be reasonably precise if we just have enough calibration objects.

But even with a very high number of calibration objects we are still not guaranteed good predictions, because there will also be some noise in the future X-measurements. This means that even with perfect estimation of $\mathbf{b}$, the predictions $\hat{y}$ may have some noise due to e.g. lack of perfect measurements of $\mathbf{x}$ (intrinsic error). An illustration is given in Chapter 4 (Figure 4.2).

The random error in the X-variables of the individual future prediction objects will be propagated through the calibration model and create additional random errors in $\hat{y}_{ij}$. How strongly this random error in $\mathbf{x}_i$ will be amplified through the model depends on how selectivity enhancement is done in the model. If the spectrum of the analyte is very different from that of the interferents, then the amplification is low. But if the analyte spectrum is similar to a linear combination of the interferents' spectra, the model has to amplify minor spectral differences in order to distinguish analyte and interferents from each other. Noise in $\mathbf{x}_i$ will then also be amplified in the prediction.

It may be desirable to assess, prior to calibration, how precise our X-data have to be to allow us to hope for a certain maximal prediction error in $\mathbf{y}_i'(y_{ij}, j = 1, 2, \ldots, j)$. A rough approach to this can be based on a *preliminary* mixture model approximation, $\mathbf{x}_i = \mathbf{K}\mathbf{y}_i + \mathbf{e}_i$. Approximate measurements of the most important constituents' spectra $\mathbf{K} = (\mathbf{k}_j, j = 1, 2, \ldots, J)$ must then be available. If the X-data are known to have a random noise with standard deviation of about σ_e, then the covariance matrix of the final predictions $\hat{y}_j, j = 1, 2, \ldots, J$ can be approximated by $\mathbf{D} = \sigma_e^2 (\mathbf{K}'\mathbf{K})^{-1}$ (sections 3.6.3 and 2.1.2).

The error standard deviation of $\hat{y}_{ij}$ will thus be at least $\sqrt{d_{jj}}$, i.e. the root of the jth diagonal element of $\mathbf{D}$. Unidentified interferents, nonlinearity effects and noise in the reference data $\mathbf{y}$ for calibration will add to this preliminary error estimate. Thus if $\sqrt{d_{jj}}$ is higher than the maximum error standard deviation that can be tolerated in future $\hat{y}_{ij}$, then something must be done to the methodology, either reducing σ_e or eliminating certain constituents that interfere with analyte j.

6.3   EXPERIMENTAL DESIGN FOR CALIBRATION

6.3.1 BASIC PRINCIPLES

First of all, it is important to note that since the X-variables in the linear regression equation and in bilinear regression are often empirical values impossible to specify in advance of an experiment (i.e. $\mathbf{X} \Leftarrow$ causal $g(\mathbf{Y})$, see section 2.2). So ordinary regression design theory as reported in e.g. Box et al. (1978) and in Deming and Morgan (1987), where exactly such control is assumed, may often be impossible to apply directly.

An important exception here is the linear mixture model where the X-variables are linear functions of the chemical constituents plus noise. In this case each $\mathbf{x}$ can be written as

$$\mathbf{x}_i = \mathbf{k}_0 + \mathbf{K}\mathbf{y}_i + \mathbf{e}_i \qquad (6.1)$$

and the combinations of y-values can be designed according to a traditional design scheme for precise estimation of $\mathbf{k}_0$ and $\mathbf{K}$ or precise predictions of future $\mathbf{y}$ (see e.g. Ott and Myers, 1968).

A simple and efficient design which can be applied in model (6.1) is a factorial design with two levels on each factor (see Figure 6.2a). This is a so-called orthogonal design making estimation of different parameters (for different constituents) statistically independent. If the number of factors to control is high, the number of calibration objects in the design can be reduced by sensible application of fractional factorial designs (see e.g. Box et al., 1978). In case of closure among the constituents as discussed in sections 3.6.1 and 3.6.2, the constituents cannot

306

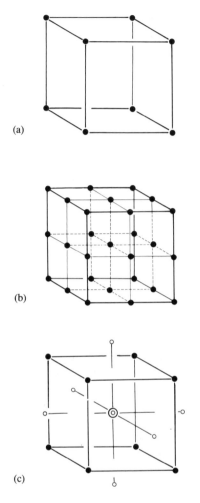

Figure 6.2 Factorial designs for three factors (analytes or known interferents). a) Factorial design with two levels for each factor. b) Factorial design with three levels for each factor. c) Central composite design

be varied independently, and modifications such as those presented in e.g. Cornell (1981) must be applied.

In general, the causal mixture model represents over-simplifications and the calibration model has to be expanded, either by using more bilinear factors than the number of known constituents (sections 3.3–3.5), or by using additional factors in $\mathbf{e}$ (see e.g. sections 3.6.2 and 3.6.3).

If a bilinear model is used, the variation in $\mathbf{X}$ is approximated by $\hat{\mathbf{T}}\hat{\mathbf{P}}'$, where $\hat{\mathbf{T}}$ spans all possible types of systematic factors; chemical concentrations, interferences and nonlinear functions and interactions of the chemical and physical factors. Some of these may be known and controllable, others not.

Experimental design for such models is complicated and is discussed in very few applications (see e.g. Næs and Isaksson, 1989). One of the problems is that the $\hat{T}$-factors are abstract and normally it is impossible to identify one single factor as e.g. protein content. Secondly, the 'dependent' Y-variable is generally part of the design and this creates problems with application of traditional regression design theory.

However, the essence of the general ideas used in the simpler cases of regression design carry over their importance and should be applied for safe calibration: The most basic point to know here is that only factors which are present and modelled (by $\hat{T}\hat{P}'$) in the calibration can be compensated for in prediction. This means that an interferent not modelled in the calibration, will create alias problems (see e.g. section 3.6.2) and therefore bad predictions. Thus, the following is of fundamental importance:

All phenomena that vary (in X) in the target population, must be spanned in the calibration set and described in the calibration model.

Each phenomenon should be spanned as well as possible, to be separated from noise and from each other. Remember, however, the dilemma in section 6.1.2 telling us to consider model fit in the larger population as well.

What then about factors in $\hat{T}$ that represent nonlinear functions of chemical constituents or interactions between chemical and physical phenomena? To be sure that all combinations of them are well represented or that all factors correponding to nonlinearities are properly spanned, each varying phenomenon should be represented at more than two levels. A general rule of thumb that can be used is that the number of levels should at least correspond to one plus the expected complexity of a polynomial approximation of the causal realationship (see e.g. Næs and Isaksson, 1989). So, if it is important to keep the number of calibration objects low, and if for instance the functional relation between **x** on one hand and the varying interferences on the other can be approximated by a second degree polynomial with interactions, then it may suffice to use 3 levels of each factor, for instance a low, an intermediate and a high value (see Figure 6.2b)

This type of multi-level multi-factor orthogonal design can give a 'combinatorial explosion', resulting in unnecessarily expensive calibration experiments. One interesting way to simplify such multi-factor factorial designs is to combine a simple 2-factorial experiment with some extra points into a central composite design (Figure 6.2c).

In case of closure one can alternatively use a type of simplex lattice design as described in e.g. Cornell(1981) (see also Figure 6.3).

Thus, factorial designs in all varying phenomena with more than two levels for each factor are reasonable candidates for BLM designs, representing a type of even spread of points over the actual region.

This even spread of design points is also important from another point of view, namely for detecting nonlinear relations between **y** and **x** and to smooth out

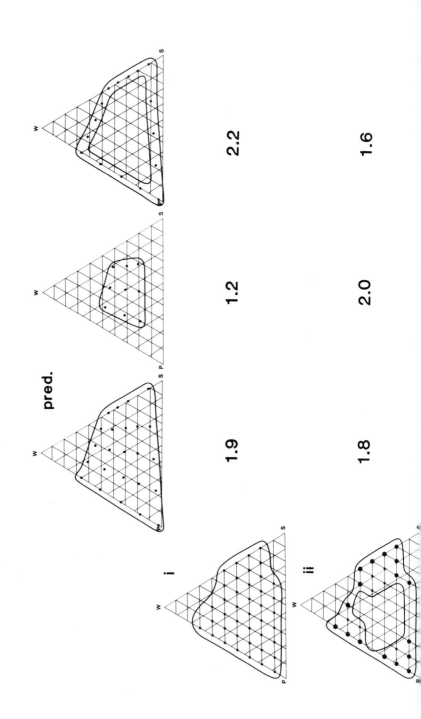

pred.

2.2

1.2

1.9

1.6

2.0

1.8

i

ii

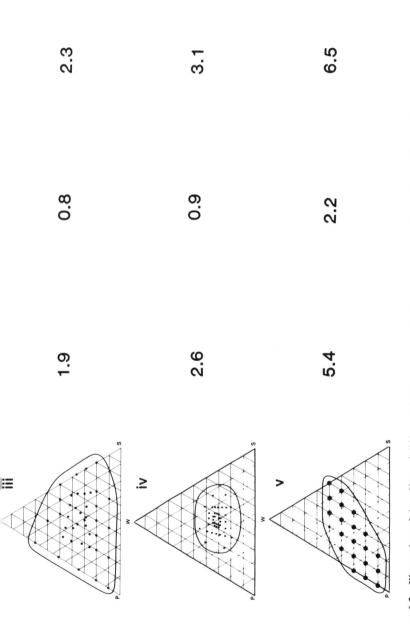

Figure 6.3 Illustration of the effect of different design strategies for the mixture experiment. Vertically are given the designs of the five calibration sets i,ii,...,v, and horizontally the design of the three test sets, illustrated in the three-phase diagram of the three constituents. The resulting prediction error for the test sets is presented in RMSEP units (see Chapter 4) for each calibration model. Solid circles indicate two points

moderate nonlinearities. This is discussed in detail in Næs and Isaksson (1989) and illustarted in section 6.3.2.

In many cases the target population has more objects near the centre than far from it. In such cases it may be natural to reflect this in the design strategy by selecting more calibration objects from near the centre. The consequences of this are not properly analysed, but an illustration is given in the next section.

Finally, remember from the above that in cases where information about interfering constituents and physical phenomena is incomplete, it may be of use to add some randomly selected objects as well.

6.3.2 ILLUSTRATION: THE IMPORTANCE OF GOOD DESIGN

The following illustration is a summary of Næs and Isaksson (1989), and concerns NIR reflectance determination of protein in mixtures of fish meal, starch and water in different proportions. A high number of mixtures were generated and measured in a Technicon InfraAlyzer 400 with 19 wavelength channels. Various subsets of calibration objects were selected from these, each set having the same number of calibration objects ($I = 41$). All the calibrations were tested on the same 25 independent prediction objects. In addition, the calibrations were tested on two different subsets of the prediction set, i.e. the central points and the border points. The calibration method used was PCR with deletion of eigenvectors corresponding to the smallest eigenvalues (see section 3.4). The data were scatter-corrected according to the multiplicative signal correction (MSC, see Chapter 7) to reduce the effect of physical phenomena on the spectral readings.

The results of the different calibrations are shown in Figure 6.3 and the results are given in RMSEP units (see Chapter 4). In each of these five designs the RMSEP corresponds to the result of PCR with optimal number of factors.

The most striking feature in the figure is the clear demonstration of the need for spanning all dimensions properly. Calibration design iv) only covers a small region of the target population and the predictions outside this region are very bad. Design v) spans the variation of the constituent protein calibrated for, but as we see this is not enough to ensure good prediction. To obtain this one obviously needs calibration objects spanning the whole space of interest (here mainly water and carbohydrates in addition to protein).

As we also see, the performance difference between the other three calibration sets are not very clear. There is, however, a tendency of better results in the middle for design i) and the 'natural population' case iii) than for design ii). As discussed in the paper by Næs and Isaksson (1989) this can be due to the 'least squares effect' of best prediction ability near the centre as discussed in section 3.6.3. and in Chapter 4, Figure 4.2. Alternatively it can be due to a nonlinearity which is not properly handled in case ii). This becomes clearer when studying the individual prediction errors as done in the original paper.

Overall it seems that the nonlinearity problem is very important in this type of application, especially when the region of interest is as large as here. To be able to handle this it is necessary that all different parts of the actual target population are covered.

6.4 CHOOSING THE CALIBRATION SET FROM AVAILABLE OBJECTS

The type of 'design' considered here was originally developed for NIR analysis, but can equally well be used in other types of data. It is based on the fact that in many situations it may be cheap and simple to obtain X-data, e.g. spectra, for a lot of different representative objects.

In such cases there is a large probability that the available object set contains the most important types of variation, but the problem is that we are not able to obtain reference data for all these objects. In other words, we can not afford to use all the available objects for calibration. It is then of great interest to select a sensible subset of objects from those with available spectrum and submit only these to chemical analysis and calibration.

This problem was first treated by Hruschka and Norris (1982). Honigs et al. (1985) presented in detail a similar approach. This procedure gave promising results in an example, but it is based on a rather complicated procedure which is difficult to evaluate from statistical principles.

In general, such searching for informative subsets of spectra can be based on two different principles:

1) Looking for representatives of *clusters* or groups of similar spectra, to cover the main variability types seen in the data.
2) Looking for representatives for *factors* or types of variation phenomena creating the dissimilarities seen in the data.

When nothing is known about the linearity structures in the $X–Y$ relationships, the cluster approach may be advisable since it makes the fewest assumptions and spreads the design points more evenly over the actual region. If, on the other hand, the $X–Y$ relationship is expected to be homogenous and linear, then the factor approach may be advisable since it can reduce the number of calibration objects the most.

Næs (1987) presented an approach based on cluster analysis (see also Zemroch, 1986). The rationale for the cluster analysis method may be put into the following sentence: It is better to use spectra (design points) covering the whole space of variation than applying many points close to each other (see Figure 6.4a). Cluster analysis will collect points which are close in the high-dimensional space; selecting only one (or a few) from each of the clusters will satisfy this requirement (Figure 6.4b). In other words, selection of objects (spectra) based on cluster analysis ensures that each design point is not too close to any other of them. A variety of clustering algorithms can be employed, ranging from hierarchical procedures (Mardia et al., 1980) to e.g. fuzzy clustering (see e.g. Gunderson et al., 1988). Application of the

312

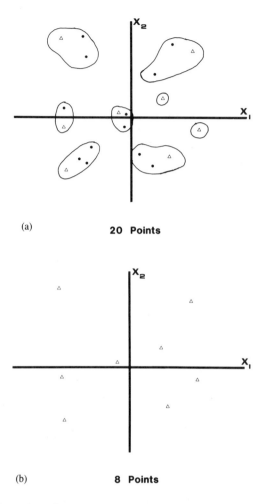

(a)   **20 Points**

(b)   **8 Points**

Figure 6.4 Illustration of the main idea behind the cluster analysis of **X** for selecting calibration set: a) The calibration objects with available X-data plotted for two X-variables, with obtained clusters encircled. b) One representative for each cluster. At least one object from each cluster is selected for the calibration set, and for these **Y** is measured. In this case the point farthest away from the centre is picked from each cluster

clustering technique is given in Figure 6.4c for the complete linkage hierachical clustering algorithm (Mardia et al., 1980).

In order to reduce the effect of very large, well understood variations like the light scattering in NIR spectra, the clustering can be based on standardized eigenvectors (the factor scores $\hat{t}_{ia}$ obtained by PCA of the X-data, section 3.3), rather than on the individual X-variables. In order to reduce the effect of random noise on the clustering, eigenvectors with eigenvalues $\mathbf{t}_a{}'\mathbf{t}_a$ close to zero are ignored.

The second approach is to search for objects to span the relevant X-space, but without necessarily representing all sub-groups of objects. This can also be based

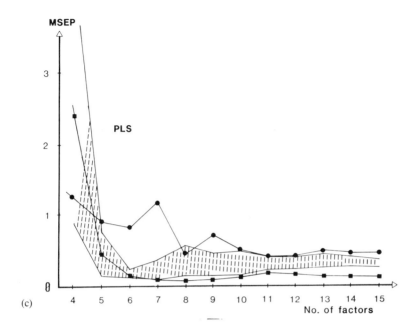

Figure 6.4 (*cont.*) Illustration of the main idea behind the cluster analysis of **X** for selecting calibration set: c) Application of the cluster algorithm for object selection. The data are from NIR determinations of fat percentage in biscuit dough and the calibration method is PLS. The shaded region represents the results from three random calibrations, i.e. MSEPs from randomly selected calibration objects. The line denoted by squares illustrates the results from the cluster analysis. The line denoted by filled circles illustrates the results from calibration objects consciously selected to cover only a sub-region of the total. The number of calibration samples were 20 in each of the calibrations. The cluster analysis and the other selections were done on a set of 40 objects

on the scores from the PC analysis of the X-data. Assuming reasonably linear or smoothly curved X–X and X–Y relationships, objects with high leverage h_i are selected. These could e.g. be two objects at high and two at low scores $\hat{t}_{ia}$ for each individual factor $a = 1, 2, \ldots, A_{\text{opt}}$. Some typical objects ($\hat{t}_{ia}$ about zero) are included to allow for possible curvatures (Martens et al., 1983a).

7   Pretreatment and Linearization

SUMMARY   Distinctions between different types of nonlinearity, i.e. curvature and multiplicative effects are given. It is shown how linear models can account for certain types of individual nonlinear relationships. Pretreatments based on transformations of Y and X and addition of extra X-variables are discussed. Multiplicative effects can be eliminated by closure, internal standards or multiplicative signal correction techniques. Other transforms such as smoothing are discussed.

7.1   WEIGHTING OF VARIABLES

Weighting of variables $x_1, x_2, \ldots, x_K$ can be important in all methods where least squares (LS) fitting (section 2.1.2.3) is used for concentrating the information in many X-variables into a lower number of 'phenomena'—bilinear factors or mixture parameters.

These a priori weights can either be applied using the weighted least squares (WLS) principle or the variables can be pre-scaled according to them. This allows us to use the simpler ordinary least squares (OLS) principle in estimation.

Here we focus on weighting of X-variables for bilinear regression, i.e. individual weighting according to

$$x_{ik} = x_{ik,\text{old}} * \text{weight}_k \tag{7.1}$$

It should be noted that this type of a priori weighting has no effect for the ordinary multiple linear regression predictor, since this is invariant to linear transformations of this type.

Weighting as defined in (7.1) has two different aspects. First of all it may be important to balance the noise of the input variables to avoid informative X-variables being overshadowed by the noise in others. Figures 7.1 and 7.2 illustrate this for spectral variables to some of which random noise has been added. The other aspect is the weighting of variables to similar total variance. This may be

314

important for plotting purposes and to prevent variables with the largest variation from dominating the rest in the modelling. This is discussed at the end of this section (see also Mardia et al., 1980).

Adjusting the X-variables to similar noise levels

In Figure 7.1a the modified O.D. spectra from six calibration objects are given at 50 wavelength channels between 400 and 800 nm. The objects represent aqueous solutions of litmus at pH 10, measured by transmission (T) with a fiber-optic spectrometer and expressed in O.D. units $\log(1/T)$. The original spectra represent the outlier-free calibration objects used in Figure 5.1—the total number

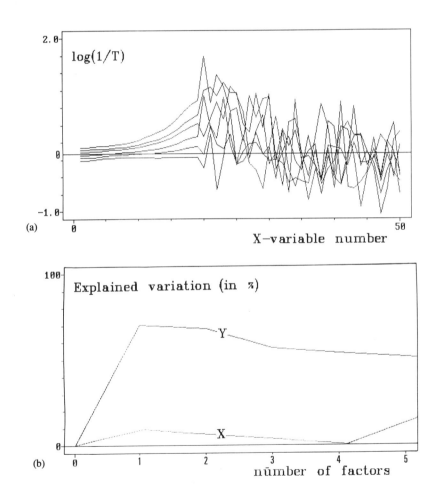

Figure 7.1 a) Six spectra with amplification of noise for wavelengths 560–800 nm, i.e. X-variables 20 to 50. b) Plot of explained variation of **Y** and **X** as a function of the number of factors

316

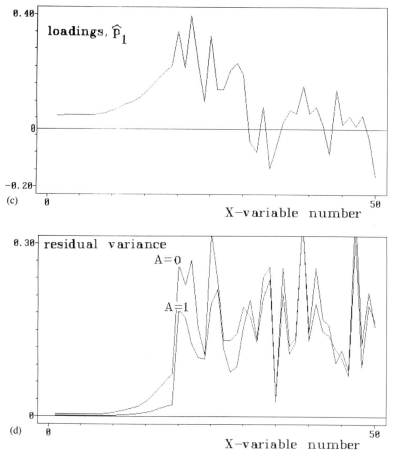

Figure 7.1 (*cont.*) c) The first loading vector, $\hat{\mathbf{p}}_1$. d) Residual X-variance for each wavelength after 0 and after 1 factor

of calibration objects is here 12. To illustrate the effect of different noise levels in **X**, the following exaggerated noise has been added to the O.D. spectra:

From wavelength channel 20 to channel 50 (560–800 nm) 3% proportional random noise and 0.4 O.D. units absolute random noise were added to the original O.D. data, simulating e.g. a monochromator working partially outside its optimal range. These data (see Figure 7.1a) represent the 50 X-variables. The litmus concentration (in the range 0–50 relative units) in the same 12 calibration objects represents variable y.

PLS regression was used as calibration method and all the X-variables were given the same a priori weight of 1.0. Figure 7.1b shows how much of the variances in **Y** and in **X** (average over the 50 wavelengths) is correctly modelled with increasing model complexity. These validation results were obtained by cross-validation, repeating the calibration 5 times. One PLS factor gives the best predictive ability— as expected since there is only one systematic type of variability in the data. But this factor only explains some 70% of the Y-variance, and only about 10% of the total X-variance.

Figure 7.1c shows the loading vector $\hat{\mathbf{p}}_1$ for this PLS factor. It shows a positive general peak in the center of the wavelength range, as expected since litmus has its maximum O.D. at about channel 25. However, while the loading vector is smooth between $k = 1$ and $k = 19$ (400–552 nm), a rather erratic pattern is evident at higher wavelengths—as expected, since these channels had excessive noise in the input data.

Figure 7.1d shows the residual variance $s(\hat{\mathbf{e}}_k)^2$, $k = 1, 2, \ldots, K$ (see Chapter 5) of the individual X-variables, after 0 and after 1 PLS factors. It shows that channels 20–50 dominate the initial variance in the data. It also shows that some of this initial variance is reduced by the 1-factor PLS regression (and hence contaminating the calibration model), while most of it remains unmodelled after the 1-factor PLS model has been subtracted. These plots indicate that channels x_{20} to x_{50} are excessively noisy, and could be eliminated altogether. However, it may be more informative just to weight them down a priori; thereby their behaviour can be studied even though their impact on the modelling has been reduced.

Figure 7.2 shows the corresponding PLS analysis after having scaled the input variables x_{20} to x_{50} by an a priori weight of 0.1, compared to weight 1.0 for x_1 to x_{19}. Figure 7.2a shows six of the 12 scaled input spectra X. These spectra were now used with the same Y-variable as in Figure 7.1 in a new PLSR modelling. Figure 7.2b gives the percentages of explained variances for Y and for X. When compared to the corresponding results in Figure 7.1b, we see how a near perfect predictive description of y is now attained after 1 factor. Even the X-data themselves are now relatively well modelled.

Figure 7.2c gives the new loading vector $\hat{\mathbf{p}}_1$. It now shows the beginning of a nice absorbance peak corresponding well to that of the analyte itself, but ends abruptly where the weights had been reduced.

The new residual variances $s(\hat{\mathbf{e}}_k)^2$, $k = 1, 2, \ldots, K$ are given in Figure 7.2d. It shows that the systematic analyte peak variation dominates the total initial variance in the peak range $k < 20$, while the noise variation has now been scaled down by the a priori weights. It also shows that the systematic analyte O.D. variation is explained by the first PLS factor, while most of the noise remains unchanged.

This example shows that excessively noisy X-variables have to be weighted down prior to the final calibration modelling. Similarly, a priori weighting of Y-variables can be important, e.g. if two or more Y-variables are to be modelled simultaneously in a PLS2 regression (see section 3.5.4). This is of course attained by:

$$y_{ij} = y_{ij,\text{old}} * \text{weight}_j \tag{7.2}$$

The a priori weight_k and weight_j can be set by prior knowledge about the different variables. This can be done in many different ways.

Adjusting the X-variables to similar noise levels The weights could for instance be proportional to the inverse of the expected noise level in the data:

318

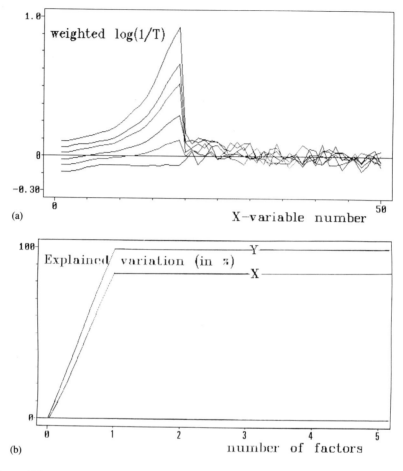

(a)

(b)

Figure 7.2 a) Same data as in Figure 7.1a, but weighted by an a priori scale constant of 0.1 for variables 20 to 50. b) Plot of explained variation of **X** and **Y** as a function of the number of factors

$$\begin{aligned} \text{weight}_k &= c/s(\hat{e}_k) \\ \text{weight}_j &= c/s(\hat{f}_j) \end{aligned} \tag{7.3}$$

where c is some constant (e.g. 1).

These a priori noise estimates can e.g. be obtained from studying independent replicate measurements. For some types of measurements the uncertainty is known to vary proportionally with the general size of that variable (e.g. '3% relative error'); $1/\bar{x}_k$ and $1/\bar{y}_j$ can then serve as weights.

In general, the noise estimates do not have to be very precise. But they should not be set too low, in order to ensure that no single variable dominates too much. Remember that the impact of the variables in the LS modelling is proportional to the square of the weights.

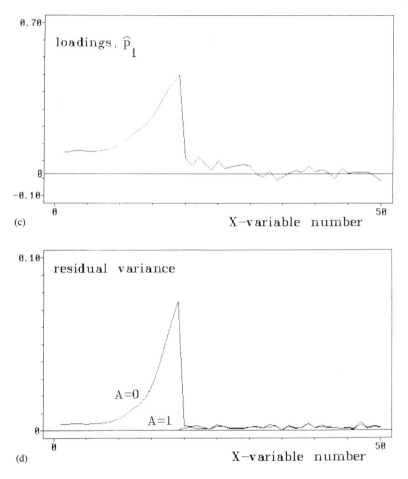

Figure 7.2 (*cont.*)   c) The first loading vector $\widehat{\mathbf{p}}_1$ for weighted data. d) Residual X-variance for each wavelength after 0 and after 1 factor

If certain X-variables are known a priori to have a higher probability of containing relevant information, these can be given correspondingly higher weights.

As the last example has shown, the a priori weights can be updated, based on a preliminary modelling: Variables yielding particularly high residuals ($s(\widehat{e}_k)$ or $\widehat{\mathrm{RMSE}}_j$) and particularly low predictive relevance (only low elements $\widehat{b}_{kj}$) after A factors, can be given reduced weights in a repeated modelling.

Variables to be completely eliminated from the modelling are simply given weights zero.

Adjusting the X-variables to similar variance   Standardization is a well-established a priori weighting in statistics. The total initial variance of every variable is then forced to 1 by dividing each variable by its total standard deviation in the calibration set:

$$\begin{aligned} \text{weight}_k &= 1/s_{xk} \\ \text{weight}_j &= 1/s_{yj} \end{aligned} \qquad (7.4)$$

This is a simple, useful, but conservative way of giving every variable an equal chance to contribute to the modelling. If this is done to both the X- and the Y-variables, it makes graphical interpretation of loading plots easier, because $\hat{\mathbf{P}}$ and $\hat{\mathbf{Q}}$ can now be compared to each other directly, and so can the elements in $\hat{\mathbf{B}}$. It is therefore particularly useful for data screening.

But such standardization can sometimes be deceptive and must be used with care: One such case is the joining of different types of variables in the same $\mathbf{X}$-matrix: Standardization usually simplifies such interdisciplinary matching, but subgroups of variables can be made to dominate too much in the modelling. Assume for instance that there are 99 X-variables of one kind (an O.D. spectrum) and one single X-variable of another kind (e.g. sample temperature) to be used together. Standardizing each of these 100 variables will give a high probability of reaching the noise in the 99 similar O.D. X-variables before the real information in the 100th X-variable has been modelled.

In that case it might be wise to give the lone X-variable temperature a higher a priori weight, at least in preliminary modelling runs. This could be done by multipying the standardized temperature variable by e.g. 5, giving it the same chance of influencing the modelling as $5^2 = 25$ of the 99 standandardized O.D. variables.

Another pitfall in standardization is that of amplifying meaningless variables. If the input level of a variable in a certain calibration set is nearly constant except for a little random noise, this random noise will be amplified to the same a priori importance as an informative variable with a very good signal/noise ratio. It may therefore be necessary to balance the weighting based on the residual variances against the weighting based on the total variances of the variables.

Pragmatic combinations of standardization and down-weighing of particularly noisy variables can be helpful. This can be attained by modifying equation (7.4) into e.g. $\text{weight}_k = 1/(s_{xk} + s(\mathbf{e}_k))$ and $\text{weight}_j = 1/(s_{yj} + s(f_j))$, where $s(\mathbf{e}_k)$ and $s(f_j)$ are a priori estimates of the general noise levels in $\mathbf{X}$ and $\mathbf{Y}$, respectively. With this weighting, variables with nothing but noise variation will be weighted down to a variance of about 0.1, while informative variables will be scaled to variances near, for instance, 1.0. Other modifications can also be used.

The weighting of individual variables, covered in section 7.1, is equivalent to the use of weighted least squares instead of ordinary least squares in ordinary regression modelling (see section 2.1.2.3). For more advanced uses, the least squares estimators can also be modified into generalized least squares (GLS) modelling (see Chapter 3.6). This allows us to reduce or eliminate the impact of X-phenomena that we know a priori to be irrelevant or unreliable. Geometrically, this can be seen as a shrinking of uninformative directions in the X-space.

But conceptually and computationally, GLS estimators can be are rather heavy. Another way to attain such enhancement of informative structures is to do the covariance weighting in the a priori weighting stage.

For instance, additional information about expected correlations between the X-variables can be incorporated in this way: Every input spectrum $\mathbf{x}_i{'}$ is post-multiplied by a non-diagonal K dimensional square weight matrix $\mathbf{D}$, and the new $\mathbf{x}_i{'}$ is then used in the subsequent multivariate calibration or prediction. To reduce the effect of known interference structures, this $\mathbf{D}$ matrix can be the inverse of the square root of the error covariance matrix or the covariance matrix of the known interferents' spectra. This generalized inverse can be obtained by submitting the interferences' covariance matrix to singular value decomposition (sections 2.1.1.13 and 3.3, ignoring X-dimensions with small eigenvalues). Conversely, particularly informative directions in X-space can be enhanced by post-multiplication by the square root of their spectra's covariance matrix.

Known intercorrelation structures between the calibration objects can likewise be handled, by pre-multiplying each X-variable in the calibration set by some non-diagonal weight matrix.

7.2   LINEARITY PROBLEMS

7.2.1 SPLITTING, NONLINEAR MODELLING OR PRETREATMENT?

If there are nonlinearities or heterogeneities in the X–X and/or X–Y relationships, the best way to obtain good linear fits to data is often to split the population of objects into more narrow sub-classes, each with sufficient linearity to let us ignore the whole problem of nonlinear calibration. The splitting can be based on a priori knowledge about the objects, or on a preliminary data analysis step (e.g. splitting on the basis of visual inspection of the score plots from a preliminary bilinear modelling, or by initial cluster analysis (Mardia et al., 1980).

It may, however, be difficult to know how to split the calibration set and there are other drawbacks as well: The splitting increases the need for calibration data, the data analysis takes more time, and having many different predictors may be confusing in practice. So, after having learned to master the linear calibration techniques, it may be useful to study the topic of nonlinearity.

In general, multivariate calibration concerns data approximation of continuous but possibly nonlinear unknown relationships. As in general Taylor series expansions, the linear simplification can give good enough description of many different types of non-linear relationships, provided it is developed over a narrow enough range. So for initial screening, for preliminary calibration results, and sometimes even for final results, linearization can be ignored. But for optimal results in terms of predictive ability and interpretation simplicity, linearity problems in data should be consciously addressed.

We refer to section 4.6 for methods to detect nonlinearity. In some cases, nonlinear calibration methods such as those treated in section 3.7 may be necessary, but in most cases a nonlinearity transformation followed by linear modelling is easier to apply and gives equally good results. This chapter is mainly devoted to such pretreatment of spectral and chemical data, intended to be used once for each data set. But it should be mentioned that a repeated iteration between nonlinearity transformation and linear modelling until convergence can be useful.

Care should be taken to avoid losing important information during the linearization process. And it is important to avoid overfitting by the linearization: Choosing the best out of many attempted alternative linearizations, or by estimating a lot of linearization parameters, may not give the best predictive ability.

7.2.2 DIFFERENT TYPES OF NONLINEARITY

The simplest and most desirable calibration situation with respect to linearity is illustrated in Figure 7.3a, i.e the instrument signal x is linearly related to analyte concentration y. There are, however, many analytical situations with different structure and two simple problems are well known.

Curvature problems: If a detector saturates at high analyte concentrations, there will be a nonlinear response of the curitvature type (Figure 7.3b). Response curvature would also arise if x represents conventional spectrophotometric data in transmission T instead of absorbance units $\log(1/T)$. These cases would traditionally be regarded as 'bad analytical practice' but it is not necessarily so: In general some curvature must be expected in all instrument responses. Thus, for instance Beer's 'Law' for spectroscopy and other conventional 'laws' are only valid under certain ideal conditions—but to limit ourselves to those ideal conditions is quite unnecessary!

Multiplicative problems: In chromatography, the response varies in proportion to the total amount of sample applied to the column, and if this varies uncontrollably, then we have a nonlinear problem of the multiplicative type (Figure 7.3c). The same type of problem arises in ordinary transmission spectroscopy if the cuvette length changes (Lambert's 'Law'). In reflectance spectroscopy of e.g. powders, changes in light scattering create a similar effect, as expressed by the Kubelka–Munk theory (see e.g. Kortum, 1969).

For the data in Figure 7.3a the variable x (error free) may be related to y according to the equation

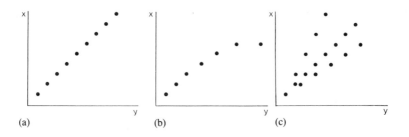

(a)　　　　　(b)　　　　　(c)

Figure 7.3　Linear and nonlinear data. a) Linear instrument response. b) Curved instrument response. c) Multiplicative problem

$$x = yk \tag{7.5}$$

For all the X-variables in a spectrum this relation can be written

$$\mathbf{x}' = y\mathbf{k}' \tag{7.6}$$

and from such an equation we can always express an inverse linear structure as

$$y = \sum_{k=1}^{k} x_k b_k = \mathbf{x}'\mathbf{b} \tag{7.7}$$

Likewise, if we expect our X-spectra to be affected by several analytes $j = 1, 2, \ldots, J$ plus various interfering constituents $a = 1, 2, \ldots, A$ in a simple additive manner, our causal structure is:

$$\mathbf{X} = \underbrace{\left(\sum_{j=1}^{J} \mathbf{y}_j \mathbf{k}_j' \right)}_{\text{analytes}} + \underbrace{\left(\sum_{a=1}^{A} \mathbf{t}_a \mathbf{p}_a' \right)}_{\text{interferents}} \tag{7.8}$$

which can be simplified (see e.g. section 3.6) to

$$\mathbf{X} = \mathbf{Y}\mathbf{K}'_{\text{analytes}} + \mathbf{T}\mathbf{P}'_{\text{interferents}} \tag{7.9}$$

Using linear algebra it is easy to see that also in this case we can find linear empirical relationship in the data for each analyte j, i.e.

$$\mathbf{y}_j = \mathbf{X}\mathbf{b}_j \tag{7.10}$$

In contrast, if the concentration data for an analyte $\mathbf{y}_j$ are known to be given in some strange non-linear unit like (mmol analyte)^2 or in a ratio to some other analyte, we would expect a nonlinear structure in the data instead of equation (7.10). Likewise, if the X-data are light absorption spectra expressed in transmittance (T) rather than absorbance units (O.D. $= \log(1/T)$), or if the X-data come from a light detector that for some wavelengths almost saturates at the highest analyte concentrations, we would expect nonlinear Y–X data structures.

However, as has been shown in practice (see e.g. Martens, 1983; Næs and Isaksson, 1988), a number of individual nonlinear relations can be approximated well by linear multivariate calibration methods—at the cost of more complicated calibration models (e.g. more bilinear factors). This is exemplified by linear calibration of NIR reflectance instruments, and will be illustrated below. But the better linearity in the input data, the simpler and better linear modelling results we usually get.

7.2.3 LINEAR METHODS CAN HANDLE SOME NONLINEAR DATA STRUCTURES

Let us look at a simple illustration to show how this can work: Assume that we want to determine a single analyte y in objects $i = 1, 2, \ldots$, without any interferences, using two detectors x_1 and x_2. These detectors can give various types of nonlinear responses due to saturation, as illustrated in Figures 7.4a and 7.4b.

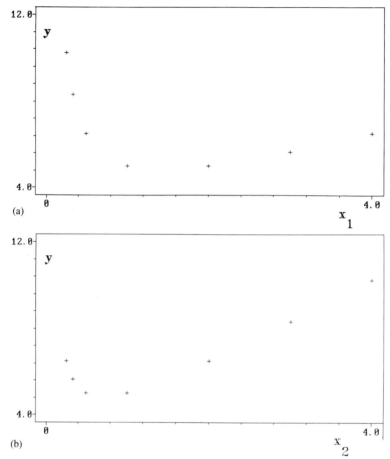

Figure 7.4 Successful linear modelling of nonlinear relations: Simplified example with two
X-variables and one Y-variable. a) Raw data: y vs. x_1 b) Raw data: y vs. x_2

Still, y is linearly related to the linear combination of them, found by regressing
y on both x_1 and x_2 (Figure 7.4c) using the model

$$y = b_0 + x_1 b_1 + x_2 b_2 + f \qquad (7.11)$$

This phenomenon is important and explains much of the reason why e.g. linear
calibration models have had such success in NIR reflectance spectroscopy: Using
for instance PCR or PLSR in this application, it is found that quite many bilinear
factors are needed (compared to the number of interferences) to obtain optimal
prediction ability (typically between 5 and 10). The interpretation of this is that
the high number of factors are needed to account for nonlinear effects between
composition, interferences and individual spectral variables. A demonstration of
this on spectroscopic data will be given later (see also Næs and Isaksson, 1988 and
Isaksson and Næs, 1988).

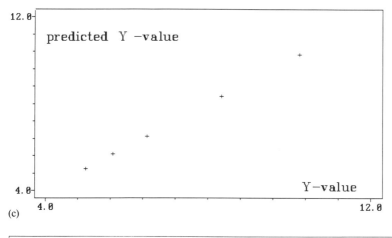

(c)

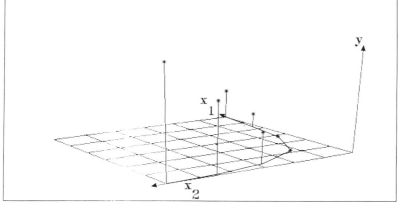

(d)

Figure 7.4 (*cont.*)   Successful linear modelling of nonlinear relations: Simplified example with two X-variables and one Y-variable. c) Prediction $\hat{y} = 2.0 + x_1 * 1.0 + x_2 * 2.0$ plotted vs y. ($\hat{\mathbf{b}}$ was here estimated using two PLSR factors). d) y vs. x_1 and x_2. The Y–X curvature forms a two-dimensional plane and can be modelled linearly by two factors, when seen from a particular angle. Both these factors can be seen in the x_1 vs. x_2 space (curved path)

But it is easy to find examples where this is not the case, as illustrated in Figure 7.5: Again y is nonlinearly related to both x_1 and to x_2 (Figures 7.5a and 7.5b), but in this case even the best linear combination of these two input X-variables gave bad nonlinear relation to y (Figure 7.5c).

In Figures 7.4d and 7.5d we see that when plotted in the three-dimensional space $y - x_1 - x_2$ the points in both examples lie along a single curved path (a 'banana'). This could be due to a single chemical phenomenon causing the variations in the X-data. The difference is in how the 'banana' is positioned relative to the X-space: In Figure 7.4d we can recognize the curvature of the 'banana' in our two-dimensional X-space (the $x_1 - x_2$ plane) itself. Thus, we can use $\mathbf{X}$ alone to estimate a linear two-factor model—a plane in which the Y–X 'banana' shape is seen as a 'boomerang' shape. This adequately describes the curved Y–X relationship.

326

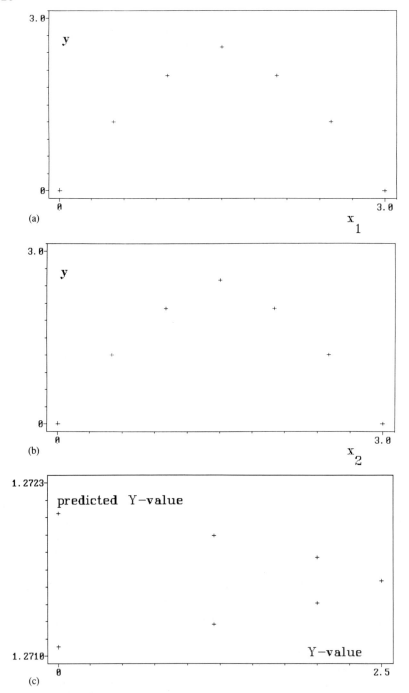

Figure 7.5 Unsuccessful linear modelling of nonlinear relations: An example with two X-variables and one Y-variable. a) Raw data: y vs. x_1. b) Raw data: y vs. x_2. c) Predicted $\widehat{y}$ $= 1.27 + x_1 * -0.002 + x_2 * 0.001$ vs y ($\mathbf{b}$ was here estimated using one PLSR factor)

In Figure 7.5d we cannot see the curvature of the 'banana' just from the X-data, because both X-variables show the same kind of nonlinear response. Hence we cannot estimate the second dimension in the 'boomerang' plane from the X-data alone, and the Y–X curvature remains unmodelled. In this case something explicit has to be done to overcome the curvature problem. How can this be done?

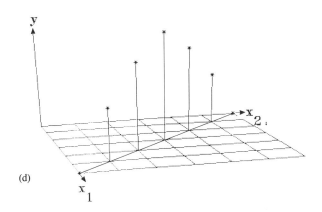

(d)

Figure 7.5 (*cont.*)   Unsuccessful linear modelling of nonlinear relations: An example with two X-variables and one Y-variable. d) y vs. x_1 and x_2. The Y–X curvature forms a two-dimensional plane. However, only one of its dimensions can be seen in the x_1 vs. x_2 space (straight path)

7.3   CURVATURE LINEARIZATION

7.3.1 TRANSFORMING THE INPUT DATA

In this book we generally advocate a separation of the linearization and the linear multivariate modelling. The reason for this is that nonlinear multivariate models can become difficult to interpret, and can easily overfit, unless the number of input variables used in the nonlinear modelling is very low. But in practice one can iterate between linearization of variables and fitting them to a linear model, so the two activities are not quite separate.

The linearizations may follow different approaches. First of all there may exist theory which indicates that a certain transformation of **x** and/or **y** will result in a more linear model. In other cases no good theory exists, and one must base the linearization on data analysis, either by using plotting techniques (see e.g. section 4.6) to indicate the transformation or by certain testing procedures.

It should be emphasized that finding the optimal transformation of variables can be complicated, but the *optimal* transformation may not be needed—a pre-processing that gives a reasonable linearity may be enough, at least in the initial stages of a calibration. In any way, this will be part of an iterative process repeating theoretical considerations, pre-processing, calibration and validation.

7.3.1.1 Modification of the Y-variables

Assume now that a monotone function $f(y)$, i.e. $\log(y)$ fits well to a linear function of $x_1, \ldots, x_k$, i.e.

$$f(y) = b_0 + \mathbf{x}'\mathbf{b} + \text{random noise} \tag{7.12}$$

In such cases, a linear predictor of $f(y)$, $\widehat{f(y)}$, can easily be found by e.g. PLSR or PCR, and y can be found by transforming $\widehat{f(y)}$ to $\hat{y}$ by the inverse

$$\hat{y} = \widehat{f^{-1}f(y)} \tag{7.13}$$

The curvature problem in Figure 7.5 could for instance be reduced by calibrating for the squared concentration, $f(y) = y^2$ and predicting the analyte concentration by the transform, $\hat{y} = (\hat{b}_0 + \mathbf{x}'\hat{\mathbf{b}})^{\frac{1}{2}}$.

A general and much used linearization procedure based on data is the so-called Box–Cox transformation or 'power transformation'. The idea behind this approach is to model a power of the goal variable y as a linear function of $\mathbf{x}$, i.e

$$(y^\lambda - 1)/\lambda \approx b_0 + \mathbf{x}'\mathbf{b} \tag{7.14}$$

and then estimate the value of λ from the available data. In this way one finds the value of λ which fits the linear function of $\mathbf{x}$ best. After estimation of λ and the regression coefficients, we can predict y by $((\hat{b}_0 + \mathbf{x}'\hat{\mathbf{b}})\hat{\lambda} + 1)^{1/\hat{\lambda}}$. If $\hat{\lambda} = 0$, it is common to define $y = ln(y)$.

A lot of theory has been developed for this approach; different estimation procedures exist, diagnostics for influence on the transformation and distributional results. Generalizations of the Box–Cox transformation are also proposed. We refer to Weisberg (1985) and Atkinson (1987) for good treatments.

7.3.1.2 Modification of the X-variables

If we knew in advance how to linearize the instrument response $\mathbf{X}$, we could replace the original nonlinear variables by new, more linear ones, by some nonliner preprocessing transformation $\mathbf{X} = \mathbf{f}(\mathbf{X})$. The bilinear model (for centered data) would then be:

$$f(\mathbf{X}) = \mathbf{TP}' + \mathbf{E}$$
$$\mathbf{Y} = \mathbf{TQ}' + \mathbf{F} \tag{7.15}$$

with the linear predictor:

$$\hat{\mathbf{Y}} = \mathbf{1}\hat{\mathbf{b}}_0' + f(\mathbf{X})\hat{\mathbf{B}} \tag{7.16}$$

for $\mathbf{Y}$ (uncentered).

A well known spectroscopic example of such analytical pre-linearization is the transformation of optical transmission data (response $= T$) into absorbance or optical density O.D.$=\log(1/T)$. This will be illustrated in the following example, which compares calibration on T and $\log(1/T)$ (see also Figure 4.2).

In Figure 7.6a the apparent transmissions T are given for 6 aqueous solutions of the blue-coloured analyte litmus at pH 10 in the wavelength range 400–800 nm (wavelength channels $k = 1, 2, \ldots, 50$). The transmission was calculated as I/I_0, where I is the Intensity of a sample, measured in a Guided Wave, model 200 fiber-optic spectrometer, and I_0 the corresponding intensity of a reference.

The transmission is seen to decrease in the middle wavelength range with increasing concentrations of the analyte dye. A total of 18 such solutions with known litmus concentrations were measured. Twelve of them were used for calibration and the rest for testing. Can these data be used for multivariate linear calibration and prediction?

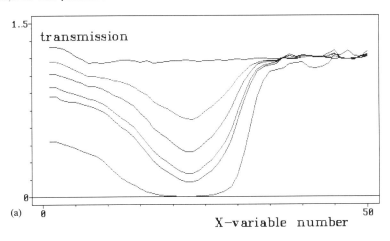

(a)

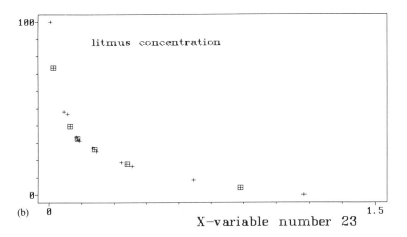

(b)

Figure 7.6 Breaking Beer's model by multi-wavelength calibration a) Transmission (T) vs. wavelength $x_k, k = 1,2,\ldots,50$ (400–800 nm) for 6 different concentrations of alkaline litmus in aqueous solution, ranging from pure water (top curve) to 100 relative concentration units (bottom curve). b) Best univariate calibration: Measured analyte concentration y vs. the transmission minimum x_{23} for the calibration and test (squares) objects

Figure 7.6b shows the best univariate calibration curve, the analyte concentration y_i vs. transmission at wavelength channel 23. As expected from Beer's model, the relationship is highly curved. Can these non-linear data be modelled by a linear calibration method, or do we have to modify the X-variables?

A PLS modelling of y vs the 50 transmission X-variables was first developed, using the mean for the 12 calibration objects as model center. This calibration model needed 2 bilinear factors to give reasonable prediction ability, and 4 factors to give optimal prediction ability, as judged by internal cross validation.

The model centre $\bar{\mathbf{x}}$ is given in Figure 7.6c. The four loading spectra corresponding to the 4 first factors are shown in Figure 7.6d; the spectra are scaled according to their relative spectral importance, $\hat{\mathbf{p}}_a \sqrt{(\hat{\mathbf{t}}_a' \hat{\mathbf{t}}_a)}$ for $a = 1, 2, 3$ and 4. The loading weights $\hat{\mathbf{w}}_a$ that convert $\mathbf{X}$ to $\hat{\mathbf{t}}_a$ have similar spectral shapes and are not

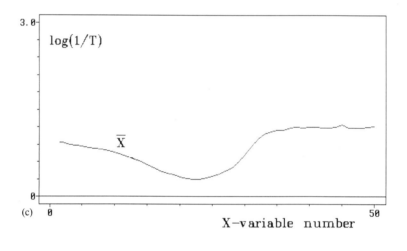

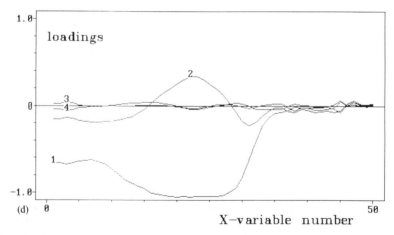

Figure 7.6 (*cont.*) Breaking Beer's model by multi-wavelength calibration c) The mean of the spectra, $\bar{\mathbf{x}}$. d) PLSR model loadings: $\hat{\mathbf{p}}_a \sqrt{(\hat{\mathbf{t}}_a' \hat{\mathbf{t}}_a)}$ for factors $a = 1,2,3$ and 4

shown here. The first two factors can be seen to describe a much higher spectral variability than the next two factors; the latter two are therefore ignored in the following. It should also be noted that factor 2 has a positive maximum where the analyte's signal is the strongest, and two negative minima on either side; this is typical for a bilinear curvature correction factor.

In Figure 7.6e the analyte concentration y_i is plotted against the two factor scores $\hat{t}_1$ and $\hat{t}_2$. Notice that a clear curvature is seen in the $\hat{t}_1 - \hat{t}_2$ horizontal plane, while the actual points lie in a flat plane in the three-dimensional space—hence the X–Y relationship can be described by a two-factor linear model.

Figure 7.6f shows that the resulting bilinear two-factor predictor yields a quite

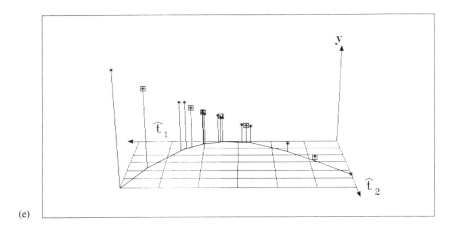

(e)

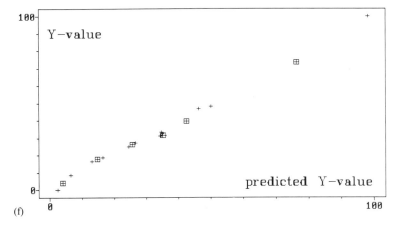

(f)

Figure 7.6 (*cont.*)   Breaking Beer's model by multi-wavelength calibration e) Measured analyte concentration y vs $\hat{t}_1$ and $\hat{t}_2$ for the calibration and test (squares) objects. A reasonably linear X–Y relation can be seen from a particular angle, in spite of the X–X curvature seen in the $\hat{t}_2$ vs $\hat{t}_1$ plane. f) Measured analyte concentration y vs predicted analyte concentration, $\hat{y} = 32.3 + \hat{t}_1 * 15.1 + \hat{t}_2 * 36.3$, for the calibration and test (squares) objects

good prediction, both for the calibration and test objects, in spite of the strong curvature in the input data!

However, since there is only one main type of variability in these data, using two or more factors in the calibration model is unnecessarily complicated. So it might be desirable to simplify the bilinear calibration model. This is shown in Figure 7.7.

The input transmission spectra (T) were here pre-linearized into O.D.=$\log(1/T)$ according to Beer's model of how photons are absorbed. Exactly the same calibration and prediction procedure was then repeated.

Figure 7.7a shows the O.D. input spectra corresponding to Figure 7.6a. Increasing

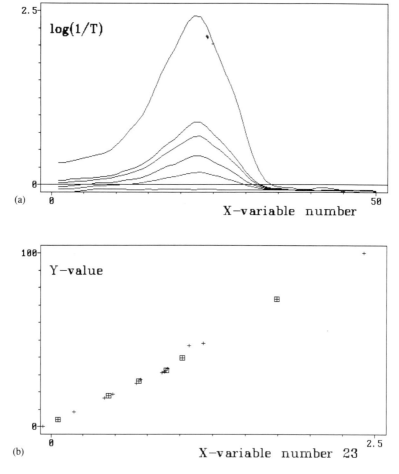

(a)

(b)

Figure 7.7 Following Beer's model in multi-wavelength calibration. a) Optical density (O.D. = $\log(1/T)$) vs wavelength $x_k, k = 1,2,...,50$ (400–800 nm) for the 6 different concentrations of alkaline litmus in aqueous solution from Figure 7.6a, ranging from pure water (bottom curve) to 100 relative concentration units (top curve). b) Best univariate calibration: Measured analyte concentration y vs. the O.D. maximum x_{23} for the calibration and test (squares) objects

intensities of the analyte absorbance peak are now evident. Figure 7.7b shows that using just the peak wavelength x_{23} now gives a very good X–Y relationship. This was not surprising, since these data are expected to contain only one systematic, linear phenomenon plus some random noise.

PLS modelling of the data indicated that very good predictive ability was now attained already after one bilinear factor, with only a minor improvement due to the second factor. In Figure 7.7c is given the model centre $\bar{x}$ and in Figure 7.7d the scaled loading vectors $\hat{\mathbf{p}}_a \sqrt{(\hat{\mathbf{t}}_a' \hat{\mathbf{t}}_a)}$ for these first two bilinear factors are given. Again the loading weights $\hat{\mathbf{w}}_a$ were quite similar to $\hat{\mathbf{p}}_a$ and are not shown. The second factor $\hat{\mathbf{p}}_2$ does display a positive–negative–positive peak pattern around the main peak in $\hat{\mathbf{p}}_1$, indicating that the $\log(1/T)$ did not linearize the X–Y and X–X relationships completely. But the second factor is very small both in X and Y, so we ignore it here.

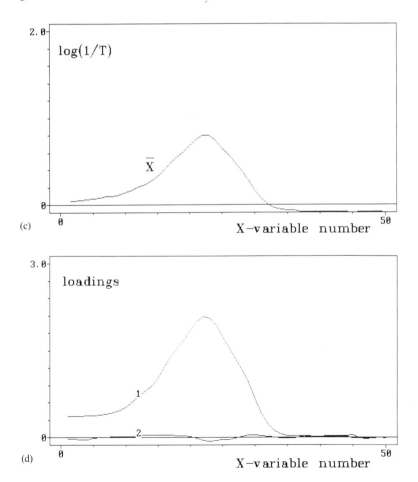

Figure 7.7 (*cont.*) Following Beer's model in multi-wavelength calibration. c) The mean of the spectra, $\bar{x}$. d) PLSR model loadings: $\hat{\mathbf{p}}_a \sqrt{(\hat{\mathbf{t}}_a' \hat{\mathbf{t}}_a)}$ for factors $a = 1$ and 2

334

Not unexpectedly, the measured and predicted analyte concentrations (using the one-factor bilinear model), are quite similar, both for the calibration and the test objects (Figure 7.7e).

In summary, this example has shown that even strongly nonlinear X–Y relationships can sometimes be modelled by bilinear methods, but at the cost of an unnecessarily complex calibration model. Proper pre-linearization simplified the subsequent calibration modelling.

A number of other pre-linearization methods are known, e.g. the Kubelka-Munk transform for diffuse reflectance data (see e.g. Kortum, 1969). In general, the input data ought to be pre-linearized whenever possible; the bilinear modelling should then be used for correcting possible deficiencies in the pre-linearization.

However, Figure 7.5, as well as some figures in section 3.5 have shown cases where an automatic linear compensation for curvatures did not work. If we do not know how to linearize the data, what can be done then?

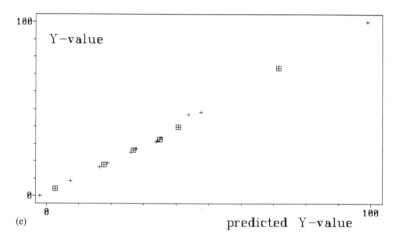

(e)

Figure 7.7 (*cont.*)   Following Beer's model in multi-wavelength calibration. e) Measured analyte concentration y vs. predicted analyte concentration, $\hat{y} = 32.3 + \hat{t}_1 * 12.8$ for the calibration and test (squares) objects

7.3.2 USING EXTRA X-VARIABLES

In some simple cases we can solve curvature problems by including some new 'artificial' X-variables that represent nonlinear transforms of the original X-variables: These extra variables could e.g. be polynomials of the original $\mathbf{X}$-variables:

$$\mathbf{x}_{\text{extra}} = (x_1^2, x_2^2, x_1 x_2 \text{ etc.}) \tag{7.17}$$

These extra X-variables will add some extra dimensions to the X-space which hopefully will be able to give adequate modelling of the nonlinearities in $\mathbf{Y}$ (see e.g. section 3.7).

The resulting linear, polynomial predictor can be written

$$\widehat{\mathbf{Y}} = \mathbf{1}\widehat{\mathbf{b}}_0{}' + \mathbf{X}\widehat{\mathbf{B}} + \mathbf{X}_{\text{extra}}\widehat{\mathbf{B}}_{\text{extra}} \tag{7.18}$$

Going back to the simple data set in Figure 7.5, the straightforward linear model could not predict y from x_1 and x_2, because the X–Y curvature was not visible in the X–X relationships: There was only one linear structure type in $\mathbf{X}$ and hence only one latent variable $\widehat{t}_1$ could be reliably estimated. But if we add an extra variable $x_3 = x_1{}^2$, then we would be able to pick up two latent variables. A linear polynomial predictor based on two PLS factors could then be obtained as

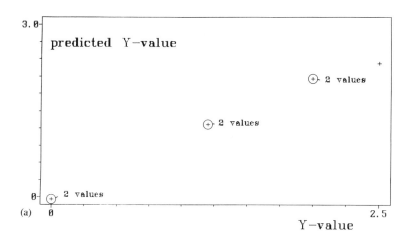

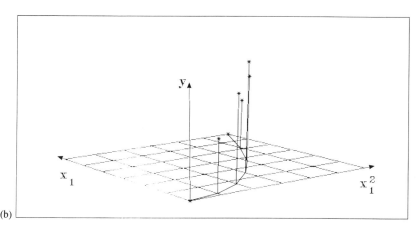

Figure 7.8 Adding extra X-variables to overcome curvature problems: The y versus x_1 and x_2 example in Figure 7.5 expanded with an extra variable, $x_3 = x_1{}^2$: a) Predicted $\widehat{y} = 4.7 + x_1 * 1.6 - x_2 * 1.6 - x_1{}^2 * 1.0$ vs y (**b** was here estimated using two PLSR factors). b) y vs. x_1 and $x_1{}^2$. The Y–X curvature forms a two-dimensional plane and can be modelled by two factors, and both of its dimensions can be seen in the X-space

$$\hat{y} = \hat{b}_0 + x_1\hat{b}_1 + x_2\hat{b}_2 + x_1{}^2\hat{b}_3 \qquad (7.19)$$

Figure 7.8a shows that the curvature in the X–Y relationship was now picked up in a two-factor bilinear model, in contrast to Figure 7.5c. Figure 7.8b shows the reason: most of the curvature can now be picked up by two dimensions in the X-space, as illustrated in this y versus x_1 and $x_1{}^2$ projection.

Care must be taken to scale the new and the old X-variables a priori so that the new ones do not dominate the bilinear estimation techniques. The scale problems can be diminished and the linearization effects somewhat amplified by using transforms like $(x_{ik} - \bar{x}_k)^2$ instead of $x_{ik}{}^2$. This breaks some of the collinearity between x_k and its square, and therefore allows e.g. PLSR regression to include more of both in fewer factors.

It should be emphasized that use of expansions of the $\mathbf{X}$-matrix as shown here should be done with care because it can make the calibration models unnecessarily complicated and it may give overfitting. Useless extra X-variables should be eliminated again after a preliminary screening.

7.4 REDUCING MULTIPLICATIVE PROBLEMS

7.4.1 LINEAR MODELS AND MULTIPLICATIVE EFFECTS

Linear (including bilinear) modelling is suitable for handling additive interferences and certain types of curvatures as described above. Other types of curvature may require explicit linearization treatment or modelling.

But there is another 'nonlinearity' problem which also has to be dealt with, namely multiplicative effects (Figure 7.1c). As mentioned in section 7.2, uncontrolled amplification problems can for instance arise in chromatography when the total amount of sample applied to the column is unknown, and in spectroscopy when the cuvette path length varies uncontrollably. Light scattering variations in diffuse spectroscopy can give similar multiplicative interferences (see e.g. Birth, 1982).

If the expected stable ('true') linear instrument response at X-variable number k, $x_{ik,\text{true}}$, is modified by the uncontrolled amplification factor β_i, then what we actually measure for object i, $x_{ik,\text{inp}}$, is:

$$x_{ik,\text{inp}} \approx x_{ik,\text{true}} * \beta_i \qquad (7.20)$$

To separate the effects of $x_{ik,\text{true}}$ and β_i in a linear way, one can of course take a log transformation. But if $x_{ik,\text{true}}$ equals e.g. $\sum y_{ij} * k_{kj}$, then the chemical information y_{ij} is nonlinearly related to $x_{ik} = \log(x_{ik,\text{inp}})$.

On the other hand, without any multiplicative correction the linear predictor would look like

$$\hat{y}_{ij} = \hat{b}_0 + \sum_{k=1}^{K} \left(x_{ik,\text{true}} * \beta_i \right) * \hat{b}_{kj} \qquad (7.21)$$

which is sensitive to the unknown factor β_i and hence unsatisfactory. Thus there is a need to normalize such input data to remove the uncontrollable scale variation β_i.

Sometimes the situation is made even more complicated by the mixing of multiplicative and additive effects:

$$x_{ik,\text{inp}} \approx \alpha_i + x_{ik,\text{true}} * \beta_i \tag{7.22}$$

where α_i represents an unknown offset deviation from some reference level and β_i is the unknown amplification relative to the reference level. The reference itself has $\alpha_i = 0$ and $\beta_i = 1$.

If we only measure one single X-variable, then it is impossible to distinguish between the 'true' analyte signal and the irrelevant scale β_i. But when several X-variables have been measured, it may sometimes be possible to estimate α_i and β_i and then correct for them.

$$x_{ik} = (x_{ik,\text{inp}} - \widehat{\alpha}_i)/\widehat{\beta}_i \tag{7.23}$$

These stabilized responses x_{ik} can then be used in subsequent bilinear calibration.

Let us assume for simplicity that α_i and β_i are about the same for all the X-variables. In spectrophotometry this could imply that the effect of light scattering was approximately the same at all wavelengths involved. The problem is then to estimate α_i and β_i in each object in such a way that we can correct the input data $x_{ik,\text{inp}}$ without loosing the X-information relevant for predicting y_{ij}!

We start with normalization based on closure. Then we proceed to internal standards, and to other multiplicative correction techniques.

7.4.2 MODELLING MULTIPLICATIVE EFFECTS

7.4.2.1 Normalization by closure

Dividing the original instrument responses x_{ik} by their sum in each object i can sometimes remove unwanted scale variations (Martens and Jensen, 1983; Hoerl et al., 1985).

For instance, assume that we want to calibrate for Y=porosity of oil-containing rocks by microscopy image analysis of rock samples. For each rock sample $i = 1, 2, \ldots, I$ the number of image pixels at various grey levels $k = 1, 2, \ldots, K$ was counted; the resulting histogram of pixel counts corresponds to the input instrument response $x_{ik,\text{inp}}$.

Depending on the size and shape of the rock samples the total number of pixels counted as representing rock would vary. It would then probably be sensible to try to use the histogram of relative (rather than absolute) frequencies of counting as X-variables for the calibration:

$$x_{ik} = x_{ik,\text{inp}} / \sum_{m=1}^{K} x_{im,\text{inp}} \tag{7.24}$$

Hence $\alpha_i = 0$ and (7.23) can be used, with

$$\widehat{\beta}_i = \sum_{m=1}^{K} x_{im,\text{inp}} \tag{7.25}$$

Such a normalization may also be sensible in some chromatography applications when the absolute amount of sample being recovered through the sample preparation and applied to the column is difficult to control. Let us look at an example concerning chromatographic amino acid analysis of proteins. In the present case we want to determine the relative concentration of a certain protein in mixtures with other proteins (interferents), from the amino acid composition of the mixtures. It could be that other ways of separating the proteins were difficult, due e.g. to immobilization due to heat treatment etc.

In the present illustration the amino acid data of the two enzymes ribonuclease (analyte) and chymotrypsinogen (interferent) are used for illustration. Their amino acid data were originally published by Gold et al. (1976) and were also adressed by Spjøtvoll et al. (1982). For the illustration, seven linear combinations of the two pure proteins' amino acid compositions were generated to represent mixtures. Each mixture was then multiplied by a different factor (ranging from 0.25 to 3) to represent the problem of non-constant total amount of material applied to the chromatograph.

Figure 7.9e shows the amino acid composition of the analyte ribonuclease, plotted as a function of amino acid no. $k = 1, 2, \ldots, 17$. On top of this is the corresponding 'unknown' amino acid composition of ribonuclease, here to be regarded as an 'unidentified' interferent. The two proteins differ at a number of amino acids, although no single amino acid is selective for any of them. We assume for simplicity that the analytical precision is the same for each amino acid, so a priori weighting is ignored.

Figure 7.9a shows the amino acid input data for the 7 protein samples, as functions of amino acid $k = 1, 2, \ldots, 17$. The most dominant characteristic is the general change in level due to the multiplicative interference—the changing amount of sample applied to the chromatograph. But some other type of differences between the amino acid patterns can also be seen.

When these raw data were used directly as X in a PLS calibration for y = ribonuclease fraction, the apparent prediction ability shown in Figure 7.9b was attained (estimated by full cross validation, using one object for testing in each of the 7 re-calculations of the model). Two factors were required to obtain any predictive ability at all. Higher factors rely on very small variations in the amino acid data to compensate further for the multiplicative problem. These effects in practice would not be valid, but due to the artificial manner in which the the present 'mixture' data were generated, they appear to be valuable here.

But the predictive ability even after 5 factors (Figure 7.9c) was not particularly impressive. And Figure 7.9d shows that this apparent prediction ability was attained by an extreme amplification of the amino acid data, with $\widehat{\mathbf{b}}$ contrasting amino acids 1, 4 and 12 against amino acids 5, 6, 9 and 10, multiplying the difference by several thousand! For real amino acid mixture data, their random noise would

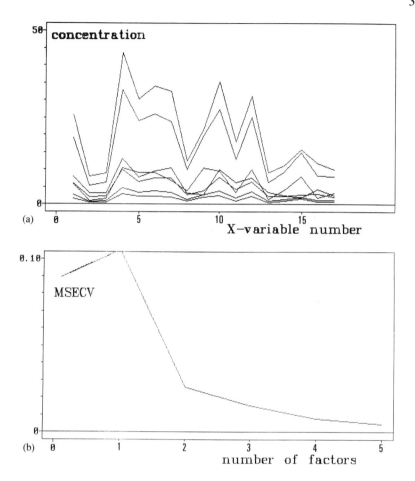

Figure 7.9 The effect of unknown scaling in chromatography: a) Concentration of amino acids $k = 1,2,\ldots,17$ for 7 mixtures of chymotrypsinogen and ribonuclease with varying unknown total protein content. b) Estimated predictive error MSECV (cross validation) for relative analyte (ribonuclease) content vs PLS factor no. $a = 1,\ldots,5$

have been correspondingly amplified and caused more prediction errors. So in this illustration the presence of an unidentified multiplicative interference was quite detrimental.

The 17 amino acids actually are the monomer 'building blocks' that make up the proteins. So dividing the raw amino acid peaks by their sum (equation (7.24)) implies a scaling to constant protein content. This solves the problem of varying, unknown total amount of protein applied to the chromatograph: The PLS calibration was then repeated for the same y, after having scaled each amino acid composition vector to 100 percent:

$$x_{ik} = x_{ik,\text{inp}} * 100 \Big/ \sum_{m=1}^{17} x_{im,\text{inp}}. \tag{7.26}$$

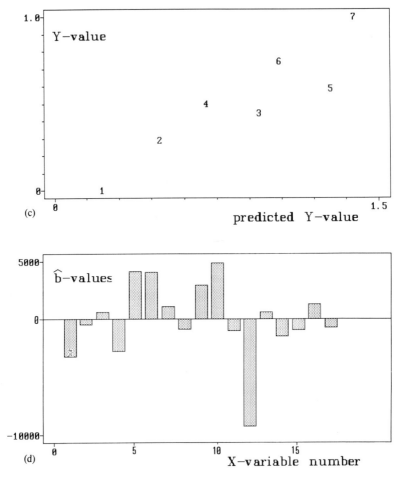

Figure 7.9 (*cont.*) The effect of unknown scaling in chromatography: c) True vs. predicted analyte content in the calibration set using 5 PLS factors. d) Estimated $\hat{b}_{kj}$ values ($k = 1,2,\ldots,17$) using 5 PLS factors

Figure 7.10a shows how much simpler the input data now look, and 7.10b shows that a one-factor PLSR model now was enough to attain a very good predictive ability. Figure 7.10c shows the predictions with this one-factor model, and Figure 7.10d the corresponding $\hat{b}$-vector.

Of course this factor now represents the difference between the analyte and the unknown interferent, since their sum is constant. The $\hat{b}$ vector is negative for those amino acids which are lower in chymotrypsinogen than in ribonuclease (nos. 9 and 15, representing the amino acids glycine and valine, see also Figure 7.9e), and vice versa.

In the present case the closure normalization worked quite well. However, in other cases closure can generate artifacts in the data in terms of part-to-whole

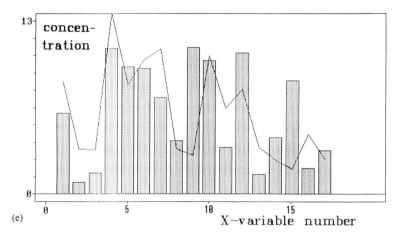

Figure 7.9 (*cont.*)   The effect of unknown scaling in chromatography: e) The amino acid compositions of the two constituents, $k_{\text{chymotrypsinogen}}$ (bars) and $k_{\text{ribonuclease}}$ (line)

correlations: If one of the variables k dominates the total sum $\sum x_{ik,\text{inp}}$, then a lot of 'artificial' intercorrelations can be generated when using this transformation. This type of artifact can in some cases be reduced by replacing the straight sum of the variables, $\sum x_{ik,\text{inp}}$ by some weighted sum of the variable in the normalization:

$$x_{ik} = x_{ik,\text{inp}} / \sum_{m=1}^{K} (d_m * x_{im,\text{inp}}) \qquad (7.27)$$

where d_m are the weights (positive, zero or negative). How to determine d_m depends on domain specific knowledge. This is somewhat related to ratioing regression as covered in section 7.4.2.4.

7.4.2.2   Internal standards

An alternative to normalization by closure is to use only one or a few of the available variables for normalization. In chromatography it is common practice to include known amounts of a foreign constituent as internal reference standard (variable g) and use its response $x_{i,g}$ in the multiplicative correction:

$$\widehat{\beta}_i = x_{i,g} / x_{\text{ref},g} \qquad (7.28)$$

where $x_{\text{ref},g}$ is the corresponding reading of some chosen reference object (alternatively, one may use $x_{\text{ref},g} = 1$).

In spectroscopy of solutions it may be useful to employ the NIR absorbance of the solvent as an internal standard if the cuvette length varies uncontrollably. Water and most organic solvents have well defined absorbance peaks e.g. near 1000 nm, which will vary in direct proportion to the optical path length.

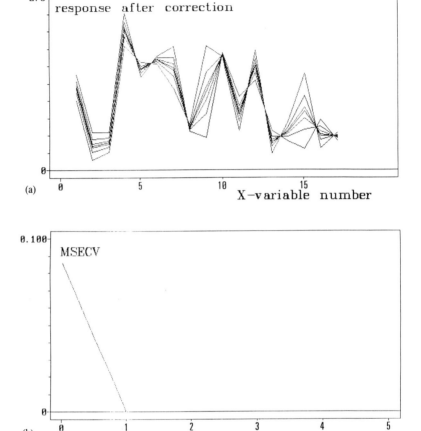

Figure 7.10 Having removed the effect of unknown multiplicative scaling in chromatography. a) Concentration of amino acids $k = 1,2,\ldots,17$ for 7 mixtures of chymotrypsinogen and ribonuclease from Figure 7.9a, after normalization to constant sum. b) Estimated predictive error MSECV (cross validation) for relative analyte (ribonuclease) content vs PLS factor no

This use of the solvent peak can also correct for other multiplicative effects, like changes in light scattering. Figure 7.11 shows the optical density of aqueous solutions of Fast Green dye at three different concentrations (0, 0.5 and 1.0 relative concentration units) in the visible and NIR wavelength range. At each dye level four different levels of light scattering were induced (using 4 different concentrations of defatted milk/water as solvent).

These spectral measurements were originally done by reflectance R, and the instrument response linearized by the so-called Kubelka-Munk transform, O.D.=$'K/S'= (1 - R)^2/2R$. Theoretically, the O.D. data in this unit should vary proportionally to the concentration of absorber (here: Fast Green dye), and inversely

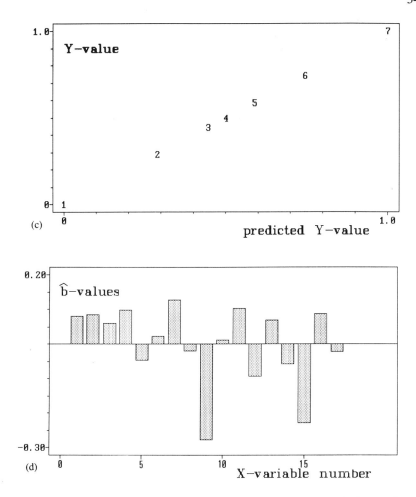

Figure 7.10 (*cont.*)   Having removed the effect of unknown multiplicative scaling in chromatography. c) True vs. predicted analyte content in the calibration set using 1 PLS factor. d) Estimated $\widehat{b}_{kj}$ values ($k = 1,2,\ldots,17$) using 1 PLS factor

proportionally to the light scattering coefficient (see also Birth, 1978, 1982).

The figure shows how the analyte peak indeed increases with analyte concentration. But unfortunately it also decreases with the milk scattering and hence cannot directly be used for predicting the analyte concentration.

However, the figure also illustrates how the O.D at the water peak near 1000 nm is unaffected by the addition of the small amounts of dye. We select one of the spectra, or the average spectrum, as reference spectrum, O.D.$_{ref}$. The solvent peak then allows us to estimate β_i by

$$\widehat{\beta}_i = \text{O.D.}_{i,975nm} / \text{O.D.}_{ref,975nm} \tag{7.29}$$

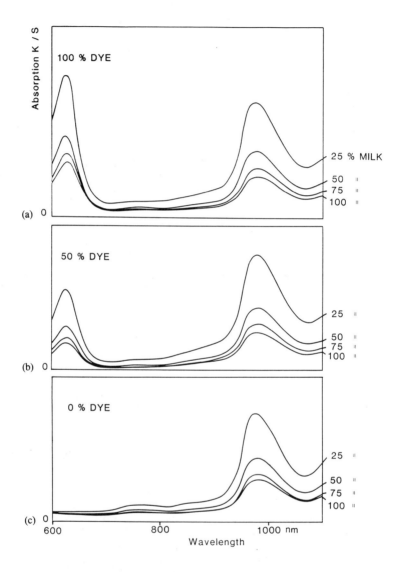

Figure 7.11 The solvent NIR peak is an internal spectroscopic reference for light scattering changes. The optical density, calculated from reflectance R according to the Kubelka Munk theory (Kortum, 1969) as 'K/S' $= (1-R)^2/2R$, plotted against wavelength for three different relative concentrations of Fast Green dye in aqueous solutions. At each dye level, spectra of samples with four different milk/water ratios are given, ranging from 25% to 100% milk defatted skim milk

7.4.2.3 Multiplicative signal correction (MSC)

This can be generalized into the so-called multiplicative signal correction (MSC). It was originally developed in order to correct for the enormous light scattering variations in reflectance spectroscopy—a variation that has a strong multiplicative component. Hence, it was originally called multiplicative scatter correction.

In diffuse reflectance spectroscopy the multiplicative problems are particularly difficult to handle, since varying light scattering levels yield enormous interference effects with both an additive and a multiplicative component effect (see e.g. (7.22)) and hence cannot just be divided away. Both offset α_i and scale β_i must be estimated and corrected for.

The basis for the original MSC was the fact that the wavelength dependency of light scattering is different from that of chemically based light absorbance. Thus, we can use data from many wavelengths to distinguish between light absorption and light scattering. This can work, even if no variables selectively express the light scattering variations like the water peak in Figure 7.11. The scatter for each sample is estimated relative to that of a reference sample, and each sample's spectrum is corrected so that all samples appear to have the same scatter level as the reference.

The MSC is a rather general technique for estimating a multiplicative interference β_i when none of the X-variables give unique information about this. What is special about the MSC is its potential for separating multiplicative (scattering) variations from the additive (chemical) information, on the basis of the same set of variables $k = 1, 2, \ldots, K$.

In spectroscopy of powders and slurries etc., for instance, there is usually no wavelength range that uniquely defines the path length or light scattering level, like we had with the solvent peak in Figure 7.11.

However, MSC is still made possible because all the major constituents absorb light in a more or less similar way. The observed X-variables therefore contain the multiplicative effect on top of the additive effects of the various analytes and interferents.

Example 1: Modelling the darkening effect of water added to white powders

This is illustrated in Figure 7.12. The example concerns calibration for protein percentage from NIR reflectance spectroscopy of mixtures of fish-flour, potato starch and water. The NIR data were obtained on a Technicon InfraAlyzer 400 at 19 fixed wavelengths. Some of the same data was used in Chapter 6 (Figure 6.3).

Figure 7.12a shows the original input data $x_{i,\text{inp}}$, in O.D. units = $\log(1/R)$, for three mixtures of fish meal and water. It shows the O.D. to increase with increasing water content. This is in part due to the increase in NIR light absorption of water at the typical water wavelengths (e.g. 1445, 1940). But if such conventional absorption had been the dominant effect, a corresponding decrease would have been expected at the typical protein and starch wavelengths. This is not the case.

Instead, a general O.D. increase is seen. This is due to increased depth of light penetration due to decreased light scattering in the samples with increasing water

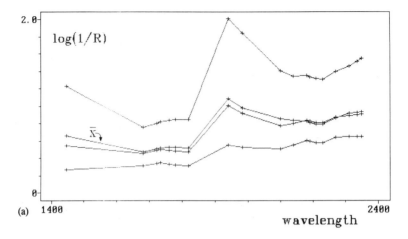

Figure 7.12 Multiplicative signal correction (MSC) a) Input spectra $x_{i,\text{inp}}$: O.D. vs. 19 NIR wavelengths in the range 1400 to 2400 nm, for 3 fish meal/water mixtures (water content: I: 10%; II: 30%; III: 60%) plus the average $\bar{\mathbf{x}}' = \mathbf{x}_{\text{ref}}$ for the 41 calibration mixtures

content (see e.g. Birth, 1978). This is analogous to the darker colour of wet paper compared to dry paper, of raw fish muscle compared to cooked fish muscle, etc. The effect of this increase in optical path length is now to be estimated. The second spectrum from the top in Figure 7.12a represents the average spectrum of the 41 calibration samples. This spectrum is chosen as MSC reference spectrum $\mathbf{x}_{\text{ref}}$.

In Figure 7.12b the spectra in 7.12a are plotted against this reference spectrum.

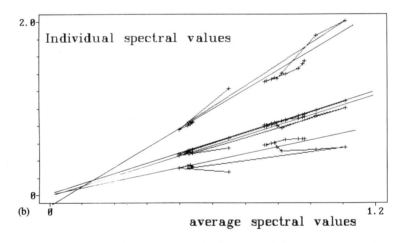

Figure 7.12 (*cont.*) Multiplicative signal correction (MSC) b) The basis for MSC: The spectra $x_{i,\text{inp}}$ in a) plotted against $\mathbf{x}_{\text{ref}}$. For instance, the uppermost curve represents $x_{ik,\text{inp}}$ for the sample with 60% water plotted against $x_{\text{ref},k}$ with consecutive wavelengths $k = 1,2,\ldots,19$ connected by straight lines. The straight lines represent the regression lines for $x_{ik,\text{inp}} = \hat{\alpha}_i + x_{\text{ref},k} * \hat{\beta}_i$. The second straight line from the top represents $x_{k,\text{ref}}$ plotted against itself

Each point in the figure thus represents one wavelength for one of the three water/fish meal samples from Figure 7.12a, $x_{ik,\text{inp}}$ vs $\bar{x}_k$. A sample with identical NIR spectrum $\mathbf{x}_{i,\text{inp}}$ as the average spectrum would give points along the diagonal line. If we for simplicity presume that the light scattering effects are the same at all the wavelengths in this plot, then we would see them as general trends affecting all the wavelengths, while more specific chemical absorbance peaks would appear as more or less erratic deviations around these general trends.

From domain-specific knowledge, and from the appearance of the curves in Figure 7.12b, it appears that the various NIR absorbance differences between water, starch and protein are sufficiently small and well distributed along the wavelength scale so that we can estimate the general trends by the model

$$x_{ik,\text{inp}} = \alpha_i + x_{\text{ref},k} * \beta_i + e_{ik,\text{inp}} \tag{7.30}$$

and estimate α_i and β_i by ordinary least squares regression of $x_{ik,\text{inp}}$ on $x_{\text{ref},k}$ over $k = 1, 2, \ldots, 19$. The resulting regression lines are shown as line extensions in Figure 7.12b. They show that increasing water contents greatly increased the relative path length, slope β_i, with some minor adjustments for relative offset, α_i for some of the curves.

The corresponding MSC corrected NIR data $x_{ik} = (x_{ik,\text{inp}} - \hat{\alpha}_i)/\hat{\beta}_i$ are shown in Figure 7.12c. Now the chemically expected variability in the spectra is the dominant remaining feature. Figure 7.12d shows that if we repeat the MSC operation on the already MSC treated data x_{ik}, no significant change takes place. The reason is that the general trends have already been removed, so mainly the apparent chemical variations remain.

Finally, Figure 7.12e compares the calibration performance of the NIR data, $\mathbf{X}$, prior to and after MSC, using a linear calibration method (PLSR) for y = protein percentage. With three major constituents in the mixtures, and their sum being 100%, two additive phenomena were expected to be seen in the NIR data. But the figure shows that the original spectra required 3 PLSR factors to give reasonable predictive ability, and as much as 10 factors to give optimal predictions. In contrast, after MSC only 2 PLSR factors were needed to give very good predictions, reaching optimality after about 5 factors.

This shows that even without MSC the additive bilinear calibration model could handle the more or less multiplicative effects quite well, but at the price of a very complicated calibration model that was very difficult to interpret and that needs many and precise data to be properly estimated. The MSC removed the great multiplicative effect, simplified the modelling and improved the predictive ability.

However, even this MSC modelling is just a data approximation; factors 3–5 are required to correct for imperfections in the preprocessing; these could be due to wavelength dependency in α_i and β_i as well as lack of linearity in the $\log(1/R)$ transform, temperature effects in the NIR spectra etc. In any case this is an illustration of how domain-specific knowledge should be built into the data analysis whenever possible, and how multivariate calibration then can be used to correct for minor imprecisions in the domain-specific assumptions.

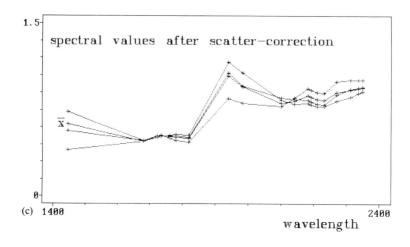

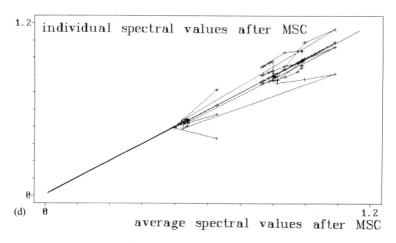

Figure 7.12 (*cont.*) Multiplicative signal correction (MSC) c) The spectra from a) plotted after MSC transformation $x_{ik} = (x_{ik,\text{inp}} - \widehat{\alpha}_i)/\beta_i$. d) The same as b), after the MSC transform

This light scattering information now mainly rests in parameters $\widehat{\alpha}_i$ and $\widehat{\beta}_i$, which can be used as extra X-variables, as illustrated in Chapter 8. These parameters can carry information about particle size, refractive index, surface treatment etc. (Geladi et al., 1985). For a comprehensive overview of light scattering effects in spectroscopy, see Birth (1978) and Birth and Hecht (1987).

Example 2: Enhancing the modelling of powders by MSC

After successful MSC one should expect the calibration to operate on simpler data structures and one could hope for more precise predictions. Table 7.1 shows a summary of the original MSC paper (Martens et al., 1983b) and concerns its use on NIR reflectance data of wheat flour fractions with widely varying particle size. The

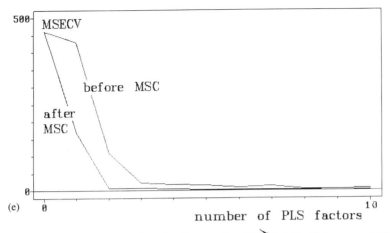

500 MSECV

before MSC

after
MSC

0

(e)   0   10
                              number of PLS factors

Figure 7.12 (*cont.*)   Multiplicative signal correction (MSC) e) Prediction error MSEP vs. no. of bilinear factors, $a = 0,1,2,..,10$ estimated for y = protein percentage by cross validation within the calibration set, before and after MSC of the X-data

NIR data X again consists of 19 wavelength channels, measured by reflectance R in a Technicon InfraAlyzer 400. The purpose of the subsequent PLS regression was to calibrate for y = protein percentage, using 23 calibration objects.

The table gives the protein prediction error RMSEP calculated for 24 independent representative test objects. The first line shows how the prediction error RMSEP is reduced from 0.33% to a more acceptable 0.24% by submitting the standard $\log(1/R)$ spectra to MSC.

The second line emphasizes the effect of MSC even more: The conventional, but pragmatic O.D. = $\log(1/R)$ transformation was here replaced by the theoretically more correct Kubelka-Munk linearization O.D. = 'K/S' = $(1 - R)^2/2R$ with Saundersson-correction for internal reflection. Ironically, this theoretically more correct response linearization gave worse prediction results (RMSEP=0.54%). However, after MSC this theoretically correct unit yielded the best results (0.20%).

The probable explanation is the following: The difference between the pragmatic $\log(1/R)$ and the more ambitious Kubelka-Munk linearization is that the latter makes the scatter effect $\approx 1/\beta_i$ purely multiplicative in the O.D. data. Purely multiplicative interferences cannot be modelled well by the additive multivariate calibration methods, so the MSC was particularly useful here.

Table 7.1   Prediction error (RMSEP in weight %) for protein in wheat flours with widely different particle size determined from NIR reflectance (R)

Response-linearization	Multiplicative correction	
	No	MSC
$\log(1/R)$	0.33	0.24
$(1-R)^2/2R$	0.54	0.20

The more pragmatic $\log(1/R)$, on the other hand, 'smears' the multiplicative scattering interference out in different ways for the different wavelengths. So, in analogy to the earlier example of implicit bilinear correction for curvatures (Figure 7.4), even the scatter information can be picked up and corrected for: The PLS regression yielded a surprisingly low prediction error (RMSEP =0.33%). But even for these $\log(1/R)$ data the MSC did have a beneficial effect.

The MSC procedure has been tested on different types of diffuse spectroscopy data (see e.g. Geladi et al., 1985; Osborne, 1988; Isaksson and Næs, 1988), and in many cases a good bilinear modelling BLM can be obtained with a lower number of factors, simplifying the interpretation. Concerning prediction ability the results after MSC are also in general somewhat better than without MSC.

Reasons for improvements by using MSC are discussed further in Isaksson and Næs (1988) along with a discussion of interpretation advantages.

The MSC approach given here—regressing each spectrum $x_i{}'$ on some reference spectrum like $\bar{x}'$ over certain X-variable intervals, can be modified and extended in a number of ways.

One such case is when the additive α_i and/or multiplicative β_i parameters are expected to vary with variable $k = 1, 2, \ldots, K$, (like the scatter coefficient's wavelength variation). Then some extra parameters can be included in the MSC model. One way is to use exponent expressions $\alpha_i{}^{ck-d}$ and/or $\beta_i{}^{ck-d}$, where c and d are nonlinearity parameters to be estimated iteratively for the whole calibration set. As a special case, if the closure restriction applies (the sum of the constituents equalling 100%) and the data are otherwise well-behaved, then a preliminary direct unmixing regression of x_i on K (the matrix of spectra of all the different constituents or groups of them, Chapter 4), can yield MSC estimates: β_i could then be equal to the sum of the preliminary constituent concentrations.

Another extension of the MSC is to regress each $x_i{}'$, not only on $\bar{x}'$, but also on a number of vectors, e.g. the spectra of the pure constituents, $k_j . j = 1, 2, \ldots, J$ Stark and Martens (1989). This improvement is implemented as Extended MSC (EMSC) in the QUIET package (1991).

7.4.2.4 Derivative ratioing regression

This merging of the multiplicative linearization and the multivariate linear additive calibration regression is done consciously in the derivative ratioing technique developed by Karl Norris (see e.g. Norris, 1983) for scanning NIR instruments. In this technique the O.D. spectra themselves are replaced by first or second 'derivatives' in the calibration regression. (Actually, what is used are differences, not true derivatives, because differences are more easily computed in practice.)

The calibration model can then be written:

$$y = b_0 + t_1 b_1 + f \tag{7.31}$$

where for instance t_1 is

$$t_1 = (x_I - x_{II})/(x_{III} - x_{IV}) \tag{7.32}$$

where (I, II) and (III, IV) represent two selected pairs of adjacent wavelengths, each pair with e.g. 10 nm wavelenth gap. The first pair is chosen at a wavelength where it can provide additive selectivity enhancement (e.g. I = analyte wavelength, II = major interferent wavelength). The second pair is positioned for optimal multiplicative scatter correction and for further selectivity enhancement.

This 'nonlinear' calibration model was designed to handle multiplicative light scatter variations, which give multiplicative effects (therefore divison) but at the same time to handle additive spectral overlaps, which require ordinary linear selectivity enhancement (therefore derivatives).

The resulting predictor then becomes:

$$\hat{y} = \hat{b}_0 + \hat{b}_1(x_I - x_{II})/(x_{III} - x_{IV}) \tag{7.33}$$

More generally this nonlinear predictor can be written:

$$\hat{y} = \hat{b}_0 + \hat{b}_1(\mathbf{x'}\hat{\mathbf{n}}/\mathbf{x'}\hat{\mathbf{d}}) \tag{7.34}$$

where $\hat{b}_0$ and $\hat{b}_1$ are ordinary linear regression parameters while vectors $\hat{\mathbf{n}} = (\hat{n}_k, k = 1, 2, \ldots, K)'$ and $\hat{\mathbf{d}} = (\hat{d}_k, k = 1, 2, \ldots, K)'$ are multivariate parameters for the numerator and denominator, respectively. In contrast to the nonlinear pretreatment methods discussed earlier, these parameters are estimated during the actual calibration regression, by testing many different alternative $(\mathbf{n}, \mathbf{d})$ sets interactively, estimating b_0 and b_1 for each of them and selecting the one with best predictive ability and/or most reasonable interpretation.

If a wavelength range can be found that displays a reasonably simple interference structure (analyte + one interferent), then the first 'derivative' numerator (vector $\mathbf{n}=(\ldots 0, 0, 0, 0, 0\ 1, -1, 0, 0, 0, 0, \ldots)'$) untangles this additive interference problem. Likewise, a second 'derivative' $\mathbf{n}=(\ldots 0, 0, 0, 0, 1, -2, 1, 0, 0, 0, \ldots)'$ can correct for two additive interferences.

The strength of this integrated approach is its simplicity, its ability to overcome curvatures by selecting the wavelength with the best combined linearity, and finally its ability to integrate the user's insight about spectroscopy in the choice of wavelengths and thereby reduce overfitting. But the method is primarily limited to scanning spectrophotometers since it usually relies on derivatives.

If there are more than three interference phenomena affecting the relevant wavelength ranges, then the first and second derivatives may not give sufficient selectivity enhancement. Higher-order derivatives can then be used, but this can amplify the random noise in the spectral measurements at the wavelengths used. The calibration regression can instead be expanded to include two or more such ratioing expressions. The danger of overfitting and losing mental overview then increases and this approach should probably be used with conscious predictive validation (see e.g. Shenk et al., 1981).

Only a few of the available wavelengths are used in the Norris regression model ((7.31) and (7.32)), because only one wavelength in principle is used to represent each phenomenon in the spectral data. In practice, various smoothing techniques are employed in order to draw information from the otherwise unused neighbouring wavelengths into the model to stabilize the calibration.

7.5 SMOOTHING

7.5.1 INTRODUCTION

Till now we have discussed systematic effects which have importance for precision of calibration results. Here we will discuss how to reduce the effect of random noise. The basic idea is that the measurements $\mathbf{X}$ contain non-systematic noise $\mathbf{E}$, i.e

$$\mathbf{X}_{measured} = \mathbf{X}_{'true'} + \mathbf{E}_{noise} \tag{7.35}$$

which can be reduced by some kind of filtering, based on various assumptions about how the 'true' data structures differ from the random noise.

The calibration methods treated earlier in this book can provide one kind of noise reduction by 'averaging' the noise over the many calibration objects and X-variables in the least squares parameter estimation (provided that the number of calibration objects I is sufficiently higher than the number of calibration factors used, $a = 1, 2, \ldots, A$).

Nevertheless, noise in the calibration data $(\mathbf{X},\mathbf{Y})$ will allways create estimation errors in the estimated calibration parameters $\hat{\mathbf{B}}$ and hence cause systematic errors in later predictions of $\mathbf{Y}$. So, improving the signal/noise ratio of input data by pre-processing will still be an advantage.

So-called smoothing of adjacent variables can be helpful in this respect, when the data come from a continuous instrument with a high enough sampling rate.

Examples of data suitable for smoothing are

* Scanning spectrophotometers
* Chromatographic or electrophoretic traces (i.e. prior to peak integration)
* Counting frequency histrograms from e.g. image summaries

Care should, however, be taken to avoid filtering out the actual information as well!

Smoothing primarily concerns how to reduce 'high-frequency' ripple noise. The corresponding 'low-frequency' noise (e.g. instrument drift during the scanning measurements) is usually more difficult to deal with and correct for, because it will often resemble the real information in the data.

7.5.2 MOVING AVERAGE FILTERS

The conceptually simplest type of smoothing out the high-frequency noise is the moving average method. The reading x_{ik} at each variable $k = 1, 2, \ldots, K$ is replaced by a weighted average of itself and its nearest neightbours from $k - D$ to $k + D$:

$$x_{ik} = \sum_{d=-D}^{+D} x_{i,k+d} u_d \tag{7.36}$$

The convolution weights u_d define the smoothing. For instance, $\mathbf{u}' = (0,0,0,1,0,0,0)$ gives no smoothing, while for instance $\mathbf{u}' = (0,1,2,4,2,1,0)/10$ does.

The denominator 10 in the latter example indicates that it is necessary to scale the filter vector so that the sum of the elements employed is equal to 1.0. At the ends of the spectrum (k near 1 or near K) full smoothing cannot be employed, and the effect at these ends will therefore be less satisfactory. The properties of moving average filters are studied in e.g. Anderson (1971) and Rabiner and Gold (1975).

7.5.3 SPLINE FILTERS

Another type of smoothing is based on so-called spline functions. The idea is that in small intervals most functions can be fitted by low-degree polynomials. One therefore divides the spectrum into pieces and fits polynomials to each of these pieces under the restriction that the resulting composite polynomial is a continuous function. Different procedures for subdivision and selection of polynomial degree exist, and we refer to Wold (1974) for a discussion of the practical aspects of the technique.

7.5.4 FOURIER ANALYSIS

Before we proceed, we note the potential confusion between the use of the word 'spectrum' in this book, referring e.g. to the 'colour spectrum' or some other input vector $\mathbf{x}_i = (x_{ik}, k = 1, 2, \ldots, K)$, and the use of the same word in Fourier analysis, where 'spectral analysis' usually refers to analysis in the transformed Fourier domain. Even in the following we shall reserve the word 'spectrum' to indicate 'colour spectrum'.

In section 3.7 we described what we called Fourier regression. In fact this idea is identical to that behind smoothing or filtering by Fourier analysis. In Fourier regression we decomposed each input spectrum $\mathbf{x}_i$ as a sum of sine and cosine patterns, and used the scores $\mathbf{T}$ of these patterns as regressors in the calibration regression for $\mathbf{Y}$. The idea was that the high frequency variation in the input spectrum was irrelevant for prediction and could be deleted, to reduce the amount of data for calibration regression and disc storage, and to attain more stable predictions.

There exist both a continuous and a discrete version of the Fourier transform, but for the type of data treated in this text the discrete one is the one of primary interest. Computationally, the fast Fourier transform (FFT) is an algorithm to perform the Fourier analysis quickly (Rabiner and Gold, 1975).

Smoothing by Fourier analysis

First we make an orthogonal transformation of the spectrum into a sum of sine and cosine shaped spectral contributions of higher and higher frequencies. The result is that instead of the spectrum itself we obtain a rotated spectrum consisting of contributions of the different types. In mathematical terms this can be written as

$$\mathbf{X} = \mathbf{T}_{\text{cosine}} \mathbf{V}'_{\text{cosine}} + \mathbf{T}_{\text{sine}} \mathbf{V}'_{\text{sine}} \tag{7.37}$$

where $\mathbf{X}$ is usually centered. and where the elements in the two Fourier loading matrices $\mathbf{V}_{\text{cosine}}$ and $\mathbf{V}_{\text{sine}}$ are as in (3.118).

By the converse transformation it is possible to regenerate $\mathbf{T}$ back into the original $\mathbf{X}$ domain. The scores $t_{ia,\text{cosine}}$ and $t_{ia,\text{sine}}$ for the first, low-frequency factors $a = 1, 2, \ldots$ will reflect the low-frequency features in the input spectrum while the scores of the last factors will primarily reflect the high-frequency 'ripple' in each spectrum (often corresponding to noise).

In Fourier filtering the input spectra $\mathbf{X}$ are first transformed into the Fourier factors $\mathbf{T}$. Undesired frequencies are then weighted down, and the inverse Fourier transform finally used for regenerating smoothed X-data. Thus the smoothing consists of multipying each of the Fourier scores $t_{ia,\text{cosine}}$ and $t_{ia,\text{sine}}$ by a weight z_a determining how much this frequency contribution should count. Then the smoothed spectra are obtained in the original domain by the back transformation.

The simplest set of smoothing weights z_a is of course the sequence $(1, 1, 1, 1, \ldots 1, 0, \ldots \ldots 0)$ where the 0's correspond to high frequencies ignored. In this way the effect of high frequency noise is eliminated and the low-frequency contributions are kept as they are. When to change from 1 to 0 corresponds exactly to how many regression factors to use in the Fourier regression (section 3.7).

More efficient weights z can also be devised, e.g. with both positive, zero and negative elements. In this way very efficient and specific smoothing 'windows' can be defined. The Savizky–Golay filter is one such example. We refer to Rabiner and Gold (1975) for more detailed treatment of digital filtering.

It can be shown that a Fourier smoothing corresponds to a convolution of the original spectrum according to (7.36) with the convolution weight vector $\mathbf{u}$ in the original domain corresponding to the weight vector z in the Fourier domain.

7.6 OTHER TRANSFORMS

Fourier deconvolution

Fourier analysis can also be used for another important linear transform, namely deconvolution: This is a way of 'sharpening' the information in the spectra, by eliminating previous implicit convolution caused by e.g. the smoothing of spectral peaks by the use of a monochromator/slit system with very wide bandwith.

Deconvolution thus is the inverse of a convolution. This process concentrates the information in many wavelengths into a few wavelengths. This can be advantageous, if the number of wavelengths to use in the modelling can be reduced without losing information, and the visual interpretation of the spectral data and the resulting models may become easier.

Fourier domain transform

The Fourier analysis can also be applied as a domain transform as such. The most well known example of this in analytical chemistry is the Fourier transform IR

(FTIR) instruments, in which the interferometer's detector response is converted from the original function of a mechanical mirror position to the desired function of wavelength.

Hadamard transform

An analogue to the Fourier transform is the Hadamard transform, which uses weights v_a', $a = 1, 2, \ldots$ consisting of orthogonal square wave functions $(-1, -1, -1, -1, 1, 1, 1, 1)$, $(-1, -1, 1, 1, -1, -1, 1, 1)$ etc. instead of cosine and sine functions. Hadamard transforms are numerically efficient and they may have a potential in multivariate calibration.

Counting transform

Converting the light intensity data of digitized images into relative pixel frequency histograms is another useful domain transform. Basically the number of pixel elements are counted at every intensity class $k = 1, 2, \ldots, K$. This can be extended into contextual histograms—counting pixel elements with various structures in their neighbouring pixels, or counting areas of various sizes ranging from individual pixels to hundreds of pixels.

This counting transform is applicable to image analysis data from e.g. electron microscopy, magnetic resonance imaging etc. in cases where the distribution of geometric shapes is interesting.

Example: In-vivo determination of chemical body composition (fat, protein and energy content and distributions) of living animals can be based on counting histograms from X-ray computer tomography images. By sacrificing some representative animals and actually measuring their body composition etc., and using these data as Y in a multivariate calibration against histrogram data X, the subsequent selection of animals for breeding progams has been greatly facilitated. This has been done for a number of different animal species (see e.g. Skjervold et al., 1981; Martens et al., 1983a). With the rapid development of other imaging techniques this type of transform will probably have an increasing value.

Position transform

The position transform concerns the conversion of a continuous input signal like an absorbance 'spectrum' into a series of peak positions. A possible example of this is the enhanced utilization of spectral shifts, like the temperature and salt shifts of water's NIR peaks: By converting e.g. the NIR absorbance peaks of water (normally around 1050 nm, 1445 nm, 1940 nm etc.) into a few variables x_k that give the wavelength (e.g. in nm) where the maximum of peak no. $k = 1, 2, 3, \ldots$ occurs. Such X-data can give simpler relationships to e.g. Y = sample temperature or salt content, since continuous sideways shifts in the original absorbance spectra are not so well suited for direct bilinear modelling. This has a potential in, for instance, thermal and sound analysis.

This is another example of how data domain transform preprocessing can bring out those aspects of the input data that are of primary concern for a given particular application and make them accessible for multivariate calibration.

8 Multivariate Calibration Illustrated: Quantifying Litmus in Dirty Samples

8.1 PROBLEM FORMULATION, DESIGN AND MEASUREMENTS

This chapter illustrates how a full calibration experiment could proceed, from problem formulation through data pre-processing to final model estimation and prediction with automatic outlier detection. It is written without any mathematical formulae—references are instead given to the technical parts of the book.

The example is based on multi-channel spectroscopy and on human assessments. But these types of measurements are intended to represent any multivariate instrumentation—chromatography, electrophoresis, NMR, image summaries etc.

The example illustrates how multivariate calibration can convert simple, but highly non-selective measurements of more or less intact samples into selective and reliable analyte quantifications. This is attained by finding informative combinations of the different instrument variables measured.

This mathematical modelling of the instrument variables $x_k, k = 1, 2, \ldots, K$ into analyte concentrations $y_j, j = 1, 2, \ldots, J$ is based on X-data from a well selected set of calibration samples with known Y-levels of the analytes of interest.

The calibration method to be used is partial least squares regression (PLSR section 3.5). This regression method compresses the information from multi-channel analytical instruments (X) into a low number of latent variables (T) that in turn define the predictions of the analytes (Y). This calibration modelling also allows us to detect outliers. In analogy, when we listen to music, we compress the sound from multi-string musical instruments into main harmonies, and can detect disharmony and imbalance.

The analyte to be determined is an organic dye called litmus (lacca musica, lacmus, tournesol). This blue colouring matter has traditionally been used as a pH indicator (red in acid, blue in basic aqueous solution). It has also been used for colouring beverages, and is now used to colour culture media for diagnostic purposes. Litmus is produced from various species of lichens.

The example illustrates how a relevant analyte can be quantified, even if it is not a well defined compound (*The Merck Index*, 10th edition, Merck & Co,, Inc. Rahway, N.J. 1983) in the chemical sense: Litmus is a mixture of several compounds, chiefly azolitmin and erythrolitmin assumed present in a fixed ratio. The reader is invited to follow the example as if being personally responsible for the chemical on-line process analysis in a chemical factory. The following computations are based on actual, empirical measurements, although their process factory setting is fictitious.

All the computations and plots in this example are performed in the UNSCRAMBLER package (Tyssø et al., 1987) on a personal computer.

The analytical problem and choice of instrumentation

Conceptually, the analytical situation is the following: A rapid and reliable on-line method is required for quantification of litmus in aqueous solution in an industrial process stream, in a certain pipeline.

Being a brightly coloured dye, litmus is considered most easily determined by spectrophotometry in the visible range (400–700 nm). A scanning spectrophotometer fitted with a fiber optic transmission probe (Guided Wave model 200) is chosen, in order to remove the sensitive instrumentation from the rather rugged process environment.

White lamp light was led from the instrument (see Figure 1.1) through a glass fiber to the process pipeline, then through the liquid in a 1 cm path length probe inserted into the process stream, and finally through another glass fiber back into the instrument's monochromator and detector system. The transmittance, representing the ratio of light intensity transmitted through sample and light intensity transmitted through air, ($T = 1$: no absorbing constituents; $T = 0$: high concentrations of absorbing constituents), was thus measured at 100 wavelength channels $x_k, k = 1, 2, \ldots, 100$ in the wavelength range from 400 nm to about 1200 nm (8 nm between each). The last three wavelength channels were discarded as being too noisy; only channels x_1 to x_{97} will thus be used for spectral information here.

The near infrared range 800–1200 nm is included in order to allow better selectivity enhancement; the spectral measurements may also be affected by other phenomena than litmus concentration.

First of all it seems that there may be chemical interference: some unidentified dye may sometimes be present in the liquid in unknown amounts. We cannot remove such interferents in the on-line spectral measurements.

Secondly there may be variability in the analyte itself, because the pH of the process is expected to vary from time to time. Consequently the colour of the analyte litmus may change uncontrollably between red and blue. The presence of possible other interferences stops us from just using the isospestic wavelength of litmus (where red and blue litmus absorb equally) for the analyte determination. Similarly the interferents themselves may vary with pH. Consequently, our process stream varies in colour hue from blue to red and even yellow! We assume that we cannot stabilize the pH in the process. The pH could be measured by a pH electrode, but for practical reasons that is considered undesirable. So we do not know the pH level at any particular moment.

Thirdly, there may be uncontrolled variations in particulate material suspended in the liquid to be analyzed. This may give varying light scattering. The process stream was sometimes nicely transparent, at other times it was milky. Such physical interference will seriously affect the spectrophotometric readings. We do not have time to remove the particulate material prior to the spectral measurements.

As a safety check and in order to avoid human alienation due to the introduction of new instrumentation and computer technology into the production plant, the operators are asked to have a look, and report how the liquid in the process stream looks to them, every time a transmission spectrum is measured. So three extra variables, measured by another spectrophotometer instrument—the human eye— were included. These simple sensory assessment variables were: Redness ($x_{98} = 1$ if the liquid is red, otherwise = 0), Blueness ($x_{99} = 1$ if the liquid is blue, otherwise = 0) and Turbidity ($x_{100} = 1$ if the liquid is turbid, otherwise = 0). It was checked in advance that the operators were not colour blind.

In analogy to the tuning of a musical instrument by a tuning fork, the reading from the fiber optic spectrophotometer was repeatedly standardized to transmittance 1 over the full wavelength range, by measurements in air. The human eye was expected to be sufficiently standardized by the brain's internal memory standards of what red and blue colours and turbidity look like.

In this analytical situation, the 'true' concentration of the analyte litmus was determined by, say, some noxious and expensive reference method too slow to be used for process control. The purpose is now to replace this slow off-line reference method **y** by the on-line spectral measurements **X** of the intact process liquid.

Experimental design: choice of calibration set

The 97 transmission variables and the three sensory assessments together form the 100 X-variables which will now be tested in multivariate calibration for y=litmus concentration. Now we have to 'teach the computer how to quantify **y** from **X**'. Since there may be interferences, we must use some type of multivariate calibration. The possible interferences are unidentified at this stage, so we cannot apply traditional 'unmixing' (direct calibration, section 3.6.1); we do not know what to unmix! So we have to rely on indirect multivariate calibration—develop the mathematical calibration model by statistical estimation based on empirical data from real samples from the process.

As discussed in sections 1.3 and 1.4, statistics is usually not the favourite for chemists and technologists.

Fortunately, multivariate calibration does not require extensive insight into classical hypothesis testing statistics. But two things are needed:

1) A minimum understanding of experimental design, and

2) A powerful multivariate calibration method implemented in a software package with suitable validation methods and graphics.

Now we need a representative, informative set of calibration samples with both X- and y-data—otherwise the computer algorithm cannot develop a good calibration model. How to obtain such calibration sample sets is treated in Chapter 6 (experimental design). In summary, it is important to choose calibration objects that span every type of X-variability that has to be modelled, be it the analyte or interferences of various types.

In the present case both the transmittance spectrum and visual appearance (**X**) in the process stream was measured at various times—and a sample was taken every time and stored for later chemical analysis of **y**.

Since the chemical reference analysis is slow and expensive, the number of calibration samples had to be limited. The many samples were inspected visually (a preliminary data analysis of the X-data could also have been used, section 6.4). Some of the selected samples were red, some were blue, some were even yellow—and they had different levels of turbidity.

On this basis, a subset of 26 samples was selected for the calibration set, and their concentration of litmus was measured by the reference method. This set of 26 samples was chosen so as to span all the apparent spectral phenomena independently of one another (different types and intensity of colours, and different levels of turbidity).

The litmus concentration (**y**) ranged from 0 to 0.15 mg/ml, and variable **y** is here expressed in percent of this maximum concentration (0 to 100%).

8.2   EXPLORATIVE DATA ANALYSIS AND DATA PRE-TREATMENTS

Figure 8.1a shows the raw X-data for some of the calibration samples. The spectral data for X-variables $k = 1$ to 97 (wavelengths 400–1176 nm) are expressed as transmittance T.

The figure shows a rather complex pattern, especially in the visible wavelength range. Some apparently random measurement noise is also apparent. The three sensory assessments Redness, Blueness and Turbidity are seen at the right side of the figure.

Without any further interpretation at this stage, is it possible to use these raw data **X** for predicting litmus concentration **Y**?

Figure 8.1b shows how traditional univariate calibration would function in this case: The transmittance at the isospestic point (x_{15}, 520 nm) is plotted vs. litmus concentration (y, in percent) for the 26 calibration samples. It shows that no calibration line in this plot would yield meaningful determination of litmus from transmittance at 520 nm. Notice for instance that the samples at zero concentration of litmus display a variety of transmittance levels. The reason is not error in the reference measurements. It appears that these samples were taken at times when

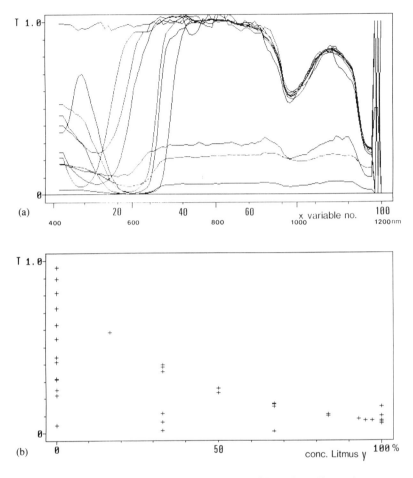

Figure 8.1 Input data and preliminary calibration (MODEL I). a) Transmittance spectra **X** of some calibration samples. The last three X-variables represent sensory redness, blueness and turbidity. b) The 'best' univariate calibration: transmittance at the isospestic point for litmus (520 nm) vs. litmus concentration (**y**) for the 26 calibration samples

some interfering substance other than litmus was present in the process stream.

The other individual X-variables gave even worse relation to y. Could multivariate calibration on the raw data (y vs the 100 transmittance-variables in **x**) do any better?

There are many calibration methods available. We here choose the bilinear partial least squares regression (PLSR) (section 3.5) due to its flexibility.

Preliminary calibration

To be useful, the PLSR requires some sort of a priori weighting of the input variables. Assuming at the outset every X-variable to have the same probability of contributing information about **y**, we for simplicity standardize all the variables to a standard deviation of 1 prior to the calibration modelling. For simplicity even **y** was standarized—this has no statistical consequence since there is only one Y-variable. But it can give nicer plots of the results. Such a priori weighting is covered in more detail in section 7.1.

The initial standard deviation of **y** was presently 42.73% litmus prior to the standardization (corresponding to an analyte concentration range from 0% to 100%). The a priori weight for **y** was therefore $1/42.73 = 0.0234$.

CALIBRATION MODEL I

The standardized X-variables and standardized **y** was submitted to PLSR modelling (section 3.5). This calibration method estimates from the X-data one abstract phenomenon ('bilinear PLS factor') after the other, subtracting their effects from the X- and Y-data. Each such abstract factor is a linear combination of the 100 X-variables, and PLSR finds those linear combinations that describe a maximum covariance remaining between **X** and **Y**. The process ideally proceeds until only irrelevant information and random noise is left in the data—in practice we choose the number of factors that appears to have the best predictive ability of **y**.

In the present data we expect the samples to be affected by 5 real phenomena (litmus conc., litmus colour, interferent conc., interference colour and turbidity). So, if the instrument response **X** had varied linearly with these real phenomena, we would expect to see 5 main phenomena in the X-data, and hence optimal predictive ability for **y** after 5 PLS factors.

Several objects and variables were tagged automatically as outliers during the calibration modelling (see Chapter 5). However, none of the outliers appeared serious enough to require any action; they were therefore accepted as valid data in this initial screening analysis.

In this preliminary analysis, the 'quick-and-dirty' validation method of leverage correction (Chapter 5) was employed, due to its computational speed (in this case $df = I - A - 1$). Figure 8.1c shows the residual variances for **y** and averaged for the 100 X-variables, as functions of the number of PLS factors. It starts at 1.0 for 0 factors (only average subtracted in **y** and **X**), since all the variables had been standardized to variance 1 prior to the PLSR.

The figure shows that most of the X-varibility is accounted for by the first few factors, reaching a minimum level after about 6–7 factors (only about 3 percent of the initial variance unexplained). The first factor explains close to 50% of the variance in litmus concentation (**y**), but a very high number of factors is needed to attain low apparent predictive error.

Not knowing much about the calibration objects or the linearity of the instruments, we want at least 4 objects per PLS factor, to avoid over-fitting (Chapter 4). With only 26 calibration samples available, we choose the 5-factor solution as the apparently 'optimal' one for this modelling. This explains about 73 percent of **y**'s initial variance (42.73^2). Figure 8.1d shows the predictor function $\widehat{\mathbf{b}}$ in the expression $\widehat{y} = \widehat{b}_0 + \mathbf{x}\widehat{\mathbf{b}}$, corresponding to the 5-factor PLSR solution (see

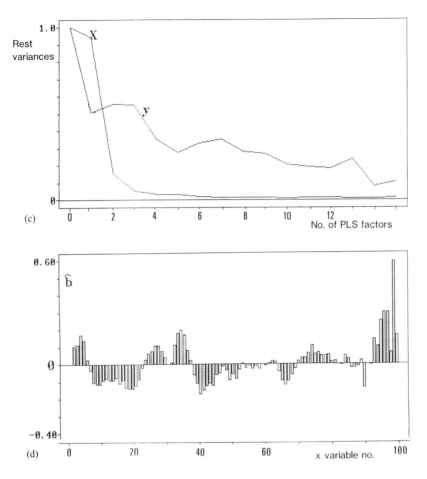

Figure 8.1 (*cont.*) Input data and preliminary calibration (MODEL I). c) Residual variance in **X** and **y** after increasing number of PLS factors a = 0,1,2,. . .,15. d) Calibration coefficient spectrum **b̂** obtained using the 5 first PLS factors

section 3.5). Not knowing much about the analytical problem, it is rather difficult to interpret the many peaks in this result.

This function leaves about 27 percent of **y**'s initial variance unexplained—corresponding to an apparent average prediction error standard deviation (RMSEP, Chapter 4) of 22% litmus.

The reduction in the litmus variability from the initial 42.73% to the residual 22% shows that there may indeed be relevant information in the *X*-data. But the residual prediction error is too high. So it seems necessary to try to optimize the calibration modelling.

The uneven and complicated way in which the prediction error of **y** in Figure 8.1c decreases with the number of factors indicates that the data contain nonlinear *X–X* and/or *X–Y* relationships.

Response linearization

In seeking a linearization of the X-data we look at the raw data (Figure 8.1b). In addition we employ domain-specific knowledge, in this case from spectroscopy: From Beer's model we know that the absorbance or optical density (O.D.) $= \log(1/T)$ is proportional to the concentrations of the analyte and chemical interferents—under ideal conditions (dilute transparent solutions). In the present case we are far from ideal conditions—the analyte concentration is relatively high and there is severe turbidity in some samples. Still, for lack of better approximation we now apply the response linearization (section 7.3) of $\log(1/T)$ for X-channels $k = 1, 2, \dots, 97$.

Figure 8.2a shows the resulting O.D. spectra for two of the calibration samples containing the same litmus concentration at pH 10. Both curves now display the characteristic absorbance peak at about x_{22} (about 580 nm). In addition, these aqueous samples display one full and one partially hidden absorbance peak at about x_{73} (984 nm) and x_{95} (1160 nm); these are due to the NIR absorption of solvent itself, water.

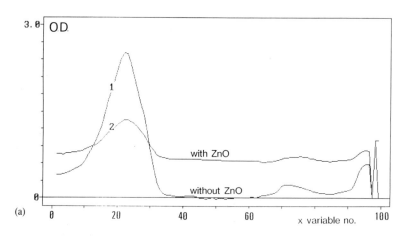

Figure 8.2 Multiplicative signal correction (MSC). a) The optical density (O.D. $= \log(1/T)$) spectra for two samples with identical litmus concentration at pH 10, but with different turbidities. Object 1 has no turbidity, object 2 has a lot of turbidity

In spite of having the same analyte concentration, the same pH and no other chemical interferents, the two curves display quite different spectra. Object 1 is a clear solution, while object 2 is a strongly turbid sample. It seems from the figure that there is a difference in baseline offset and in amplification of the O.D. signal.

Such multiplicative amplification interferences are difficult to correct for with additive linear calibration models like PLSR. But it can be removed by so-called multiplicative signal correction (MSC, section 7.4) prior to the analysis in the following way.

Multiplicative Signal Correction

Figure 8.2b shows the same spectra for objects 1 and 2 plotted against the spectrum for object 1. The diagonal hence represents object 1's spectrum plotted against itself; each point represents one wavelength channel $k = 1, 2, \ldots, 97$. The second line shows that the spectrum of object 2 is a linear function of that of object 1, with a positive offset and a slope lower than 1. If the second line is moved vertically till its ordinate offset is 0.0, and then rotated up till its slope is 1.0, then it would stongly resemble the diagonal.

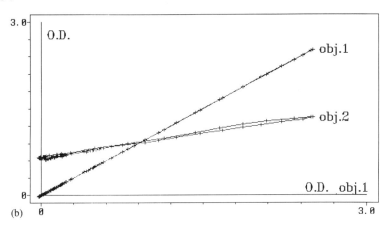

(b)

Figure 8.2 (*cont.*)   Multiplicative signal correction (MSC). b) The basis for MSC: Diagonal curve: O.D. spectrum of object 1 plotted against itself for wavelength channels $k = 1,2,\ldots,95$. Second curve: O.D. spectrum of object 2 plotted against = O.D. spectrum of object 1 for wavelength channels $k = 1,2,\ldots,95$

The figure indicates more or less identical offset and slope effects both for the litmus peak in the visible wavelength range and for the two NIR water peaks. The O.D. differences between samples in this latter wavelength region primarily reflect turbidity differences and not differences in dye concentrations or pH. How much to move the offset and how much to change the slope in order to correct the spectrum of each individual object can for instance be assessed from the shape of its baseline and solvent peaks in the NIR range, x_{40} to x_{95} (720 to 1160 nm).

Thus, we can now use the NIR wavelength range to assess the turbidity, and then scatter-correct the whole spectrum to a common 'average turbidity level'. All the corrected spectra will seem to represent samples with the same average turbidity level. This is discussed in more detail in section 7.4.

8.3   CALIBRATION

The linearized and scatter-corrected spectra can now be submitted to a renewed calibration effort. Figure 8.3a shows the scatter corrected O.D. data for the same

366

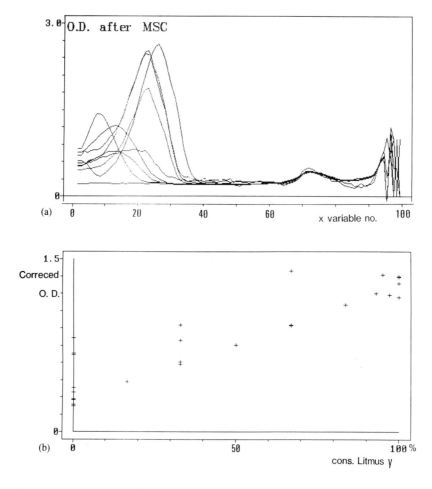

Figure 8.3 Pre-processed spectral data. a) Scatter-corrected O.D.=log(1/T) spectra **X** of some calibration samples. The last three X-variables represent sensory redness, blueness and turbidity. b) The 'best' univariate calibration: Corrected O.D. at the isospestic point for litmus (520 nm) vs. litmus concentration (**y**) for the 26 calibration samples

objects that were given in Figure 8.1a. Figure 8.3b now repeats the best univariate calibration. Compared to the transmittance data in Figure 8.1b, the linearization and scatter correction has improved the quality of the spectral readings. But the selectivity is still not satisfactory when only one wavelength variable is used: the readings for samples with no litmus present vary a lot, and even at intermediate and high litmus levels deviations are apparent.

Let us now repeat a multivariate calibration, based on the PLSR method. In order not to lose the turbidity information entirely, the MSC offset and slope were inserted as x_{96} and x_{97} respectively, instead of the corrected O.D. values which were deemed unnecessary. As before, x_{98}, x_{99} and x_{100} represent the sensory assessments redness, blueness and turbidity, with levels 0 or 1.

To optimize the PLSR calibration modelling, the a priori weighting needs more attention: We have 3 different types of X-variables; 95 O.D. variables, 2 MSC parameters and 3 sensory assessments, and there is a need to balance these against each other. But many of the baseline X-variables now have very little variance, and most of this is noise, so it is not a good idea to standardize all the X-variables like we did in MODEL I; that would amplify the baseline noise needlessly.

After having studied a preliminary run on unweighted variables, it was decided to standardize the last 5 X-variables, to multiply X-variables 4 to 95 by a factor of 10 and to drop the first 3 X-variables as unreliable. This a priori weighting was expected to give comparable residual levels for all the X-variables.

The y-variable is still the concentration of litmus (in percent).

CALIBRATION MODEL II

When submitted to PLSR, these linearized and scatter-corrected spectral X-data improved the quality of the calibration (MODEL II), compared to the transmittance input spectra (MODEL I). A more formal and conservative, but computationally time-consuming validation method (cross validation, Chapter 5) was now used, since a usable calibration was now sought. The validation showed that the best calibration model was obtained after 6 PLS factors. This solution was named MODEL II.

With this 6-factor calibration model, 98 percent of the inital variance in y was correctly predicted (as opposed to 73 percent in MODEL I). The remaining 2 percent represents an average absolute prediction standard deviation of 5.8% litmus.

So, in spite of the unidentified interfering dye, the pH variations and the turbidity variations, litmus predictions could now apparently be performed on-line in the concentration range from 0% (pure water) to 100% (1.25 mg/ml), to a precision of 5.8% (0.0725 mg/ml).

For monitoring the analyte concentration in the chemical process, this result was deemed satisfactory for process control, since analytical speed was regarded as more important than analytical precision. Thus, based on calibration MODEL II the analysis was tentatively put on-line: The calibration produced valuable results even before we had full understanding of what the interference problems were. But we retained the human colour assessments in the modelling as a safety check.

However, it was considered necessary to get a better understanding of how the calibration model developed. Therefore the calibration samples were submitted to more detailed chemical analysis off-line.

In addition to the already known litmus concentration (now termed y_1), the following additional measurements were taken: Concentration (by weight) of dispersed solid material (y_2) and of the unidentified interferent (y_5) and pH. Variables y_2 and y_5 were expressed in percent of some reference level. Since pH was not expected to vary linearly with the other constituents, it was expressed in terms of two design-variables: $y_3 = 1$ if pH<5, otherwise $y_3 = 0$, and $y_4 = 1$ if pH>8, otherwise $y_4 = 0$.

The last calibration regression was now repeated, this time calibrating for all 5 Y-variables instead of only y_1 = litmus concentration. The a priori weighting of the X-variables are now the same as in MODEL II; the Y-variables were now standardized prior to the PLS regression.

CALIBRATION MODEL III

The PLS2 algorithm was now used in the regression (section 3.5.4.1), since there are more than one Y-variable. Cross validation was again used for finding the optimal number of factors to use in the calibration model.

A few objects were tagged as outliers during the calibration modelling. Each of them was inspected, and it was concluded that they all represented 'good outliers', i.e. particularly informative objects. Therefore no objects had to be eliminated from the calibration set in this case.

But certain 'good outliers' were submitted to further analyses. This showed that the unidentified interferent y_5 was actually bromcresol green (actually another pH indicator) and that the dispersed solids (y_2) giving turbidity was white ZnO powder ('zinc white').

Figure 8.4a summarizes how much of the initial variance of the 5 Y-variables was correctly predicted from the X-variables in the cross validation, as the PLSR MODEL III develops in complexity. It shows that most of the predictive ability was attained with a 4-factor model, with a slight further predictive improvement by using 6 factors.

What does this 6-factor model mean? Let us first study how the different Y-variables are described by the PLS factors $a = 1, 2, \ldots, 6$. Figure 8.4b shows the residual variance $s(\widehat{f})^2$ (section 4.2.4) of each of the 5 Y-variables after 0,1,2,3,4,5 and 6 factors. The five Y-variables have been joined by straight line segments for each factor, to improve the visualization. The variance explained by each factor is denoted by that factor no. (large numbers); the variances explained by factors 2 and 5 have been shaded for illustration.

Since all the variables were standardized prior to the analysis, all 5 variables have an initial variance of 1.0, after having only subtracted the average ('0 factors'). This is the total variance that the calibration model is intended to describe.

The first factor then accounts for some variance in y_1 (litmus) and a lot of variance in y_3 (pH<5?) and y_4 (pH>8?). The second factor accounts for a lot of variance in y_1 plus a little in y_5 (the unidentified interferent), which is extensively described by the third factor instead. The fourth factor accounts primarily for y_2 (suspended material), the fifth for y_5, the sixth for y_1 and y_3. After 6 factors, there is only a little Y-variance unexplained in all the variables ('rest').

Let us look at this from another angle. Figure 8.5 shows the obtained X-parameters that define the calibration model that describes the data in Figure 8.3a:

Figure 8.5a gives the average spectrum $\bar{x}$, around which the calibration model is to be developed. Figure 8.5b, c and d give the X-loading spectra $\widehat{p}_a$ of factors $a = 1$

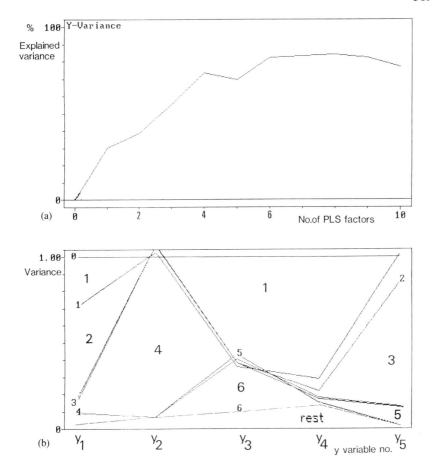

Figure 8.4 MODEL III: Joint calibration of 5 Y-variables from corrected X-data. a) Percentage correctly predicted variance, averaged over the 5 Y-variables, vs. no. of PLS factors. This predictive ability was estimated by cross validation. b) Residual variance plotted for variables y_j, $j = 1,2,...,5$, after $0,1,2,...,6$ PLS factors (small numbers). The residual variances for adjacent Y-variables have been joined by straight lines. Major variances explained by a factor have been marked with that factor no (large numbers); in addition, the variances explained by factors 2 and 5 have been symbolized by shaded patterns

and 2, 3 and 4, and 5 and 6. The first two factors are apparently dominated by the fiber optics instrument's corrected O.D. data (x_1 to x_{95}), while the MSC parameters and the visual assessments play increasingly important roles in the subsequent factors.

Interpreting individual PLS factors in multi-factor solutions can be quite difficult. The reason is that the individual factors do not necessarily represent individual chemical or physical phenomena—they usually represent combinations of these real phenomena. But for instance factor 4, which in Figure 8.4b was seen primarily to model the amount of undissolved, scattering material (y_2), does show some

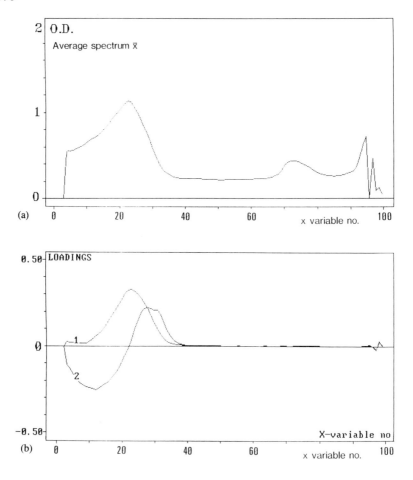

Figure 8.5 Calibration model parameters for X-data in Figure 8.3a. a) Average spectrum $\bar{\mathbf{x}}$. b) Loading spectra $\hat{\mathbf{p}}_1$, $\hat{\mathbf{p}}_2$

weak and erratic spectral features in Figure 8.5c, but its highest loadings shows a relationship to the two MSC parameters (x_{96} and x_{97}) and to the visual assessment of turbidity (x_{100}).

Figure 8.6 shows the model results in a more comprehensive way: In Figure 8.6a the X-loadings of factors 1 and 2 are plotted against each other, with adjacent wavelength channels joined by line segments. Variables far from the origin in the plot represent variables being modelled by these factors. A major spectral loop, from X-variables 3, via 10 to 22 and on to 27, 31 and down to 35 is evident.

Letters A, B, C, D and E represents $\hat{q}_{ja}$, the loadings of $y_j, j = 1, 2, \ldots, 5$ for the same two factors. This shows that litmus concentration (y_1,A) points in the same direction as X-variable no. 15, relative to the origin. Interestingly, x_{15} represents the isospestic point of litmus, 520 nm.

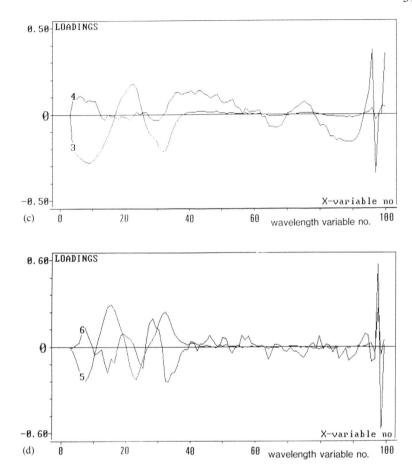

Figure 8.5 (*cont.*) Calibration model parameters for X-data in Figure 8.3a. c) Loading spectra $\hat{\mathbf{p}}_3$, $\hat{\mathbf{p}}_4$. d) Loading spectra $\hat{\mathbf{p}}_5$, $\hat{\mathbf{p}}_6$

The region of this plot near zero is amplified in Figure 8.6b. It shows that y_3 (pH<5?, C) is related to x_{98} (sensory redness) and oppositely related to y_4 (pH>8?, D) and x_{99} (sensory blueness).

Figure 8.6c shows that the concentration of the 'unidentified interferent' bromcresol green (y_5,E) is related positively to wavelengths around x_8 and x_{31} and negatively to wavelengths around x_{22} along factor 3. In addition, a turbidity phenomenon is evident in a combination of factors 4 and 3: The concentration of suspended ZnO material (y_2,B) is here very strongly positively related to the MSC baseline offset (x_{96}, here also called 'a') and to sensory assessment of turbidity (x_{100}), and negatively related to the multiplicative MSC parameter (x_{97}, also called 'b').

In Figure 8.6d one can observe some spectral phenomenon along factor 5 correlating positively with sensory redness (x_{98}) and negatively to the concentration

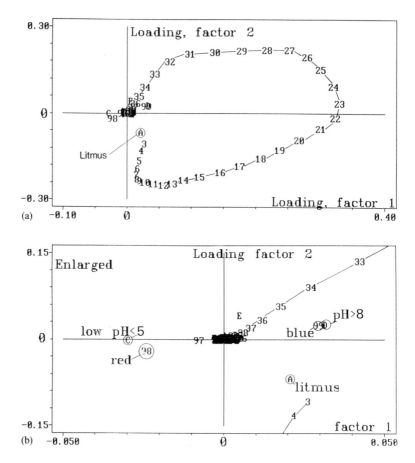

Figure 8.6 MODEL III: Loadings for **X** and **Y**. Numbers 1,2,...,100 represents the 100 X-variables, while letters A, B, C, D and E represent Y-variables 1, 2, 3, 4 and 5. X-variables no. 96 and 97, representing the MSC parameters, are denoted 'a' and 'b'. a) The X-loadings $\hat{p}_2$ vs $\hat{p}_1$ with adjacent wavelength channels joined by line segments. b) Enlargened window from a); for $-.05 < \hat{p}_{k1} < .05$ and $-.15 < \hat{p}_{k2} < .15$

of the 'unidentified interferent' bromcresol green (E). This could be a compensation for the fact that the interferent varies between yellow and blue with pH shifts, instead of red and blue as for litmus. In addition, the figure shows a tendency for negative relationship between sensory blueness (x_{99}) and the Y-variable pH<5? (y_3,C). The latter at the same time shows some degree of correlation with sensory redness.

This graphical inspection of pairs of factors' loadings reveal how the X- and Y-variables relate to the latent variables $a = 1, 2, \ldots, 6$ obtained in the modelling. Parallel to that, plots of the different calibration objects along these same factors were also studied—the so-called 'score plots'. This will not be illustrated here, but is extensively covered in sections 3.4 and 3.5.

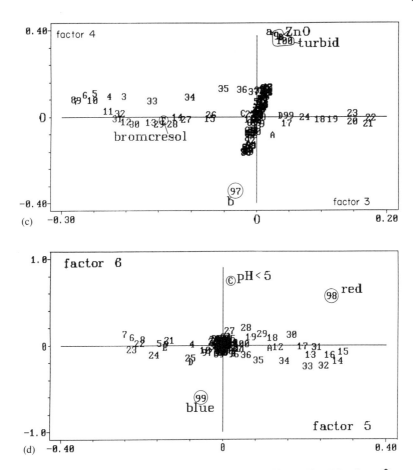

Figure 8.6 (*cont.*)   MODEL III: Loadings for **X** and **Y**. c) The *X*-loadings $\hat{\mathbf{p}}_4$ vs $\hat{\mathbf{p}}_3$.
d) The *X*-loadings $\hat{\mathbf{p}}_6$ vs $\hat{\mathbf{p}}_5$ *for full caption see p.372*

Interpreting high-dimensional calibration models in this way can be somewhat
time-consuming and difficult. (Still, looking at these 6 factors is much simpler than
looking at the 105 input variables!) Studying only pairs of factors can be a little
dangerous: In this case the experimental design of the calibration set was such that
the different chemical and physical phenomena came out more or less in separate
PLS factors. But in other cases it could be that the most readily understandable
information appeared for instance in the factor 1 vs. factor 3 plane. So other
factor combinations should also be plotted and studied; with computerized 2- and
3-dimensional colour graphics this is not too difficult.

Still, there is a need for summarizing the 6-factor calibration model. One way
to do that for each *Y*-variable is to collapse the 6-factor models onto the one
dimension that shows how this *Y*-variable is best predicted from the collection of
X-variables: $\hat{y}_j = \hat{b}_{0,j} + \mathbf{x}\hat{\mathbf{b}}_j$. Figure 8.7 shows the estimated **b**-vectors for each of
the 5 *Y*-variables.

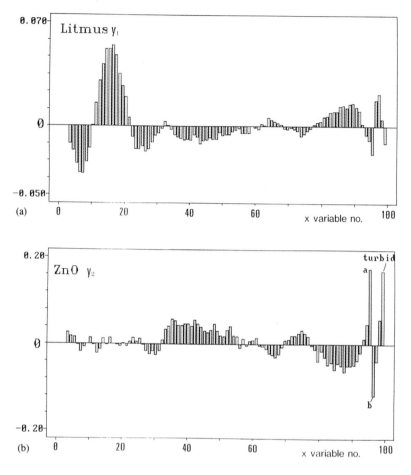

Figure 8.7 MODEL III: The resulting predictors $\widehat{\mathbf{b}}_j$ The vectors $\widehat{\mathbf{b}}_j = (\widehat{b}_{kj}, k = 1,2,\ldots,100)$ obtained with the 6-factor PLSR solution. Outstanding X-variables are marked explicitly. a) Predictor for the original analyte litmus, y_1. b) Predictor for ZnO powder, y_2

Litmus (y_1) is seen (Figure 8.7a) to be predicted by combinations of positive O.D.s around its isospestic wavelength (x_{15}), but with strong negative side peaks that probably compensate for the presence of the interferent bromcresol green.

The amount of dispersed ZnO powder (y_2) is here (Figure 8.7b) primarily predicted by a combination of the spectral MSC parameters x_{96} and x_{97} ('a' and 'b') and the visual turbidity assessment.

Acid state (pH<5?, y_3) is primarily predicted (Figure 8.7c) from the contrast between red and blue colour assessment ('red and not blue') Conversely, alkaline state (pH>8?, y_4) is primarily predicted (Figure 8.7d) from the contrast between blue and red colour assessment ('blue and not red'). Hence, it appears that the human ability to summarize information in relevant ways make the visual

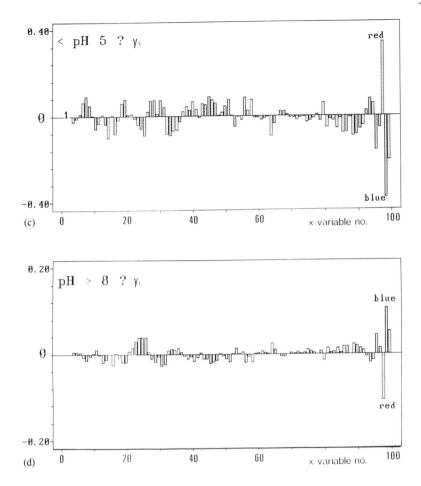

Figure 8.7 (*cont.*)   c) Predictor for pH variable 'pH<5?', y_3. d) Predictor for pH variable 'pH>8?', y_4 (*for full caption see p.374*)

assessments more relevant than the spectrophotometric data for predicting pH state in this way. The reason could be the nonlinear character of these pH and colour perception scales.

Finally, Figure 8.7e shows that the interferent bromcresol green (y_5) was primarily predicted by spectral combinations resembling the predictor for litmus, but with opposite sign. In addition, being 'not red' also helps predict the interferent (whose colours are either yellow or blue).

In summary, this joint calibration for 5 Y-variables provided insight into what was really going on in the samples and in the measurements. The predictive ability for litmus, using $\hat{\mathbf{b}}_1$ (Figure 8.7a) from this 6-factor MODEL III solution appeared similar to, but somewhat less satisfactory than that obtained in MODEL II, calibrating for litmus alone (prediction error 8.1% litmus in MODEL III, vs. 5.8% in MODEL II). Thus the price of the increased overview was a slight

376

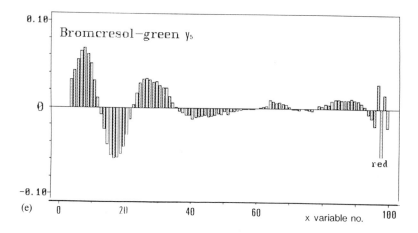

Figure 8.7 (*cont.*)   e) Predictor for the interferent bromcresol green, y_5 *for full caption see*
p.374

decrease in predictive ability for the main analyte. So, for subsequent litmus predictions during the process control the optimized litmus predictor (MODEL II) might still be the one to use.

However, it could well be that even more optimized predictors can be developed for particular sample types.

8.4   SPECIALIZED CALIBRATIONS

Till now we have attempted to make broad calibration models, applicable for the whole range of sample qualities expected to be encountered (the whole statistical population of future samples, see sections 1.3 and 2.1.2). But multivariate calibration is a question of local linear approximation of some unknown, basically non-linear X–Y and X–X data structure. In principle, an approximation over a narrow range of sample qualitites can give better local prediction results than an approximation covering a broad range. This is discussed in more detail in Chapter 6.

It was decided that two particular types of sample qualities passing the fiber optics detector needed particularly precise and reliable predictions. One sample type was pure solutions of litmus at alkaline pH, without any ZnO or bromcresol green present. The other was pure bromcresol green at various pH-levels, but without any ZnO or litmus present.

CALIBRATION MODEL IV

The set of 26 calibration samples contained 8 pure alkaline solutions of litmus at various concentrations (including pure water, $y_1 = 0$). Based on these 8 samples, a separate PLSR calibration model was estimated, using only y_1 as Y-variable, and using the same a priori X-weights as in MODELs II and III. A 1-factor PLSR model was attained, as expected (there is only one chemical phenomenon affecting these X-data—varying litmus levels).

(In addition a certain degree of non-linear instrument response was expected. In section 3.5 such effects were modelled by a second PLS factor. But there were presently too few samples to model this statistically.)

CALIBRATION MODEL V

The set of 26 calibration samples contained 6 pure solutions of pure bromscresol green (including pure water), although at varying pH levels. A separate PLS calibration model was estimated, using only y_5 as Y-variable, and employing the same a priori X-weights as above. A 2-factor PLSR model was required (apparently corresponding well to the yellow and blue states of bromcresol green with varying pH).

Now a two-step prediction scheme can be used: The calibration model from MODEL III can be used as a general, informative 'gate keeper' predictor, detecting serious outliers and giving general information about pH, ZnO level and litmus and bromcresol green concentrations. Based on these preliminary predictions of sample qualities, it can be decided if an unkown object should be passed on to the more specialized, and hopefully more precise, predictors (calibration models from MODEL IV or V).

8.5 PREDICTION, OUTLIER DETECTION AND CLASSIFICATION

An independent set of 10 quite different samples of the same general type had been obtained. They were measured by the spectrophotometer and assessed visually, and their spectra linearized and scatter corrected, just like calibration samples—and just like the thousands of subsequent samples to come. In addition, the 5 Y-variables were measured explicitly for these 10 test samples, to check how the calibration model behaved, and to assess the need for further refinements of the calibration models.

Figure 8.8 shows their final X-data (scatter-corrected O.D spectra, two MSC parameters and three sensory assessments).

Figure 8.9 compares the true vs the predicted litmus concentrations (using the broad 6-factor predictor from MODEL III), for the 26 calibration samples as well as the 10 test samples (squares). Most of the test samples gave very precise litmus predictions even with this 'overview' predictor. But two of the test samples (9 and 10) showed strong prediction errors. However, these samples were automatically detected as outliers by MODEL III, based on their data $\mathbf{x}_i$.

The basis for this outlier detection is shown in Figure 8.10a. For each of the 10 test samples the residual X-variance $s(\hat{\mathbf{e}}_i)^2$ is plotted against its so-called 'leverage', h_i (Chapter 5). The residual X-variance reveals if the X-data of a sample do not fit the calibration model (in music: disharmony!). The leverage instead reveals if its

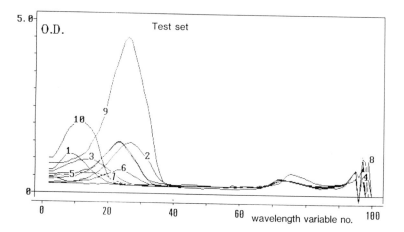

Figure 8.8 Test data. *X*-data (scatter-corrected O.D. spectra, MSC parameters and sensory assessments) for the 10 test samples

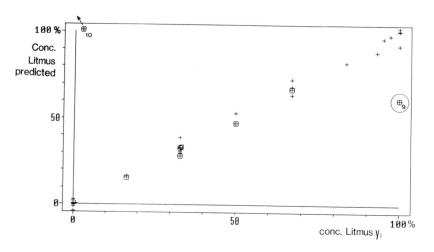

Figure 8.9 Multivariate calibration and prediction. MODEL III, 6 PLS factors ($\hat{\mathbf{b}}_j$, for analyte $j = 1$, litmus (Figure 8.7a). Abscissa: True litmus concentration. Ordinate: Predicted litmus concentrations. Crosses: Calibration samples. Squares: 10 test samples not used in developing the calibration model. Test objects 9 and 10 are marked explicitly because of their erroneous predictions. Both were automatically tagged as outliers on the basis of their *X*-data

X-data seem to reflect abnormal levels of the phenomena being described in the calibration model ('too much of a good thing'—in music: imbalance). The arrows in Figure 8.10a show the approximate detection limits obtained from the calibration set (4 times the average residual *X*-variance and 2 times the average leverage in the calibration set). New objects falling outside the dotted box will be defined as outliers.

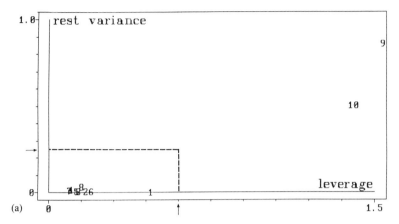

Figure 8.10 Outlier detection The test objects' X-data $\mathbf{x}_i, i = 1, 2, \ldots, 10$ examined for two types of extremeness. Abscissa: Leverage h_i. Ordinate: Residual X-variance $s(\hat{e}_i)^2$. The arrows and dotted box represent the detection limits from the calibration modelling. a) MODEL III: The widest calibration model

Figure 8.10a shows that test objects 1–8 are well within the error limits of MODEL III, while samples 9 and 10 are well outside. Upon closer inspection it was found that object 9 represented an operator error with the spectrophotometer—the wavelength range had unfortunately been shifted about 25 nm. This is evident in the spectra in Figure 8.8, but was overlooked in the routine operation. Object 10 was found to contain an unexpected interferent—(methyl orange at alkaline pH).

Since such errors can now be automatically detected in the outlier test, their erroneous predictions can be stopped from creating mistakes in the process control.

Table 8.1 shows the predicted and true values for all 5 Y-variables for the 10 individual test samples, predicted with MODEL III. On the basis of these predicted Y-values it could be decided which of the new objects should have their final chemical predictions from the more specialized calibrations (MODELs IV and V) instead of the general calibrations (MODEL III). The results of this classification based on Y-predictions is given in column 2 in Table 8.2. Column 3 shows the results of the outlier detection based on MODEL III.

The classification of new objects into different calibration models can also be done on the basis of outlier detection based on the X-data: An object not being an outlier with respect to a certain calibration model can normally be assumed to give valid prediction of $\mathbf{Y}$ with this model.

Figure 8.10b shows how the 10 test samples behaved relative to the specialized calibration for clear samples of blue litmus solutions only (MODEL IV). The dotted lines represent the detection limits from the calibration establishing MODEL IV. Likewise, Figure 8.10c shows the test samples' behaviour relative to the calibration for only bromcresol green (MODEL V). Columns 4 and 5 in Table 8.2 summarize these classifications.

Table 8.1 Predicted (and measured) values of the 5 variables, rounded off to the nearest integer. The predictions were based on calibration MODEL III, using 6 PLS factors. The percentages are relative to some reference concentration; the binary variables are defined as 1=yes, 0=no. Symbol m represents missing data (used for pure water, since it has low buffer capacity)

Y no. Object	1 Litmus	2 ZnO	3 pH<5?	4 pH>8?	5 Bromcresol green	Truth
1	<0 (0)	0 (0)	1 (1)	0 (0)	74 (75)	alk. bromcres.
2	1 (0)	5 (0)	0 (0)	1 (1)	53 (50)	acid ,,
3	67 (67)	<0 (0)	1 (1)	0 (0)	3 (0)	acid litmus
4	0 (0)	8 (0)	1 (m)	0 (m)	2 (0)	pure water
5	48 (50)	<0 (0)	0 (0)	1 (1)	5 (0)	alk. litmus
6	16 (16.5)	2 (0)	0 (0)	1 (1)	5 (0)	,,
7	34 (33)	<0 (0)	1 (1)	0 (0)	1 (0)	acid ,,
8	34 (33)	37 (30)	0 (0)	1 (1)	7 (0)	alk. litmus, turbid
9	41 (100)	<0 (0)	<0 (0)	2 (1)	163 (0)	operator mistake
10	91 (0)	3 (0)	0 (0)	0 (1)	75 (0)	methyl orange
Unit:	%	%	yes/no	yes/no	%	

Table 8.2 Classification of the 10 test samples with respect to predicted variables $\hat{Y}$ from Table 8.1, and with respect to outlier detection in MODELs III, IV and V

Object	Classified from $\hat{Y}$ into model no.	Fitting this model? MODEL III	MODEL IV	MODEL V
1	V	+	no	+
2	V	+	no	+
3	III	+	no	no
4	IV,V	+	+	+
5	IV	+	+	no
6	IV	+	+	no
7	III	+	no	no
8	III	+	no	no
9	—	no	no	no
10	—	no	no	no

Thus it seems that objects 4, 5 and 6 could preferably have their litmus concentrations predicted using the specialized litmus model IV, objects 1,2 and 4 their bromcresol concentrations using model V, while the rest of the objects should have their Y-predictions made in model III. (Notice that pure water seems to be well predicted in all the models. This was of course expected, since all these models extend to pure water.)

Compared to a wide 'over-view' calibration, the specialized, narrow calibrations should theoretically give better linear approximations and hence better predictions. But it should be remembered that in the present case very few calibration objects were available for estimating the specialized models IV and V. So the cost of

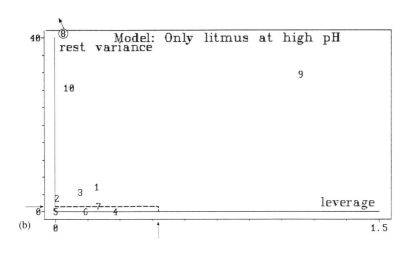

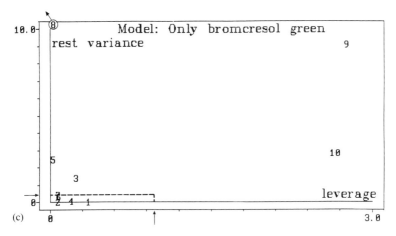

Figure 8.10 (*cont.*) b) MODEL IV: Only clear alkaline litmus solutions. c) MODEL V: Only bromcresol green at various pH levels *for full caption see p.379*

statistical imprecision in practice could be greater than the benefit of better local approximation. Therefore it would be advisable to include the test samples into the respective calibration sub-sets and re-calibrate to get better statistical estimates. Then this more complicated, but tentially better calibration scheme could be implemented instead of MODEL II, which was running in the process control while we performed this detailed model interpretation and optimization.

Updating the calibration models

It would be advisable to check the performance of these new calibrations by determining the analyte concentrations (litmus and bromcresol green) by the traditional reference method once in a while. As more control data (**X,Y**) thus accumulate, these can also be used for further updating of the calibration models. By successively eliminating the oldest calibration data we can thus adapt the calibration to modifications in the chemical process and/or drift in the analytical instrumentation.

In contrast to properly done sensory analyses by trained panelists (see e.g. Martens M. et al., 1986), the present colour assessments are not quantitative. In addition, human assessments may not be practical in on-line process control. Therefore, after having proven that the fiber-optic spectrophotometer in fact performed as expected, we may want to drop the extra control provided by the human sensory assessments (redness, blueness and turbidity, x_{98}, x_{99} and x_{100}). Figure 8.7a indicates that such a variable reduction would have little consequence for the determination of the main analyte, litmus; the sensory contributions to its $\hat{\mathbf{b}}$-vector were small.

For determination of ZnO powder they could probably also be eliminated: Although the sensory turbidity measurement x_{100} gave strong contributions in Figure 8.7b, Figure 8.6c showed the two MSC parameters x_{96} and x_{97} ('a' and 'b') to yield almost the same information.

But the two pH assessments predicted (y_3 and y_4, Figure 8.7c and d) could possibly deteriorate if the colour assessments were dropped. The determination of bromcresol green could likewise be slightly affected, although the strong spectral features in the visible range in Figure 8.7e indicate this problem to be minor.

So, going back to the original purpose of the analysis, namely determining litmus, the sensory assessments may now safely be dropped. The number of X-variables could be reduced further, if so is desired. Redundant or irrelevant variables could be dropped or at least merged together. For instance, one could average many of the channels in the range x_{40} to x_{97} into one or two baseline variables and one or two water absoprtion variables. If a simplified instrument with fixed wavelength bands were to replace the scanning spectrophotometer, as few as 6 filters could possibly suffice, representing the 6 PLS factors in model III.

8.6   CONCLUSION

This example has shown how multivariate calibration can convert highly non-selective high-speed measurements of intact samples into selective information, thereby replacing slower, more expensive or noxious traditional analytical techniques.

It has also shown that multivariate data analysis can provide new insight into complicated analytical problems.

Finally, it has demonstrated how unexpected instrument errors or sample qualities, which might have passed unnoticed and created grave problems with the traditional 'calibration line', were automatically detected as outliers in the multivariate modelling.

The present calibration was based on a causal relationship of the extended reverse type (section 2.2); the spectrophotometric responses are caused by the analyte and interferents. This is the most common causal basis for analytical instruments: **X** is caused by **Y**. But other examples in the book have illustrated other valid bases for multivariate calibration.

One such case is the forward causal basis (**Y** is caused by **X**), represented by the quantification of individual protein 'polymers' in mixtures from the mixtures' amino acid 'monomer' profiles (section 7.4.2). Another case is the common-cause for **X** and **Y**. This was examplified by the high-sensitivity determination of dioxins in smoke, from simpler chromatographic measurements of various halogenated benzenes and phenols presumably generated by the same burning process as the dioxin (section 1.1.4).

One and the same calibration approach can successfully solve all these different causal situations. Such an approach was illustrated in the present chapter. It consists of a 'soft modelling' regression method (PLSR) with extensive graphics and outlier warnings in its calibration and prediction stages. This is employed interactively, and it is optimized by response linearization and other data preprocessing, plus a rough a priori weighting of the different variables to balance their noise levels.

So the actual computations in multivariate calibration are reasonably simple, once the use of a calibration program package has been learned.

A minimum of chemometric understanding is required, as described by Level 1 in this book, with respect to experimental design, a priori weighting, model validation and outlier detection. The more advanced statistical insight described by Level 2 is not necessary, although it may allow further optimization of the calibration.

What is critical in multivariate calibration is to guard against mistaken modelling of nonsense. Successful calibration must be based on valid, stable, enduring relationships between **X** and **Y** for the type of samples (the 'population of objects') in question. It should not reflect incidental indirect correlations that lack predictive validity.

To ensure predictive validity, the analyst must apply whatever background knowledge is available,- for selecting sufficiently informative variables and objects, for assessing the calibration results, and for optimizing these. And the analyst should update this knowledge by studying the resulting calibration models and outlier warnings—calibration is a learning process. But complete insight is not required in advance—good empirical data on how the samples actually behave can be more valuable than theoretical speculations on how they ought to behave.

With multivariate calibration properly used, the analyst can do a better job and have more fun.

References

Aastveit, A. and Martens, H. (1984) ANOVA interactions by partial least squares regression. *Biometrics*, **42**, 829–844.

Anderson, T.W. (1958) *An introduction to multivariate statistical analysis*. John Wiley and Sons, New York.

Anderson, T.W. (1971) *The statistical analysis of time series*. John Wiley and Sons, New York.

Akaike, H. (1974) A new look at the statistical model identification. *IEEE Transactions on Automatic Control*, **AC-19**, 716–723.

Atkinson, A.C. (1987) *Plots, transformations and regression*. Oxford Science Publications, Oxford.

Basilevski, A. (1983) *Applied matrix algebra in the statistical sciences*. North Holland Publishing Company, New York.

Beckman, R.J. and Cook, R.D. (1983) Outlier....s, *Technometrics*, **25**, 119–163.

Belsley, D.A., Kuh, E. and Welsch, R.E. (1980) *Regression diagnostics. Identifying influential data and sources of collinearity*. John Wiley and Sons, New York.

Bergmann, G., von Oepen, B. and Zinn, P. (1987) Improvement in the definitions of sensitivity and selectivity. *Analytical Chemistry*, **59**(20), 2522—2526.

Berk, K.N. (1984) Validating regression procedures with new data. *Technometrics*, **26**, 331–338.

Berkson, J. (1969) Estimation of a linear function for a calibration line. Considerations of a recent proposal. *Technometrics*, **11**, 644–660.

Bibby, J. and Toutenburg, H. (1977) *Prediction and improved estimation in linear models*. John Wiley and Sons, Chichester.

Birth, G.S. (1978) The light scattering properties of foods. *J. Food. Sci.*, **43**, 916–925.

Birth, G.S. (1982) Diffuse thickness as a measure of light scattering. *Appl. Spec.*, **36**, 675–682.

Birth, G. and Hecht, H.G. (1987) The physics of near-infrared reflectance. In *Near-infrared technology in agricultural and food industries* (ed. P.C. Williams and K. Norris). Am. Assoc. Cereal Chem., St. Paul, Minnesota, 1–15.

Bjørsvik, H.R. and Martens, H. (1989) Data analysis: PLS calibration of NIR instruments by PLS regression. In D.A. Burns, ed: *Near-infrared analysis*. Marcel Dekker (in press).

Borgen, O.S. and Kowalski, B.R. (1985) An extension of the multivariate component-resolution method to three components. *Analytica Chimica Acta*, **174**, 1–26.

Box, G.E.P., Hunter, W.G. and Hunter, J.S. (1978) *Statistics for experimenters. An introduction to design, data analysis and model building*. John Wiley and Sons, New York.

Breiman, L, Friedman, J.H., Olshen, R.A. and Stone, C.J. (1984) *Classification and regression trees*. Wadsworth International Group, Belmont, California.

Breiman, L. and Friedman, J.H. (1985) Estimating optimal correlations for multiple regression and correlation. *Journal of Amer. Stat. Assoc.*, **80**, 580–598.

Brown, G. (1979) An optimization criterion for linear inverse estimation. *Technometrics*, **21**, 575–579.

Brown, P.J. (1982) Multivariate calibration (with discussion). *Journal of the Roy. Stat. Soc.*, Ser. B., **44**, 287–321.

Brown, P.J. and Sundberg, R. (1985) Confidence and conflict in multivariate calibration. *Technical Report* no. 140, Stockholm University, Box 6701, 11385 Stockholm.

Brown, P.J. and Sundberg, R. (1987) Confidence and conflict in multivariate calibration. *Journal of the Roy. Stat. Soc.*, Ser. B., **49**, 46–57.

Carroll, J.D. and Chang, J.J. (1970) Analysis of individual differences in multidimensional scaling via an N-way generalization of "Eckart-Young" decomposition. *Psychometrika*, **35**, 283–319.

Cook, R.D. (1977) Detection of influential observations in linear regression. *Technometrics*, **19**, 15–18.

Cook, R.D. and Weisberg, S. (1982) *Residuals and influence in regression*. Chapman and Hall, London.

Cornell, J.A. (1981) *Experiments with mixtures: Designs, models and the analysis of mixtures*. John Wiley and Sons, New York.

Cowe, I. and McNicol, J.W. (1985) The use of principal components in the analysis of near-infrared spectra. *Applied Spectroscopy*, **39**, 257–266.

Cowe, I. A., McNicol, J.W. and Cuthbertson, D.C. (1985) A designed experiment for the examination of techniques used in the analysis of near infrared spectra. *Analyst*, **110**, 1227-1240.

Critchley, F. (1985) Influence in principal component analysis. *Biometrika*, **72**, 627–636.

Davies, A.M.C. and McClure, W.F. (1985) Near infrared analysis in the Fourier domain with special reference to process control. *Anal. Proc. (London)*, **22**, 321–322.

Deming, S.N. and Morgan, S.L. (1987) *Experimental design: A chemometric approach*. Elsevier Publishing Company, New York.

Dempster, A.P., Schatzoff, M. and Wermuth, N. (1977) A simulation study of alternatives to ordinary least squares. *J. Amer. Stat. Ass.*, **72**, 77–91.

Devaux, M-F., Bertrand, D., Robert, P. and Morat, J-L. (1987) Extraction of near Infra-red spectral information by fast Fourier transform and principal component analysis. Application to the discrimination of baking quality of wheat flours. *Journal of Chemometrics*, **1**, 103–110.

Eastment, H.T and Krzanowski, W.J. (1982) Cross-validatory choice of the number of components from a principal component analysis. *Technometrics*, **24**, 73–77.

Efron, B. (1982) *The jackknife, the bootstrap and other resampling techniques*. Society for Industrial and Applied Mathematics, Philadelphia, Pennsylvania. ISBN 0-89871-179-7.

Efron, B. and Gong, G. (1983) A leisurely look at the Bootstrap, the jacknife and cross-validation. *The Amer. Stat.*, **37**, 36–48.

Eisenhart, C. (1939) Interpretation of certain regression methods and their use in biological and industrial research. *Ann. Math. Stat.*, **10**, 162–186.

Esbensen, K.H. and Wold, S. (1983) SIMCA, SELPS, GDAM, SPACE and UNFOLD: The ways towards regionalized principal component analysis and some constrained N-way decomposition with geological illustrations. *Proc. Nordic. Symp. on Applied Statistics*, Stokkand Forlag Publ., Stavanger, Norway.

Esbensen, K.H., Wold, S. and Geladi, P. (1988) Relationships between higher-order data array configurations and problem formulations in multivariate data analysis. *J. of Chemometrics*, **3**, 33–48.

Esbensen, K.H. and Martens, H. (1987) Predicting oil-well permeability and porosity from wire-line petrophysical logs—a feasibility study using partial least squares regression. *Chemometrics and Intelligent Laboratory Systems*, **2**, 221–232.

Fearn, T. (1983) Misuse of ridge regression in the calibration of a near infrared reflectance instruments. *Applied statistics*, **32**, 73–79.

Fearn, T. (1986) Some statistical comments on the errors in NIR calibration. *Analytical Proceedings*, **23**. 123–125.

Frank, I. (1987) Intermediate least squares regression method. *J. Chemometrics*, **1**, 233–242.

Friedman, J.H. and Stuetzle, W. (1981) Projection pursuit regression. *J. Am. Stat. Ass.*, **76**, 817–823.

Fujikoshi, Y. and Nishii, R. (1984) On the distribution of a statistic in multivariate inverse regression analysis. *Hiroshima Math. J*, **14**, 215–225.

Geladi, P., McDougall, D. and Martens, H. (1985) Linearization and scatter-correction for near-infrared reflectance spectra of meat. *Applied Spectroscopy*, **39**, 491–500.

Geladi, P. and Kowalski, B.R. (1986) Partial least squares regression: A Tutorial. *Analytica Chimica Acta*, **185**, 1–17.

Gold, R.J.M., Tenenhouse, H.S. and Adler, L.S. (1976) A method for calculating the relative protein content of the major keratin components from their amino acid composition. *Biochemical Journal*, **159**, 157–160.

Golub, G.H. and Kahn, W. (1965) Calculating the singular values and pseudo-inverse of a matrix. *SIAM Journal of Numerical Analysis*, **13**, 205–244.

Golub, G. and Van Loan, C.F. (1983) *Matrix computations*. The Johns Hopkins University Press, Baltimore, Maryland.

Gunderson, R.W., Thrane, K.R. and Nilson, R.D. (1988) A false-colour technique for display and anlysis of multivariate chemometric data. *Chemometrics and Intelligent Laboratory Systems*, **3**, 119–131.

Gunst, R.F. and Mason, R.L. (1977) Biased estimation in regression. An evaluation using mean squared error. *Journal of American Statistical Association*, **72**, 616–628.

Gunst, R.F. and Mason, R.L. (1979) Some considerations in the evaluation of alternate prediction equations. *Technometrics*, **21**, 55–63.

Gunst, R.F. and Mason, R.L. (1980) *Regression analysis and its applications*. Marcel Dekker, Inc., New York.

Haaland, D.M. (1988) Quantitative infrared analysis of borophosphosilicate films using multivariate statistical methods. *Analytical Chemistry*, **60**, 1208–1217.

Haaland, D.M. and Easterling, R.G. (1982) Application of new least-squares methods for the quantitative infrared analysis of multicomponent samples. *Applied Spectroscopy*, **36**, 665–673.

Haaland, D.M. and Thomas, E.V. (1988a) Partial least-squares method for spectral analyses. I: Relation to other quantitative calibration methods and the extraction of qualitative information. *Analytical Chemistry*, **60**, 1193–1202.

Haaland, D.M. and Thomas, E.V. (1988b) Partial least-squares method for spectral analyses. II: Applications to simulated and glass spectral data. *Analytical Chemistry*, **60**, 1202–1208.

Halperin, M. (1970) On inverse estimation in linear regression. *Technometrics*, **12**, 727–736.

Hampden-Turner, C. (1987) *Ways of the mind. Charts and concepts of the mind and its labyrinths*. McMillan, New York.

Harman, H.H. (1967) *Modern factor analysis*. University of Chicago Press, Chicago.

Harshman, R.A. (1970) Foundations of the PARAFAC procedure: models and conditions for an "explanatory" multi-mode factor analysis. *UCLA working papers on phonetics*, **16**, 1–84.

Helland, I.S. (1987) On the interpretation and use of R^2 in regression analysis. *Biometrics*, **43**, 61–70.

Helland, I.S. (1988) On the structure of partial least squares regression. *Communications in statistics (sim and comp)*, **17**, 58–607.

Henrici, P. (1964) *Elements of numerical analysis*. John Wiley and Sons, New York.

388

Hildrum, K.I., Martens, M. and Martens, H. (1983) Research on analysis of food quality. International Symp. 'Control of food quality and food analysis' at Reading University, England, March 22–24.

Ho, C-N., Christian, G.D. and Davidson, E.R. (1978) Application of the method of rank annihilation for quantitative analyses of multi-component fluorescence data from the video fluorometer. *Anal. Chem.*, **50**, 1108–1113.

Hoadley, B. (1970) A Bayesian look at inverse linear regression, *J. Amer. Stat. Ass.*, **65**, 356–369.

Hocking, R.R. (1976) The analysis and selection of variables in linear regression. *Biometrics*, **32**. 1–49.

Hoerl, A.E. and Kennard, R.W. (1970) Ridge regression, biased estimation for nonorthogonal problems. *Technometrics*, **12**, 55–67.

Hoerl, A.E., Kennard, R.W. and Hoerl, R.W. (1985) Practical use of ridge regression: A challenge met. *Appl. Stat.*, **34**, 114–120.

Honigs, D.E. (1984) Improvements of the near-infrared diffuse-reflectance technique. PhD Thesis, Indiana University, Dept. of Chemistry, Bloomington, IN 47405.

Honigs, D.E., Hieftje, G.M. and Hirschfeld, T. (1984) A new method for obtaining individual component spectra from those of complex mixtures. *Appl. Spec.* **38**, 317–322.

Honigs, D.E and Hieftje, G.M., Mark, H.C. and Hirschfeld, T.B. (1985) Unique sample selection via near-infrared spectral subtraction. *Anal. Chem.*, **57**, 2299–2303.

Hotelling, H. (1933) Analysis of a complex of statistical variables into principal components. *J. Educ. Psychol.*, **24**, 417–441, 498–520.

Hruschka, W.R. (1987) Data analysis: wavelength selection methods. In *Near infrared reflectance spectroscopy* (P.C. Williams and Norris, K., eds.) American Cereal Association, St. Paul, Minnesota, 35–55.

Hruschka, W.R. and Norris, K. (1982) Least squares curve fitting of near infrared spectra predicts protein and moisture content of ground wheat. *Applied Spectroscopy*, **36**, 261–265.

Huber, P.J. (1981) *Robust statistics*. John Wiley and Sons, New York.

Hunter, R.S. (1975) *The measurement of appearance*. John Wiley and Sons, New York.

Høskuldsson, A. (1988) PLS regression methods. *J. of Chemometrics*, **2**, 211–228.

Isaksson, T. and Næs, T. (1987) A comparative study of different multivariate calibration methods on NIR data. Proceedings, Eurofood, Loen, Norway, June 1987.

Isaksson, T. and Næs, T. (1988) The effect of multiplicative scatter correction (MSC) and linearity transformation in NIR spectroscopy. *Applied Spectroscopy*, **42**, 1273–1284.

Jensen, S.Å., Munck, L. and Martens, H. (1982) The botanical constituents of wheat and wheat milling fractions. I. Quantification by autofluorescence. *Cereal Chemistry*, **59**(6), 477–484.

Jochum, C., Jochum, P. and Kowalski, B.R. (1981) Error propagation and optimal performance in multicomponent analysis. *Anal. Chem.*, **53**, 85–92.

Johnson-Laird, P.N. (1983) *Mental models*. Cambridge University Press, Cambridge.

Joiner, B.L. (1981) Lurking variables: Some examples. *The American Statistician*, **35**(4), 227–233.

Joliffe, I.T. (1982) A note on the use of principal components in regression. *Applied Statistics*, **31**, 300–303.

Joliffe, I.T. (1986) *Principal component analysis*. Springer Verlag, New York.

Jöreskog, K.G. and Wold, H. (1981) *Systems under indirect observation, causality-structure-prediction*, Vols. I and II. North Holland, Amsterdam.

Kalivas, J.H. and Kowalski, B.R. (1982) Compensation for drift and interferences in multicomponent analysis. *Anal. Chemistry*, **54**, 560–565.

Knafl, G., Spiegelman, C., Sacks, J. and Ylvisaker, D. (1984) Nonparametric calibration. *Technometrics*, **26**, 233–241.

Kortum, G. (1969) *Reflectance spectroscopy: Principles, methods, applications.* Springer Verlag, New York.

Kowalski. B.R. (ed.) (1984) *Chemometrics: Mathematics and statistics in chemistry.* D.Reidel, Dordrecht, The Netherlands.

Krutchkoff, R.G. (1967) Classical and inverse regression methods of calibration. *Technometrics*, **9**, 425–439.

Kvalheim, O.M. (1987a) Oil-source correlation by the combined use of principal component modelling, analysis of variance and a coefficient of congruence. *Chemometrics and Intelligent Laboratory Systems*, **2**, 127–136.

Kvalheim, O.M. (1987b) Latent structure decompositions (projections of multivariate data). *Chemometrics and Intelligent Laboratory Systems*, **2**, 283–290.

Kvalheim, O.M. (1988) A partial least squares approach to interpretative analysis of multivariate data. *Chemometrics and Intelligent Laboratory Systems*, **3**, 189–197.

Lanczos, C. (1950) An iteration method for the solution of the eigenvalue problem of linear differential and integral operations. *J. of Research of the National Bureau of Standards*, **45**, 255–282.

Lawton, W.H. and Sylvestre, E.A. (1971) Self modelling curve resolution. *Technometrics*, **13**, 617–633.

Lawton, W.H., Sylvestre, E.A. and Maggio, M.S. (1972) Self modelling nonlinear regression. *Technometrics*, **14**, 513–532.

Lea, P., Martens, H., Mielnik M. and Slinde, E. (1983) Analysis of mixtures: Haemoglobin and myoglobin in various molecular states determined from visible light spectra. *Proc. Nordic. Symp. on Applied Statistics*, Stokkand Forlag Publ., Stavanger, Norway, 165–183,

Lieftinck-Koeijers, C.A.J. (1988) Multivariate calibration: a generalization of the classical estimation. *Journal of Multivariate Analysis*, **25**, 31–44.

Lindberg, W., Persson, J.Å. and Wold, S. (1983) Partial least squares method for spectrofluorimetric analysis of mixtures of humic acid and ligninsulfonate. *Analytical Chemistry*, **55**, 643–648.

Lakoff, G. (1987) *Women, fire and dangerous things. What categories reveal about the mind.* The University of Chicago Press, Chicago.

Lodder, R.A. and Hieftje, G.M. (1988) Analysis of intact tablets by near-infrared reflectance spectrometry. *Applied Spectroscopy*, **42**(4), 556–558.

Lorber, A., Wangen, L. and Kowalski, B.R. (1987) A theoretical foundation for PLS. *J. of Chemometrics*, **1**, 19–31.

Lorber, A. and Kowalski, B.R. (1988) The effect of interference and calibration design on accuracy. Implications for sensor and sample selection. *J. of Chemometrics*, **2**, 67–80.

Lwin, T. and Maritz, J.S. (1980) A note on the problem of statistical calibration. *Applied Statistics*, **29**, 135–149.

Lwin, T. and Maritz, J.S. (1982) An analysis of the linear-calibration controversy from the perspective of compound estimation, *Technometrics*, **24**, 235–242.

Lwin, T. and Spiegelman, C.H. (1986) Calibration with working standards. *Applied Statistics*, **35**, 256–261.

Malinowski, E.R. and Howery, D.G. (1980) *Factor analysis in chemistry.* John Wiley and Sons, New York.

Mallows, C.L. (1973) Some comments on C_p. *Technometrics*, **15**, 661–675.

Mandel, J. (1971) A new analysis of varaince model for non-additive data. *Technometrics*, **13**, 1–18.

Mandel, J. (1982) Use of the singular value decomposition in regression analysis. *The American Statistician*, **36**, 15–24.

Manne, R. (1987) Analysis of two partial-least-squares algorithms for multivariate calibration. *Chemometrics and Intelligent Laboratory Systems*, **2**, 187–197.

Mansfield, E.R., Webster, J.T. and Gunst, R.F. (1977) An analytic variable selection technique for principal component regression. *Applied Statistics*, **26**. 34–40.

Mardia, K.V., Kent, J.T. and Bibby, J.M. (1980) *Multivariate analysis*. Academic Press, London.

Martens, H. (1979) Factor analysis of chemical mixtures. Non-negative factor solutions for cereal amino acid data. *Analytica Chimica Acta*, **112**, 423–441.

Martens, H. (1980) On the calibration of a multivariate instrument for quantitative estmation of individual components in a mixture. Proc. Symp. in Applied Statistics, Lyngby, Denmark, Jan 1980, NEUCC, DTH, DK-2800 Denmark, 393–414.

Martens, H. (1983) Understanding food research data. In *Food research and data analysis* (Martens, H. and Russwurm, H. jr, eds.) Applied Science Publ., London, 5–38.

Martens, H. (1985) Multivariate calibration. Quantitative interpretation of non-selective chemical data. Dr. techn. thesis, University of Trondheim (NTH). ISBN 82–90394–10–1.

Martens, H. (1987) A general partial least squares calibration algorithm. NCC Note STAT/35/87 from the Norwegian Computing Center, P.O.Box 114, N-0314 Oslo 3, Norway.

Martens, H. (1989) Soft modelling of multivariate chaos. *Order and Chaos. The Interdisciplinary Journal of Non-linear Dynamics*, (in press).

Martens, H. and Bach-Knutsen, K.E. (1980) Fractioning barley proteins by computer factor analysis. *Cereal Chemistry*, **57**, 97–105.

Martens, H. and Jensen, S.Å. (1983) Partial least squares regression: A new two-stage NIR calibration method. *Proc. 7th World Cereal and Bread Congress*. Prague June 1982 (Holas and Kratochvil, eds.) Elsevier Publ., Amsterdam, 607–647.

Martens, H. and Næs, T. (1983) Calibration as a practical problem. *Proc. Nordic Symp. on Applied Statistics*, Stavanger, Norway, Stokkand Forlag Publ., 113–135. Reprinted in Martens (1985).

Martens, H. and Næs, T. (1984) Multivariate calibration. I. Concepts and distinctions. *Trends in Analytical Chemistry*, **3**, 204–210.

Martens, H. and Næs, T. (1987) Multivariate calibration by data compression. In *Near-infrared technology in agricultural and food industries*. (ed. P.C. Williams. and K.Norris) Am. Assoc. Cereal Chem., St. Paul, Minnesota, 57–87.

Martens, H., Vangen, O. and Sandberg, E. (1983a) Multivariate calibration of an X-ray computer tomograph by smoothed PLS regression. *Proc. Nordic Symp. on Applied Statistics*. Stokkand Forlag Publ., Stavanger, Norway, 235–268. Reprinted in Martens (1985).

Martens, H. Jensen, S.Å. and Geladi, P. (1983b) Multivariate linearity transformation for near-infrared reflectance spectrometry. *Proc. Nordic Symp. on Applied Statistics* (O.H.J Christie, ed.) June 12–14, 1983. Stokkand Forlag Publishers, Skagenkaien 12, N-4000 Stavanger, Norway. ISBN 82–90496–02–8 , 208–234. Reprinted in Martens (1985).

Martens, H., Wold, S. and Martens, M. (1983c) A layman's guide to multivariate data analysis. In Martens H. and Russwurm, H. jr. (eds), *Food Reserch and Data Analysis*. Applied Science Publishers, London, 473–492.

Martens, H., Karstang, T. and Næs, T. (1987) Improved selectivity in spectroscopy by multivariate calibration. *Journal of Chemometrics*, **1**, 201–219.

Martens, H., Rødboffen, M., Martens, M., Risvik, E. and Russwurm, J.H. (1988) Dissimilarities in cognition of flavour terms related to various sensory laboratories in a multivariate study. *Journal of Sensory Studies*, **3**, 123–135.

Martens, M. (1985) Sensory and chemical quality criteria for white cabbage studied by multivariate data analysis. *Lebensmittel Wissenschaft u. Technol.*, **18**, 100–104.

Martens, M. (ed.) (1987) Data-approximation by PLS methods. *Proceedings Symposium May 19 1987*. Report no. 800 from the Norwegian Computing Center, P.O.Box 114, N-0314 Oslo 3, Norway.

Martens, M. and Martens, H. (1986a) Partial least squares regression. In *Statistical procedures in food research* (ed. J.R. Piggott), Elsevier Appl. Sci. Publ., London.

Martens, M. and Martens, H. (1986b) Near infrared reflectance determination of sensory quality of peas. *Applied Spectroscopy*, **40**, 303–310.

Massart, D.L., Vandeginste, B.G.M, Deming, S.N., Michotte, Y. and Kaufman, L. (1988) *Chemometrics: a textbook*. Elsevier Science Publishers, Amsterdam.

McClure, W.F. (1984) Fourier analysis of near infrared spectra. *Proc. Symp. on NIR spectroscopy*, Melbourne, Victoria, Australia, 15–16 Oct 1984. Royal Austr. Chem. Institute.

McClure, W.F., Hamid, A., Giesbrecht, F.G. and Weeks, W.W. (1984) Fourier analysis enhances NIR diffuse reflectance spectroscopy. *Applied Spectroscopy*, **38**, 322–329.

Meuzelaar, H.I.C. and Isenhour, T.L. (eds) (1987) *Computer-enhanced analytical spectroscopy*. Plenum Press, New York.

Minkkinen, P. (1987) Evaluation of the fundamental sampling error in the sampling of particulate solids. *Analytica Chimica Acta*, **196**, 237–245.

Norman, D.A. (1983) Some observations on mental models. In: Genter, D. and Stevens. A.L., *Mental models*. Lawrence Elrbaum Ass. Publ., London, 5–14.

Norris, K.H. (1983) Extracting information from spectrophotometric curves. Predicting chemical composition from visible and near-infrared spectra. *Proc. IUFost Symp. Food Research and Data Analysis*, Sept. 1982, Oslo, Norway (Martens and Russwurm, eds). Applied Science Publ., 95–113.

Norris, K.H., Barnes, R.F., Moore, J.E. and Shenk, J.S. (1976) Predicting forage quality by infrared reflectance spectroscopy. *Journal of Animal Science*, **43**, 889–897.

Nyden, M.R., Forney, G.O. and Chittur, K. (1988) Spectroscopic quantitative analysis of strongly interacting systems: Human plasma protein mixtures. *Applied Spectroscopy*, **42**(4), 588–594.

Næs, T. (1985a) Comparison of approaches to multivariate linear calibration, *Biometrical Journal*, **27**, 265–275.

Næs, T. (1985b) Multivariate calibration when the error covariance matrix is structured. *Technometrics*, **27**, 301–311.

Næs, T. (1985c) Inverse versus classical calibration in practice. International Statistical Institute (ISI) meeting, 12–22 August, Amsterdam 1985, *Proceedings*, 75–76.

Næs, T. (1986a) Multivariate calibration by covariance adjustment. *Biometrical Journal*, **28**, 99–107.

Næs, T. (1986b) Detection of multivariate outliers in linear mixed models. *Communications in Statistics (Theory and Methods)*, **15**, 33–47.

Næs, T. (1987a) Leverage and influence measures related to principal component regression. *Chemometrics and Intelligent Laboratory Systems*, **5**, 155–168.

Næs, T. (1987b) PLS versus some other satatistical calibration methods. *Proc. Seminar on PLS Data Approximation*, 19 May, Norwegian Computing Centre (ed. M. Martens).

Næs, T. (1989) The design of calibration in near infra-red reflectance analysis by clustering. *Journal of Chemometrics*, **1**, 121–134.

Næs, T. and Isaksson, T. (1988) Multivariate calibration in near-infrared reflectance spectroscopy—some statistical aspects. (Unpublished manuscript).

Næs, T. and Isaksson, T. (1989) Selection of samples for calibration in near-infrared spectroscopy. I. General principles illustrated by example. *Applied spec.*, **43**, 328–335.

Næs, T. and Martens, H. (1984) Multivariate calibration II. Chemometric methods. *Trends in Analytical Chemistry*, **3**, 266–271.

Næs, T. and Martens, H. (1985) Comparison of prediction methods for collinear data. *Communications in Statistics (Sim. and Comp.)*, **14**, 545–576.

Næs, T. and Martens, H. (1987a) Multivariate calibration. Quantification of harmonies and disharmonies in analytical data. *Computer-enhanced analytical spectroscopy* (eds. H. Meuzelaar and T.L Isenhour), Plenum Publ. Comp., 121–141.

Næs, T. and Martens, H. (1987b) Testing adequacy of linear random models. *Mathematische operationsforschung und statistik (ser. Statistics)*, **18**, 323–331.

Næs, T. and Martens, H. (1988) Principal component regression in NIR analysis. *Journal of Chemometrics*, **2**, 155–167.

of Chemometrics, **2**, 155–167.

Næs, T., Irgens, C. and Martens, H. (1986) Comparison of linear satatistical methods for calibration of NIR instruments. *Applied Statistics*, **35**, 195–206.

Oden, A. (1973) Simultaneous confidence intervals in inverse linear regression. *Biometrika*, **60**, 339–343.

Oman, S.D. (1984) Analyzing residuals in calibration problems. *Technometrics*, **26**, 347–353.

Oman, S.D. (1985) An exact formula for the mean squared error of the inverse estimator in the linear calibration problem. *Journal of Stat. Planning and Inference*, **11**, 189–196.

Oman, S. (1988) Confidence-regions in multivariate calibration. *Annals of Statistics*, **16**, 174–187.

Oman, S.D. and Wax, Y. (1984) Estimating fetal age by ultrasound measurements: An example of multivariate calibration. *Biometrics*, **40**, 947–960.

Osborne, B.G. (1988) A comparative study of methods of linearization and scatter correction in near infrared reflectance spectroscopy. *Analyst*, **113**, 263–267.

Osborne, B.G. and Fearn, T. (1986) *Near infrared spectroscopy in food analysis*. Longman Science and Technical, Great Britain.

Osten, D.W. (1988) Selection of optimal regression models via cross-validation. *Journal of Chemometrics*, **2**, 39–48.

Osten, D.W. and Kowalski, B.R. (1985) Background detection and correction in multicomponent analysis. *Anal. Chem.*, **57**, 908–917.

Ott, R.L. and Myers, R.H. (1968) Optimal experimental designs for estimating the independent variable in regression. *Technometrics*, **10**, 811–823.

Ottestad, P. (1975) Component analysis. An alternative system. *Int. Stat. Rev.*, **43**, 83–108.

Otto, M. and Wegscheider, W. (1985) Spectrophotometric multicomponent analysis applied to trace element determinations, *Anal. Chem.*, **57**, 63–69.

Pedersen, B. and Martens, H. (1988) Multivariate calibration of fluorescence data. In L. Munck (ed.), *Fluorescence spectroscopy*, Longman.

Popper, R., Risvik, E. Martens, H. and Martens, M. (1988) A comparison of multivariate approaches to sensory analysis and the prediction of acceptability. *Proc. U. R.-SCI Int. Symp. on Food Acceptability*. Reading, Sept. 1987.

QUIET—Quantitative Inference Engine Toolbox. Version 1.0 User's guide (1991). Consensus Analysis AS, P.O.Box 1391, N-1401 SKI, Norway.

Rabiner, L.R. and Gold, B. (1975) *Theory and application of digital signal processing*. Prentice-Hall, Englewood Cliffs, NJ.

Rao, C.R. (1959) Some problems involving linear hypotheses in multivariate analysis. *Biometrika*, **46**, 49–58.

Rao, C.R. (1965) The use and interpretation of principal component aanalysis in applied research. *Sankhya*, A, 329–358.

Ricker, N.L. (1988) The use of biased least-squares estimators for parameters in discrete-time impulse–reponse models. *Ind. Eng. Chem. Res.*, **27**, 343–350.

Sanchez, E. and Kowalski, B.R. (1986) Generalized standard addition method. *Analytical Chemistry*, **58**, 496–499.

Saxberg, B.E.H. and Kowalski, B.R. (1979) Generalized standard addition method. *Analytical Chemistry*, **51**, 1034–1038.

Searle, S.R. (1971) *Linear models*. John Wiley and Sons, New York.

Searle, S.R (1974) Prediction, mixed models and variance components. In *Reliability and biometry. Statistical analysis of lifelengths*. Philadelphia, Society for Industrial and Applied Mathematics, 229–266.

Searle, S.R. (1982) *Matrix algebra useful for statistics*. John Wiley and Sons, New York.

Scheffe, H. (1973) A statistical theory of calibration. *Annals of Statistics*, **1**, 1–37.

Sharaf, M.A., Illman, D.L. and Kowalski, B.R. (1986) *Chemometrics*. John Wiley and Sons, New York.

Shenk, J.S., Landa, I., Hoover, M.R. and Westerhaus, M.O. (1981). Description and evaluation of a near infrared reflectance spectrocomputer for forage and grain analysis.

Crop Sci., **21**, 355–358.

Shukla, G.K. (1972) On the problem of calibration. *Technometrics*, **14**, 547–553.

Skjervold, H., Grønseth, K., Vangen, O. and Evensen, A. (1981) Z. *Tierzuchtung und Zuchtungsbiologie*, **98**, 77–79.

Snee, R.D. (1976) Validation of regression models: Methods and examples. *Technometrics*, **19**, 415–428.

Solopchenko, G.N. (1987) Inverse problems in measurement. *Measurement (J. Int. Measurement Confederation)*, **5**(1), 10–19.

Spiegelman, C.H. (1980) Design aspects of Scheffe's calibration theory using linear splines. *Journal of National Bureau of Standards*, **85**, 295–304.

Spjøtvoll, E., Martens, H. and Volden, R. (1982) Restricted least squares estimation of the spectra and concentrations of two unknown constituents available in mixtures. *Technometrics*, **24**, 173–180.

Stark, E. and Martens, H. (1989) *Extended multiplicative signal correction*. (In press).

Stark, E., Luchter, K. and Margoshes, M. (1986) Near-infrared analysis (NIRA): A technology for quantitative and qualitative analysis. *Applied Spec. Reviews*, **22**, 335–399.

Stewart, G.H. (1987) Collinearity and least squares regression. *Statistical Science*, **2**, 68–100.

Stone, M. (1974) Cross-validatory choice and assessment of statistical prediction. *J. Roy. Stat. Soc.*, B, 111–133.

Sundberg, R. (1985) When is the inverse regression estimator MSE-superior to the standard regression estimator in multivariate controlled calibration situations? *Stat. and Prob. Letters*, **3**, 75–79.

Sundberg, R. and Brown, P.J. (1987) Multivariate calibration with more variables than observations. Royal Institute of Technology, S-10044 Stockholm 70, Sweden.

Tyssø, V., Esbensen, K. and Martens, H. (1987) UNSCRAMBLER, an interactive program for multivariate calibration and prediction. *Chemometrics and Intelligent Laboratory Systems*, **2**, 239–243.

UNSCRAMBLER *users guide*, Version 2 (1987), Version 3 (1991). Programme package for multivariate calibration. Marketed by CAMO A/S, Trondheim, Norway.

Valberg, A. and Seim, T. (1983) Quantification of sensory colour differences from physical measurement. Implications for food appearance. In *Food research and data analysis* (Martens, H. and Russwurm, H.jr., eds). Applied Science Publishers, London, 321–342.

Vandeginste, B.G.M., Leyten, F., Gerritsen; M., Noor, J.W. and Kateman, G. (1987) Evaluation of curve resolution and iterative target transformation factor analysis in quantitative analysis by liquid chromatography. *Journal of Chemometrics*, **1**, 57–71.

Velleman, P.F. and Welsch, R.E. (1981) Efficient computing of regression diagnostics. *The Amer. Statistician*, **35**, 234–242.

Webster, J.T., Gunst, R.F. and Mason, R.L. (1974) Latent root regression analysis. *Technometrics*, **16**, 513–522.

Weisberg, S. (1985) *Applied linear regression*. John Wiley and Sons, New York.

Wetzel, D.L. (1983). Near infrared analysis. *Anal. Chem.*, **55**, 1165A–1176A.

Williams, E.J. (1969) A note on regression methods in calibration. *Technometrics*, **11**, 189–192.

Williams, P.C. (1987a) Commercial near-infrared reflectance analyzers. In *Near-infrared technology in the agricultural and food industries* (eds. P.C. Williams and K.H. Norris), Amer. Cereal Assoc., 107–142.

Williams, P.C. (1987b) Variables affecting near-infrared reflectance spectroscopic analysis. In *Near-infrared technology in the agricultural and food industries* (eds. P.C. Williams and K.H. Norris). Amer. Cereal Assoc., 143–167.

Williams, P.C. and Norris, K.H. (1987c) *Near-infrared technology in the agricultural and food industries*. Amer. Cereal Assoc., St. Paul, Minnesota.

Wold, H. (1966) Estimation of principal components and related models by iterative least squares. In *Multivariate analysis* (ed. P.R. Krishnaiah), Academic Press, New York.

Wold, H. (1981) Soft modelling: The basic design and some extensions. In *Systems under indirect observation, causality-structure-prediction* (eds. K.G. Jöreskog and H. Wold), North Holland, Amsterdam.

Wold. H. (1983) In *Food research and data analysis* (eds. H. Martens and H. Russwurm), Applied Science Publ., London.

Wold, S. (1974) Spline functions in data analysis. *Technometrics*, **16**, 1–11.

Wold, S. (1978) Cross-validatory estimation of the number of components in factor analysis and principal components models. *Technometrics*, **20**, 397–406.

Wold, S., Martens, H. and Wold, H. (1983a) The multivariate calibration problem in chemistry solved by the PLS method. *Proc. Conf. Matrix pencils* (A. Ruhe, B. Kågström, eds), March 1982, *Lecture Notes in Mathematics*, Springer Verlag, Heidelberg, 286–293.

Wold, S., Ruhe, A., Wold, H. and Dunn III, W.J. (1984) The collinearity problem in linear regression. The partial least squares (PLS) approach to generalized inverses. *SIAM Journal of Science and Statistical Computations*, **5**, 735–743.

Wold, S., Geladi, P., Esbensen, K.H. and Öhman, J. (1987) Multi-way principal components and PLS-analysis. *J. of Chemometrics*, **1**, 41–56.

Wold, S., Albano, C., Dunn II, W.J., Esbensen, K., Hellberg, S., Johansson, E. and Sjöström, M. (1983b) Pattern recognition: Finding and using regularities in multi-variate data. In: Martens, H. and Russwurm, H.jr (eds), *Food research and data analysis*. Applied Science Publ., London, 147–188.

Young, F. (1981) Quantitative analysis of qualitative data. *Psychometrika*, **46**, 357–388.

Yum, B.J. (1987) Statistical calibration when both measurements are subject to error, A simulation study. *Computers in Engng.*, **12**, 57–65.

Zemroch, P.J. (1986) Cluster analysis as an experimental design generator, with application to gasoline blending experiments. *Technometrics*, **28**, 39–49.

Öberg, T. and Bergström, J.G.T. (1987) Emission and chlorination pattern of PCDD/PCDF prediction from indicator parameters. *Chemosphere*, **16**, 1221–1230.

Öberg, T. and Bergström, J.G.T. (1988) *Indicator parameters for PCDD/PCDF*. Report no. MKS-87/109 from Environmental consultants at Studsvik AB, S-61182, Nyköping, Sweden. T. Öberg Konsult AB, Gamla Brovägen 13, S-37160 Lyckeby, Sweden.

Symbols and Abbreviations

Numbers refer to page of definition or example

398

Index and Cross Reference List

Main entry points are in italics